PLANT AGRICULTURE

Readings from
SCIENTIFIC AMERICAN

PLANT AGRICULTURE

Selected and Introduced by

Jules Janick
Purdue University

Robert W. Schery
The Lawn Institute

Frank W. Woods
University of Tennessee

Vernon W. Ruttan
University of Minnesota

W. H. Freeman and Company
San Francisco

Most of the SCIENTIFIC AMERICAN articles in
Plant Agriculture are available as separate Offprints.
For a complete list of approximately 700 articles now
available as Offprints, write to W. H. Freeman and
Company, 660 Market Street, San Francisco,
California 94104.

PREFACE

Plant agriculture, the production of useful plants, refers to a series of complex and related technologies involving particular species forms known as crops. The study of crop plants and their culture is of far more than academic interest, since life as we know it would be inconceivable without them. Although crop plants constitute but a small percentage of the hundreds of thousands of plant species with which man shares this planet, they bear him a very special relationship. These plants, many of which do not exist apart from cultivation, were selected in prehistory by our forbears, and have developed along with our civilization. They now appear irreplaceable, for in contrast to the fantastic changes that have taken place in our technology and social organization since we began to record our history little replacement has occurred in the basic crops that sustain us. Our domesticated plants represent a unique achievement of early mankind.

The management of our crops varies greatly—from selective harvest to the careful manipulation of each phase of growth. Our level of civilization depends largely upon our ability to produce an abundance of plant products efficiently and dependably. The techniques for the production of food, fiber, spices, and medicinals have played a key role in the story of mankind. The present mounting pressure of world population on world food supply underscores our dependency on the efficient management of plant life.

A collection of readings must necessarily be a smorgasbörd. In this one, however, the selections have been organized into five parts, each of which emphasizes a different approach to the study of plant agriculture. The first part, Agricultural Beginnings, explores the origins of agriculture in both the Old and the New World. The physiology of living plants is the general subject of the articles that make up Part II, Plant Growth and Development. Part III, Plant Environment consists of articles on light, soil, water, and climate. Part IV, Production Technology is concerned with land use, plant breeding, fertilizers, pest control, harvesting, and processing. Part V, Food Needs and Potentials, deals with the dynamics of the world food problems. The twenty-five articles that these parts comprise appeared in *Scientific American* between 1950 and 1969.

January 1970

JULES JANICK
ROBERT W. SCHERY
FRANK W. WOODS
VERNON W. RUTTAN

CONTENTS

Note on cross-references: Cross-references within the articles are of three kinds. A reference to an article included in this book is noted by the title of the article and the page on which it begins; a reference to an article that is available as an offprint but is not included here is noted by the article's title and offprint number; a reference to a SCIENTIFIC AMERICAN article that is not available as an offprint is noted by the title of the article and the month and year of its publication.

PLANT AGRICULTURE

I

AGRICULTURAL BEGINNINGS

I

AGRICULTURAL BEGINNINGS

INTRODUCTION

Civilizations have begun independently in various parts of the world, notably in the Middle East, the Far East, and southern Mexico. In certain favorable environments within these parts of the world nomadic peoples who lived by hunting and gathering became dependent upon the plants and animals they gradually learned to domesticate, and settled down to life in villages. Research generally confirms the centers of origin for crop plants proposed nearly half a century ago by the great Russian plant scientist N. I. Vavilov. The rewards of recent botanical-archeological investigations have been rich and exciting as the details of the genesis of agriculture emerged.

Western scholars have, for various reasons, been more interested in the agricultural beginnings of the Near East and Latin American than in those of the Far East. Consequently, we have less detailed information about the Far Eastern centers of agricultural origin than about the others, although it is well known that many of our domesticated plants originated in China, India, and parts of Southeast Asia.

In the Middle East, where Western civilization had its birth, several centers of origin are known. In "The Agricultural Revolution," Robert J. Braidwood examines the archeological record of ancient settlements in the foothills of the Zagros Mountains, where diggings have resulted in rich finds of artifacts, and even coprolites. Evidence that a number of cultivated plants were in existence there at least ten millennia ago is found in charred seeds of wheat, barley, and several legumes—all common in the diggings. Possession of these domesticated plants, plus nearly all of the important domestic animals, provided "vast new dimensions for cultural evolution." When crop plants are no longer able to self-seed and animals to fend for themselves, man becomes captive of his possessions, and hunter transforms to farmer. But the foothills of the Zagros support fewer people today than they did in ancient times, a tragic consequence of deforestation, soil erosion and exhaustion—a warning to modern man.

In the Western Hemisphere, pre-Columbian Indians succeeded in transforming a wild grass into maize, one of the world's most important food grains. In search of the origin of this staple food, Richard S. Mac-Neish, author of "The Origins of New World Civilization," and several co-workers turned in 1960 to the Tehuacán Valley of Mexico after unsuccessful attempts in Honduras and Guatemala. Through the skillful integration of the expertise of specialists in many disciplines, their project has produced such a wealth of information that it is already regarded as a classic, revealing in detail a very gradual progression "from savagery to civilization" as completely as any project in the Middle East, where archeological exploration has gone on for more than a century. Radio-carbon dates prove that the transition from food gathering to production took place in the Tehuacán Valley between five and nine millennia ago and indicate that there were multiple centers of plant domestication.

In many parts of the world, the transition from hunting and gathering to food production took place rather abruptly, perhaps because of the migration of peoples (or at least the influx of ideas) from more advanced cultures. That this could have been the way in which agriculture came to Western Europe is the conclusion reached by Johannes Iversen in "Forest Clearance in the Stone Age." The hypothesis set forth in his article is based upon the fossil pollen record preserved in lake beds where tools used by Neolithic farmers have also been preserved. Iversen and his co-workers were able to use the fossil pollen record to date a fascinating sequence of changes in the flora of certain areas. Their data indicate that soon after man appeared in the forests that covered much of Europe following the Ice Age, large areas of forested land were suddenly cleared, as evidenced by a rapid increase in the amount of pollen from exotic weeds and types of shrubs associated with forest clearance and burning. In some areas the record indicates that the forest reconquered, in much the same way that forests do today in parts of the world where shifting agriculture is still practiced. Proof that Iversen's hypothesis is at least possible is given by his description of the changes in flora that took place when he and his fellow workers experimentally duplicated what Stone Age man may have done with his flint tools and fire.

Agricultural practices have always been adapted to local conditions. In "The Chinampas of Mexico," Michael D. Coe explores part of the legacy of the Aztec's chinampas farming, which is still practiced in the Mexico City area. In this method of intensive cultivation, raised beds built along a lake margin are held in place by peripheral trees and stakes, and are renewed by the addition of organic matter and mud dredged from the shallows of the lake. The soil is rich, and water is of course plentiful; about seven crops are taken annually from each cultivated rectangle. This method of cultivation, which appears to go back nearly two thousand years, was so productive that it supplied the Aztecs with enough surplus food to provision an army that conquered all of the then-civilized parts of Mexico. Remnants of the system are still seen in the "floating gardens" of Xochimilco, which, of course, are actually fixed in place.

1

THE AGRICULTURAL REVOLUTION

ROBERT J. BRAIDWOOD
September 1960

Tool-making was initiated by pre-*sapiens* man. The first comparable achievement of our species was the agricultural revolution. No doubt a small human population could have persisted on the sustenance secured by the hunting and food-gathering technology that had been handed down and slowly improved upon over the 500 to 1,000 millennia of pre-human and pre-*sapiens* experience. With the domestication of plants and animals, however, vast new dimensions for cultural evolution suddenly became possible. The achievement of an effective food-producing technology did not, perhaps, predetermine subsequent developments, but they followed swiftly: the first urban societies in a few thousand years and contemporary industrial civilization in less than 10,000 years.

The first successful experiment in food production took place in southwestern Asia, on the hilly flanks of the "fertile crescent." Later experiments in agriculture occurred (possibly independently) in China and (certainly independently) in the New World. The multiple occurrence of the agricultural revolution suggests that it was a highly probable outcome of the prior cultural evolution of mankind and a peculiar combination of environmental circumstances. It is in the record of culture, therefore, that the origin of agriculture must be sought.

About 250,000 years ago wide-wandering bands of ancient men began to make remarkably standardized stone hand-axes and flake tools which archeologists have found throughout a vast area of the African and western Eurasian continents, from London to Capetown to Madras. Cultures producing somewhat different tools spread over all of eastern Asia. Apparently the creators of these artifacts employed general, non-specialized techniques in gathering and preparing food. As time went on, the record shows, specialization set in within these major traditions, or "genera," of tools, giving rise to roughly regional "species" of tool types. By about 75,000 years ago the tools became sufficiently specialized to suggest that they corresponded to the conditions of food-getting in broad regional environments. As technological competence increased, it became possible to extract more food from a given environment; or, to put the matter the other way around, increased "living into" a given environment stimulated technological adaptation to it.

Perhaps 50,000 years ago the mod-

AIR VIEW OF JARMO shows 3.2-acre site and surroundings. About one third of original area has eroded away. Archeologists dug the square holes in effort to trace village plan.

JARMO IN IRAQI KURDISTAN is the site of the earliest village-farming community yet discovered. This photograph of an upper level of excavation shows foundation and paving stones. Site was occupied for perhaps 300 years somewhere around 6750 B.C.

EXCAVATION AT KARIM SHAHIR contained confused scatter of rocks brought there by ancient men and disturbed by modern plowing. This prefarming site had no clear evidence of permanent houses, but did have skillfully chipped flints and other artifacts.

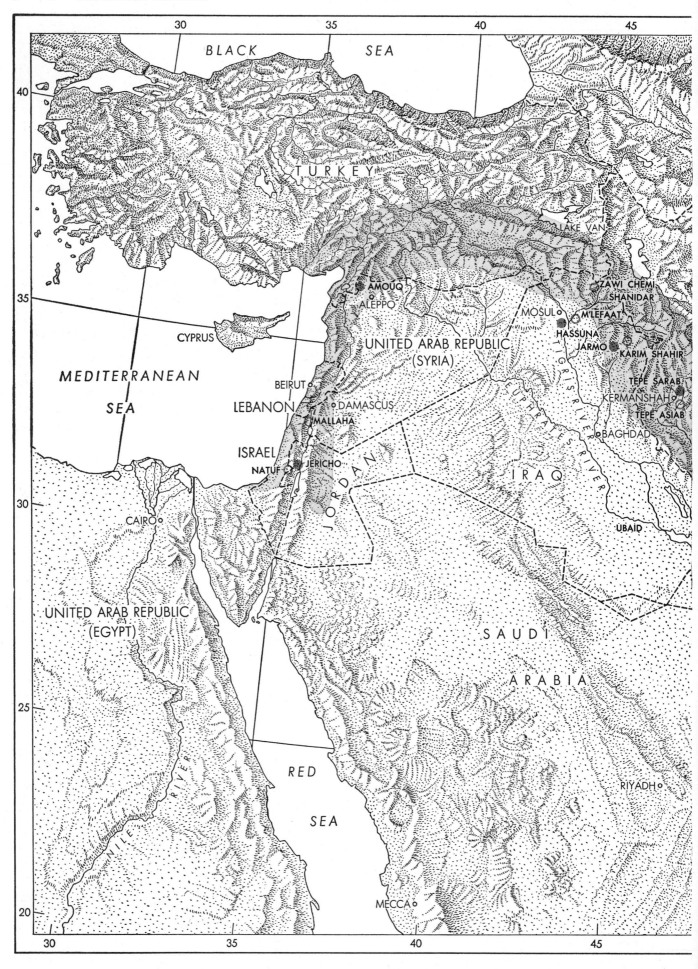

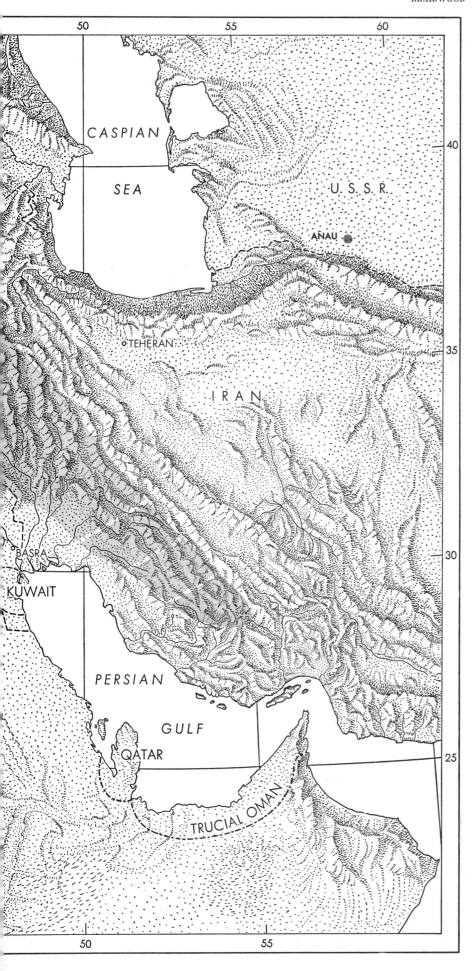

ern physical type of man appeared. **The record** shows concurrently the first appearance of a new genera of tools: **the blade tools** which incorporate a qualitatively higher degree of usefulness and skill in fabrication. The new type of **man** using the new tools substituted more systematic food-collection and organized hunting of large beasts for the simple gathering and scavenging of his predecessors. As time passed, the human population increased and men were able to adjust themselves to environmental niches as diverse as the tropical jungle and the arctic tundra. By perhaps 30,000 years ago they spread to the New World. The successful adaptation of human communities to their different environments brought on still greater cultural complexity and differentiation. Finally, between 11,000 and 9,000 years ago some of these communities arrived at the threshold of food production.

In certain regions scattered throughout the world this period (the Mesolithic in northwestern Europe and the Archaic in North America) was characterized by intensified food-collection: the archeological record of the era is the first that abounds in the remains of small, fleet animals, of water birds and fish, of snails and mussels. In a few places signs of plant foods have been preserved, or at least we archeologists have learned to pay attention to them. All of these remains show that human groups had learned to live into their environment to a high degree, achieving an intimate familiarity with every element in it. Most of the peoples of this era of intensified food-collecting changed just enough so that they did not need to change. There are today still a few relict groups of intensified food-collectors—the Eskimos, for example—and there were many more only a century or two ago. But on the grassy and forested uplands bordering the fertile crescent a real change was under way. Here in a climate that provided generous winter and spring rainfall, the intensified food-collectors had been accumulating a rich lore of experience with wild wheat, barley and other food plants, as well as with wild dogs,

HILLS FLANKING fertile crescent, where agricultural revolution occurred, are indicated in color. Hatched areas are probably parts of this "nuclear" zone of food-producing revolution. Sites discussed in this article are indicated by large circles. Open circles are prefarming sites; solid circles indicate that food production was known there.

goats, sheep, pigs, cattle and horses. It was here that man first began to control the production of his food.

Not long ago the proponents of environmental determinism argued that the agricultural revolution was a response to the great changes in climate which accompanied the retreat of the last glaciation about 10,000 years ago. However, the climate had altered in equally dramatic fashion on other occasions in the past 75,000 years, and the potentially domesticable plants and animals were surely available to the bands of food-gatherers who lived in southwestern Asia and similar habitats in various parts of the globe. Moreover, recent studies have revealed that the climate did not change radically where farming began in the hills that flank the fertile crescent. Environmental determinists have also argued from the "theory of propinquity" that the isolation of men along with appropriate plants and animals in desert oases started the process of domestication. Kathleen M. Kenyon of the University of London, for example, advances the lowland oasis of Jericho as a primary site of the agricultural revolution [see "Ancient Jericho," by Kathleen M. Kenyon; SCIENTIFIC AMERICAN, April, 1954].

In my opinion there is no need to complicate the story with extraneous "causes." The food-producing revolution seems to have occurred as the culmination of the ever increasing cultural differentiation and specialization of human communities. Around 8000 B.C. the inhabitants of the hills around the fertile crescent had come to know their habitat so well that they were beginning to domesticate the plants and animals they had been collecting and hunting. At slightly later times human cultures reached the corresponding level in Central America and perhaps in the Andes, in southeastern Asia and in China. From these "nuclear" zones cultural diffusion spread the new way of life to the rest of the world.

In order to study the agricultural revolution in southwestern Asia I have since 1948 led several expeditions, sponsored by the Oriental Institute of the University of Chicago, to the hills of Kurdistan north of the fertile crescent in Iraq and Iran. The work of these expeditions has been enriched by the collaboration of botanists, zoologists and geologists, who have alerted the archeologists among us to entirely new kinds of evidence. So much remains to be done, however, that we can describe in only a tentative and quite incomplete fashion how food production began. In part, I must freely admit, my reconstruction depends upon extrapolation backward from what I know had been achieved soon after 9,000 years ago in southwestern Asia.

The earliest clues come from sites of the so-called Natufian culture in Palestine, from the Kurdistan site of Zawi Chemi Shanidar, recently excavated by Ralph S. Solecki of the Smithsonian Institution, from our older excavations at Karim Shahir and M'lefaat in Iraq, and from our current excavations at Tepe Asiab in Iran [see map on preceding two pages]. In these places men appear to have moved out of caves, although perhaps not for the first time, to live in at least semipermanent communities. Flint sickle-blades occur in such Natufian locations as Mallaha, and both the Palestine and Kurdistan sites have yielded milling and pounding stones—strong indications that the people reaped and ground wild cereals and other plant foods. The artifacts do not necessarily establish the existence of anything more than intensified or specialized food-collecting. But these people were at home in a landscape in which the grains grew wild, and they may have begun to cultivate them in open meadows. Excavations of later village-farming communities, which have definitely been identified as such, reveal versions of the same artifacts that are only slightly more developed than those from Karim Shahir and other earlier sites. We are constantly finding additional evidence that will eventually make the picture clearer. For example, just this spring at Tepe Asiab we found many coprolites (fossilized excrement) that appear to be of human origin. They contain abundant impressions of plant and animal foods, and when analyzed in the laboratory they promise to be a gold mine of clues to the diet of the Tepe

SICKLE BLADES FROM JARMO are made of chipped flint. They are shown here approximately actual size. When used for harvesting grain, several were mounted in a haft of wood or bone. Other Jarmo flint tools show little advance over those found at earlier sites.

Asiab people. The nature of these "antiquities" suggests how the study of the agricultural revolution differs from the archeology of ancient cities and tombs.

The two earliest indisputable village-farming communities we have so far excavated were apparently inhabited between 7000 and 6500 B.C. They are on the inward slopes of the Zagros mountain crescent in Kurdistan. We have been digging at Jarmo in Iraq since 1948 [see "From Cave to Village," by Robert J. Braidwood; SCIENTIFIC AMERICAN, October, 1952], and we started our investigations at Tepe Sarab in Iran only last spring. We think there are many sites of the same age in the hilly-flanks zone, but these two are the only ones we have so far been able to excavate. Work should also be done in this zone in southern Turkey, but the present interpretation of the Turkish antiquities law discourages our type of "problem-oriented" research, in which the investigator must take most of the ancient materials back to his laboratory. I believe that these northern parts of the basins of the Tigris and Euphrates rivers and the Cilician area of Turkey will one day yield valuable information.

Although Jarmo and Tepe Sarab are 120 miles apart and in different drainage systems, they contain artifacts that are remarkably alike. Tepe Sarab may have been occupied only seasonally, but Jarmo was a permanent, year-round settlement with about two dozen mud-walled houses that were repaired and rebuilt frequently, creating about a dozen distinct levels of occupancy. We have identified there the remains of two-row barley (cultivated barley today has six rows of grains on a spike) and two forms of domesticated wheat. Goats and dogs, and possibly sheep, were domesticated. The bones of wild animals, quantities of snail shells and acorns and pistachio nuts indicate that the people still hunted and collected a substantial amount of food. They enjoyed a varied, adequate and well-balanced diet which was possibly superior to that of the people living in the same area today. The teeth of the Jarmo people show even milling and no marginal enamel fractures. Thanks apparently to the use of querns and rubbing stones and stone mortars and pestles, there were no coarse particles in the diet that would cause excessive dental erosion. We have calculated that approximately 150 people lived in Jarmo. The archeological evidence from the area indicates a population density of 27 people per square mile, about the same as today. Deforestation, soil deteriora-

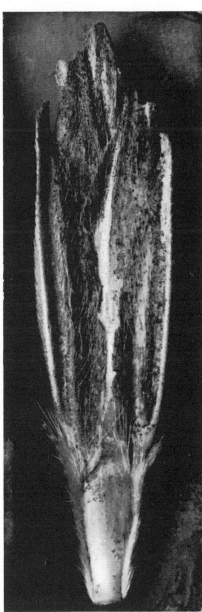

JARMO WHEAT made imprint upon clay. Cast of imprint (*left*) resembles spikelet of present-day wild wheat *Triticum dicoccoides* (*right*). Specimens are enlarged seven times.

tion and erosion, the results of 10,000 years of human habitation, tend to offset whatever advantages of modern tools and techniques are available to the present population.

Stone vessels of fine craftsmanship appear throughout all levels at Jarmo, but portable, locally made pottery vessels occur only in the uppermost levels. A few impressions on dried mud indicate that the people possessed woven baskets or rugs. The chipped flint tools of Jarmo and Tepe Sarab, in both normal and microlithic sizes, are direct and not very distant descendants of those at Karim Shahir and the earlier communities. But the two farming villages exhibit a geo-

metric increase in the variety of materials of other types in the archeological catalogue. Great numbers of little clay figurines of animals and pregnant women (the "fertility goddesses") hint at the growing nonutilitarian dimensions of life. In both communities the people for the first time had tools of obsidian, a volcanic glass with a cutting edge much sharper and harder than stone. The obsidian suggests commerce, because the nearest source is at Lake Van in Turkey, some 200 miles from Jarmo. The sites have also yielded decorative shells that could have come only from the Persian Gulf.

For an explanation of how plants and animals might have been domesticated

KERNELS OF JARMO WHEAT were carbonized in fires of ancient village. They resemble kernels of wild wheat growing in area today. They are enlarged approximately four times.

Wild grasses have to scatter their seeds over a large area, and consequently the seed-holding spike of wild wheat and barley becomes brittle when the plant ripens. The grains thus drop off easily. A few wild plants, however, exhibit a recessive gene that produces tough spikes that do not become brittle. The grains hang on, and these plants do not reproduce well in nature. A man harvesting wild wheat and barley would necessarily reap plants with tough spikes and intact heads. When he finally did sow seeds, he would naturally have on hand a large proportion of grains from tough-spike plants—exactly the kind he needed for farming. Helbaek points out that early farmers must soon have found it advantageous to move the wheat down from the mountain slopes, from 2,000 to 4,300 feet above sea level (where it occurs in nature), to more level ground near a reliable water supply and other accommodations for human habitation. Still, the plant had to be kept in an area with adequate winter and spring rainfall. The piedmont of the fertile crescent provides even today precisely these conditions. Since the environment there differs from the native one, wheat plants with mutations and recessive characteristics, as well as hybrids and other freaks, that were ill adapted to the uplands would have had a chance to survive. Those that increased the adaptation of wheat to the new environment would have made valuable contributions to the gene pool. Domesticated wheat, having lost the ability to disperse its seeds, became totally dependent upon man. In turn, as Helbaek emphasizes, man became the servant of his plants, since much of his routine of life now depended upon the steady and ample supply of vegetable food from his fields.

between the time of Karim Shahir and of Jarmo, we have turned to our colleagues in the biological sciences. As the first botanist on our archeological team, Hans Helbaek of the Danish National Museum has studied the carbonized remains of plants and the imprints of grains, seeds and other plant parts on baked clay and adobe at Jarmo and other sites. He believes that the first farmers, who grew both wheat and barley, could

only have lived in the highlands around the fertile crescent, because that is the only place where both plants grew wild. The region is the endemic home of wild wheat. Wild barley, on the other hand, is widely scattered from central Asia to the Atlantic, but no early agriculture was based upon barley alone.

Helbaek surmises that from the beginning man was unintentionally breeding the kind of crop plants he needed.

The traces and impressions of the grains at Jarmo indicate that the process of domestication was already advanced at that place and time, even though human selection of the best seed had not yet been carried far. Carbonized field peas, lentils and vetchling have also been found at Jarmo, but it is not certain that these plants were under cultivation.

Apparently farming and a settled community life were cultural prerequisites for the domestication of animals. Charles A. Reed, zoologist from the University of Illinois, has participated in the Oriental Institute expeditions to Iraq and Iran and has studied animal skeletons we have excavated. He believes that animal domestication first occurred in this area, because wild goats, sheep, cattle, pigs, horses, asses and dogs were all present

CARBONIZED BARLEY KERNELS from Jarmo, enlarged four times, are from two-row grain. The internodes attached to kernels at right indicate tough spikes of cultivated barley.

there, and settled agricultural communities had already been established. The wild goat (*Capra hircus aegagrus*, or pasang) and sheep (*Ovis orientalis*), as well as the wild ass (onager) still persist in the highlands of southwestern Asia. Whether the dog was the offspring of a hypothetical wild dog, of the pariah dog or of the wolf is still uncertain, but it was undoubtedly the first animal to be domesticated. Reed has not been able to identify any dog remains at Jarmo, but doglike figurines, with tails upcurled, show almost certainly that dogs were established in the domestic scene. The first food animal to be domesticated was the goat; the shape of goat horns found at Jarmo departs sufficiently from that of the wild animal to certify generations of domestic breeding. On the other hand, the scarcity of remains of cattle at Jarmo indicates that these animals had not yet been domesticated; the wild cattle in the vicinity were probably too fierce to submit to captivity.

No one who has seriously considered the question believes that food needs motivated the first steps in the domestication of animals. The human proclivity for keeping pets suggests itself as a much simpler and more plausible explanation. Very young animals living in the environment may have attached themselves to people as a result of "imprinting," which is the tendency of the animal to follow the first living thing it sees and hears during a critically impressionable period in its infancy [see "'Imprinting' in Animals," by Eckhard H. Hess; SCIENTIFIC AMERICAN Offprint 416].

Young animals were undoubtedly also captured for use as decoys on the hunt. Some young animals may have had human wet nurses—a practice in some primitive tribes even today. After goats were domesticated, their milk would have been available for orphaned wild calves, colts and other creatures. Adult wild animals, particularly goats and sheep, which sometimes approach human beings in search of food, might also have been tamed.

Reed defines the domesticated animal as one whose reproduction is controlled by man. In his view the animals that were domesticated were already physiologically and psychologically preadapted to being tamed without loss of their ability to reproduce. The individual animals that bred well in captivity would have contributed heavily to the gene pool of each succeeding generation. When the nucleus of a herd was established, man would have automatically selected against the aggressive and un-

CLAY FIGURES from Sarab, shown half size, include boar's head (*top*), what seems to be lion (*upper left*), two-headed beast (*upper right*), sheep (*bottom left*) and boar.

"FERTILITY GODDESS" or "Venus" from Tepe Sarab is clay figure shown actual size. Artist emphasized parts of body suggesting fertility. Grooves in leg indicate musculature.

STONE PALETTES from Jarmo show that the men who lived there were highly skilled in working stone. The site has also yielded many beautifully shaped stone bowls and mortars.

POTTERY MADE AT JARMO, in contrast to the stonework, is simple. It is handmade, vegetable-tempered, buff or orange-buff in color. It shows considerable technical competence.

manageable individuals, eventually producing a race of submissive creatures. This type of unplanned breeding no doubt long preceded the purposeful artificial selection that created different breeds within domesticated species. It is apparent that goats, sheep and cattle were first husbanded as producers of meat and hides; wild cattle give little milk, and wild sheep are not woolly but hairy. Only much later did the milk- and wool-producing strains emerge.

As the agricultural revolution began to spread, the trend toward ever increasing specialization of the intensified food-collecting way of life began to reverse itself. The new techniques were capable of wide application, given suitable adaptation, in diverse environments. Archeological remains at Hassuna, a site near the Tigris River somewhat later than Jarmo, show that the people were exchanging ideas on the manufacture of pottery and of flint and obsidian projectile points with people in the region of the Amouq in Syro-Cilicia. The basic elements of the food-producing complex —wheat, barley, sheep, goats and probably cattle—in this period moved west beyond the bounds of their native habitat to occupy the whole eastern end of the Mediterranean. They also traveled as

far east as Anau, east of the Caspian Sea. Localized cultural differences still existed, but people were adopting and adapting more and more cultural traits from other areas. Eventually the new way of life traveled to the Aegean and beyond into Europe, moving slowly up such great river valley systems as the Dnieper, the Danube and the Rhone, as well as along the coasts. The intensified food-gatherers of Europe accepted the new way of life, but, as V. Gordon Childe has pointed out, they "were not slavish imitators: they adapted the gifts from the East . . . into a new and organic whole capable of developing on its own original lines." Among other things, the Europeans appear to have domesticated rye and oats that were first imported to the European continent as weed plants contaminating the seed of wheat and barley. In the comparable diffusion of agriculture from Central America, some of the peoples to the north appear to have rejected the new ways, at least temporarily.

By about 5000 B.C. the village-farming way of life seems to have been fingering down the valleys toward the alluvial bottom lands of the Tigris and Euphrates. Robert M. Adams believes that there may have been people living in the

lowlands who were expert in collecting food from the rivers. They would have taken up the idea of farming from people who came down from the higher areas. In the bottom lands a very different climate, seasonal flooding of the land and small-scale irrigation led agriculture through a significant new technological transformation. By about 4000 B.C. the people of southern Mesopotamia had achieved such increases in productivity that their farms were beginning to support an urban civilization. The ancient site at Ubaid is typical of this period [see "The Origin of Cities," by Robert M. Adams, SCIENTIFIC AMERICAN Offprint 606].

Thus in 3,000 or 4,000 years the life of man had changed more radically than in all of the preceding 250,000 years. Before the agricultural revolution most men must have spent their waking moments seeking their next meal, except when they could gorge following a great kill. As man learned to produce food, instead of gathering, hunting or collecting it, and to store it in the grain bin and on the hoof, he was compelled as well as enabled to settle in larger communities. With human energy released for a whole spectrum of new activities, there came the development of specialized nonagricultural crafts. It is no accident that such innovations as the discovery of the basic mechanical principles, weaving, the plow, the wheel and metallurgy soon appeared.

No prehistorian worth his salt may end or begin such a discussion without acknowledging the present incompleteness of the archeological record. There is the disintegration of the perishable materials that were primary substances of technology at every stage. There is the factor of chance in archeological discovery, of vast areas of the world still incompletely explored archeologically, and of inadequate field techniques and interpretations by excavators. There are the vagaries of establishing a reliable chronology, of the whimsical degree to which "geobiochemical" contamination seems to have affected our radioactive-carbon age determinations. There is the fact that studies of human paleo-environments by qualified natural historians are only now becoming available. Writing in the field, in the midst of an exciting season of excavation, I would not be surprised if the picture I have presented here needs to be altered somewhat by the time that this article has appeared in print.

2

THE ORIGINS OF NEW WORLD CIVILIZATION

RICHARD S. MACNEISH
November 1964

Perhaps the most significant single occurrence in human history was the development of agriculture and animal husbandry. It has been assumed that this transition from food-gathering to food production took place between 10,000 and 16,000 years ago at a number of places in the highlands of the Middle East. In point of fact the archaeological evidence for the transition, particularly the evidence for domesticated plants, is extremely meager. It is nonetheless widely accepted that the transition represented a "Neolithic Revolution," in which abundant food, a sedentary way of life and an expanding population provided the foundations on which today's high civilizations are built.

The shift from food-gathering to food production did not, however, happen only once. Until comparatively recent times the Old World was for the most part isolated from the New World. Significant contact was confined to a largely one-way migration of culturally primitive Asiatic hunting bands across the Bering Strait. In spite of this almost total absence of traffic between the hemispheres the European adventurers who reached the New World in the 16th century encountered a series of cultures almost as advanced (except in metallurgy and pyrotechnics) and quite as barbarous as their own. Indeed, some of the civilizations from Mexico to Peru possessed a larger variety of domesticated plants than did their European

conquerors and had made agricultural advances far beyond those of the Old World.

At some time, then, the transition from food-gathering to food production occurred in the New World as it had in the Old. In recent years one of the major problems for New World prehistorians has been to test the hypothesis of a Neolithic Revolution against native archaeological evidence and at the same time to document the American stage of man's initial domestication of plants (which remains almost unknown in both hemispheres).

The differences between the ways in which Old World and New World men achieved independence from the nomadic life of the hunter and gatherer are more striking than the similarities. The principal difference lies in the fact that the peoples of the Old World domesticated many animals and comparatively few plants, whereas in the New World the opposite was the case. The abundant and various herds that gave the peoples of Europe, Africa and Asia meat, milk, wool and beasts of burden were matched in the pre-Columbian New World only by a half-domesticated group of Andean cameloids: the llama, the alpaca and the vicuña. The Andean guinea pig can be considered an inferior equivalent of the Old World's domesticated rabbits and hares; elsewhere in the Americas the turkey was an equally inferior counterpart of the Eastern Hemisphere's many

varieties of barnyard fowl. In both the Old World and the New, dogs presumably predated all other domestic animals; in both beekeepers harvested honey and wax. Beyond this the New World list of domestic animals dwindles to nothing. All the cultures of the Americas, high and low alike, depended on their hunters' skill for most of their animal produce: meat and hides, furs and feathers, teeth and claws.

In contrast, the American Indian domesticated a remarkable number of plants. Except for cotton, the "water bottle" gourd, the yam and possibly the coconut (which may have been domesticated independently in each hemisphere), the kinds of crops grown in the Old World and the New were quite different. Both the white and the sweet potato, cultivated in a number of varieties, were unique to the New World. For seasoning, in place of the pepper and mustard of the Old World, the peoples of the New World raised vanilla and at least two kinds of chili. For edible seeds they grew amaranth, chive, panic grass, sunflower, quinoa, apazote, chocolate, the peanut, the common bean and four other kinds of beans: lima, summer, tepary and jack.

In addition to potatoes the Indians cultivated other root crops, including manioc, oca and more than a dozen other South American plants. In place of the Old World melons, the related plants brought to domestication in the New World were the pumpkin, the

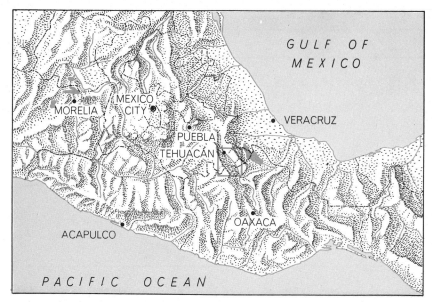

TEHUACÁN VALLEY is a narrow desert zone in the mountains on the boundary between the states of Puebla and Oaxaca. It is one of the three areas in southern Mexico selected during the search for early corn on the grounds of dryness (which helps to preserve ancient plant materials) and highland location (corn originally having been a wild highland grass).

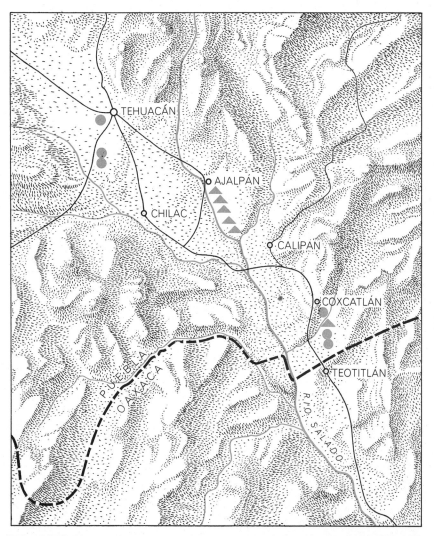

SIX CAVES (dots) and six open-air sites (triangles) have been investigated in detail by the author and his colleagues. Coxcatlán cave (top dot at right), where early corn was found in 1960, has the longest habitation record: from well before 7000 B.C. until A.D. 1500.

gourd, the chayote and three or four distinct species of what we call squash. Fruits brought under cultivation in the Americas included the tomato, avocado, pineapple, guava, elderberry and papaya. The pioneering use of tobacco—smoked in pipes, in the form of cigars and even in the form of cane cigarettes, some of which had one end stuffed with fibers to serve as a filter—must also be credited to the Indians.

Above all of these stood Indian corn, *Zea mays,* the only important wild grass in the New World to be transformed into a food grain as the peoples of the Old World had transformed their native grasses into wheat, barley, rye, oats and millet. From Chile to the valley of the St. Lawrence in Canada, one or another of 150 varieties of Indian corn was the staple diet of the pre-Columbian peoples. As a food grain or as fodder, corn remains the most important single crop in the Americas today (and the third largest in the world). Because of its dominant position in New World agriculture, prehistorians have long been confident that if they could find out when and where corn was first domesticated, they might also uncover the origins of New World civilization.

Until little more than a generation ago investigators of this question were beset by twin difficulties. First, research in both Central America and South America had failed to show that any New World high culture significantly predated the Christian era. Second, botanical studies of the varieties of corn and its wild relatives had led more to conflict than to clarity in regard to the domesticated plant's most probable wild predecessor [see "The Mystery of Corn," by Paul C. Mangelsdorf; SCIENTIFIC AMERICAN Offprint 26]. Today, thanks to close cooperation between botanists and archaeologists, both difficulties have almost vanished. At least one starting point for New World agricultural activity has been securely established as being between 5,000 and 9,000 years ago. At the same time botanical analysis of fossil corn ears, grains and pollen, together with plain dirt archaeology, have solved a number of the mysteries concerning the wild origin and domestic evolution of corn. What follows is a review of the recent developments that have done so much to increase our understanding of this key period in New World prehistory.

The interest of botanists in the history of corn is largely practical: they study the genetics of corn in order to produce improved hybrids. After the wild ancestors of corn had been sought for nearly a century the search had narrowed to two tassel-bearing New World grasses—teosinte and *Tripsacum*—that had features resembling the domesticated plant. On the basis of crossbreeding experiments and other genetic studies, however, Paul C. Mangelsdorf of Harvard University and other investigators concluded in the 1940's that neither of these plants could be the original ancestor of corn. Instead teosinte appeared to be the product of the accidental crossbreeding of true corn and *Tripsacum*. Mangelsdorf advanced the hypothesis that the wild progenitor of corn was none other than corn itself—probably a popcorn with its kernels encased in pods.

Between 1948 and 1960 a number of discoveries proved Mangelsdorf's contention to be correct. I shall present these discoveries not in their strict chronological order but rather in their order of importance. First in importance, then, were analyses of pollen found in "cores" obtained in 1953 by drilling into the lake beds on which Mexico City is built. At levels that were estimated to be about 80,000 years old—perhaps 50,000 years older than the earliest known human remains in the New World—were found grains of corn

pollen. There could be no doubt that the pollen was from wild corn, and thus two aspects of the ancestry of corn were clarified. First, a form of wild corn has been in existence for 80,000 years, so that corn can indeed be descended from itself. Second, wild corn had flourished in the highlands of Mexico. As related archaeological discoveries will make plain, this geographical fact helped to narrow the potential range—from the southwestern U.S. to Peru—within which corn was probably first domesticated.

The rest of the key discoveries, involving the close cooperation of archaeologist and botanist, all belong to the realm of paleobotany. In the summer of 1948, for example, Herbert Dick, a graduate student in anthropology who had been working with Mangelsdorf, explored a dry rock-shelter in New Mexico called Bat Cave. Digging down through six feet of accumulated deposits, he and his colleagues found numerous remains of ancient corn, culminating in some tiny corncobs at the lowest level. Carbon-14 dating indicated that these cobs were between 4,000 and 5,000 years old. A few months later, exploring the La Perra cave in the state of Tamaulipas far to the north of Mexico City, I found similar corncobs that proved to be about 4,500 years old. The oldest cobs at both sites came close

to fitting the description Mangelsdorf had given of a hypothetical ancestor of the pod-popcorn type. The cobs, however, were clearly those of domesticated corn.

These two finds provided the basis for intensified archaeological efforts to find sites where the first evidences of corn would be even older. The logic was simple: A site old enough should have a level of wild corn remains older than the most ancient domesticated cobs. I continued my explorations near the La Perra cave and excavated a number of other sites in northeastern Mexico. In them I found more samples of ancient corn, but they were no older than those that had already been discovered. Robert Lister, another of Mangelsdorf's co-workers, also found primitive corn in a cave called Swallow's Nest in the Mexican state of Chihuahua, northwest of where I was working, but his finds were no older than mine.

If nothing older than domesticated corn of about 3000 B.C. could be found to the north of Mexico City, it seemed logical to try to the south. In 1958 I went off to look for dry caves and early corn in Guatemala and Honduras. The 1958 diggings produced nothing useful, so in 1959 I moved northward into Chiapas, Mexico's southernmost state. There were no corncobs to be found,

EXCAVATION of Coxcatlán cave required the removal of one-meter squares of cave floor over an area 25 meters long by six meters wide until bedrock was reached at a depth of almost five meters. In this way 28 occupation levels, attributable to seven distinctive culture phases, were discovered. Inhabitants of the three lowest levels lived by hunting and by collecting wild-plant foods.

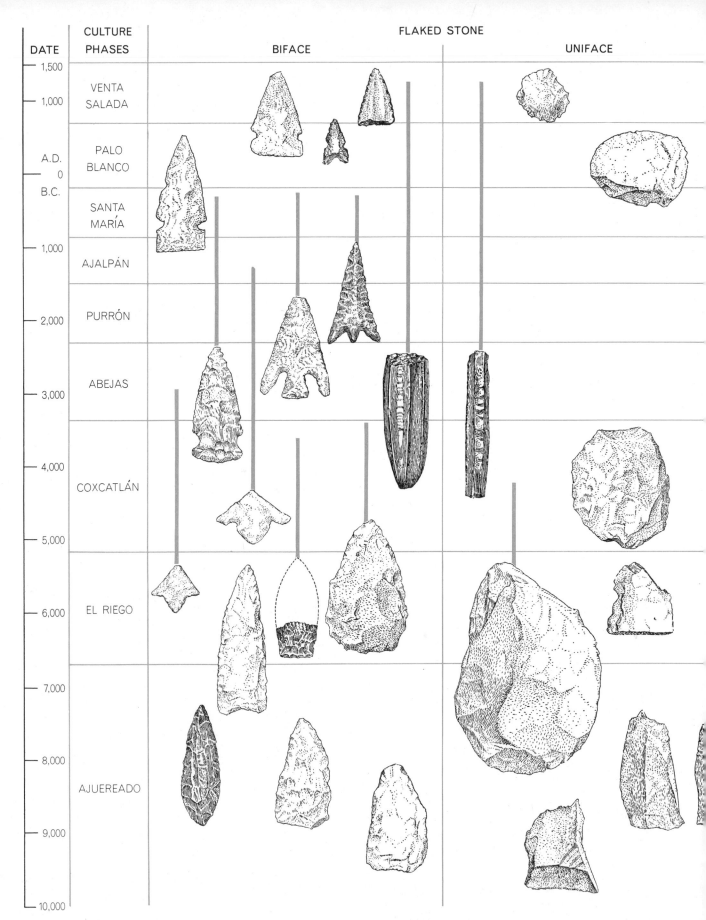

DATE	CULTURE PHASES	FLAKED STONE	
		BIFACE	UNIFACE

DATE

— 1,500
— 1,000

A.D.
— 0
B.C.

— 1,000

— 2,000

— 3,000

— 4,000

— 5,000

— 6,000

— 7,000

— 8,000

— 9,000

— 10,000

CULTURE PHASES

VENTA SALADA

PALO BLANCO

SANTA MARÍA

AJALPÁN

PURRÓN

ABEJAS

COXCATLÁN

EL RIEGO

AJUEREADO

STONE ARTIFACTS from various Tehuacán sites are arrayed in two major categories: those shaped by chipping and flaking (*left*) and those shaped by grinding and pecking (*right*). Implements that have been chipped on one face only are separated from those that show bifacial workmanship; both groups are reproduced at half their natural size. The ground stone objects are not drawn to a common scale. The horizontal lines define the nine culture phases thus far distinguished in the valley. Vertical lines (*color*) indicate the extent to which the related artifact is known in cultures other than the one in which it is placed. At Tehuacán the evolution of civilization failed to follow the classic pattern established by the Neolithic Revolution in the Old World. For instance, the mortars,

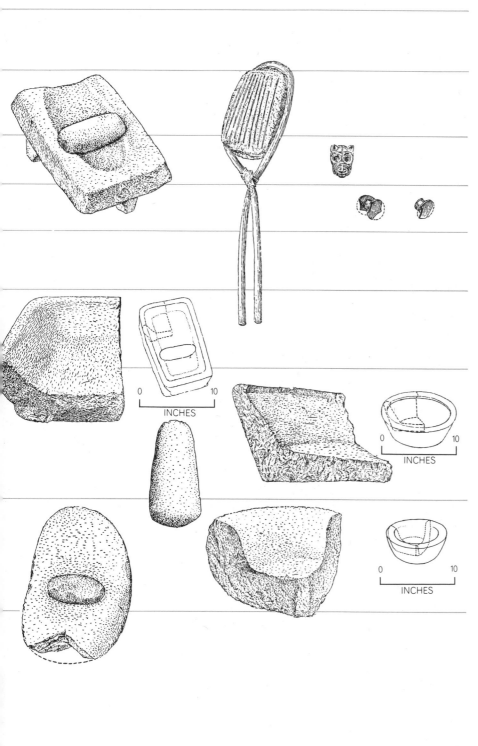

but one cave yielded corn pollen that also dated only to about 3000 B.C. The clues provided by paleobotany now appeared plain. Both to the north of Mexico City and in Mexico City itself (as indicated by the pollen of domesticated corn in the upper levels of the drill cores) the oldest evidence of domesticated corn was no more ancient than about 3000 B.C. Well to the south of Mexico City the oldest date was the same. The area that called for further search should therefore lie south of Mexico City but north of Chiapas.

Two additional considerations enabled me to narrow the area of search even more. First, experience had shown that dry locations offered the best chance of finding preserved specimens of corn. Second, the genetic studies of Mangelsdorf and other investigators indicated that wild corn was originally a highland grass, very possibly able to survive the rigorous climate of highland desert areas. Poring over the map of southern Mexico, I singled out three large highland desert areas: one in the southern part of the state of Oaxaca, one in Guerrero and one in southern Puebla.

Oaxaca yielded nothing of interest, so I moved on to Puebla to explore a dry highland valley known as Tehuacán. My local guides and I scrambled in and out of 38 caves and finally struck pay dirt in the 39th. This was a small rock-shelter near the village of Coxcatlán in the southern part of the valley of Tehuacán. On February 21, 1960, we dug up six corncobs, three of which looked more primitive and older than any I had seen before. Analysis in the carbon-14 laboratory at the University of Michigan confirmed my guess by dating these cobs as 5,600 years old—a good 500 years older than any yet found in the New World.

With this find the time seemed ripe for a large-scale, systematic search. If we had indeed arrived at a place where corn had been domesticated and New World civilization had first stirred, the closing stages of the search would require the special knowledge of many experts. Our primary need was to obtain the sponsorship of an institution interested and experienced in such research, and we were fortunate enough to enlist exactly the right sponsor: the Robert S. Peabody Foundation for Archaeology of Andover, Mass. Funds for the project were supplied by the National Science Foundation and by the agricultural branch of the Rockefeller

pestles and other ground stone implements that first appear in the El Riego culture phase antedate the first domestication of corn by 1,500 years or more. Not until the Abejas phase, nearly 2,000 years later (marked by sizable obsidian cores and blades and by grinding implements that closely resemble the modern mano and metate), do the earliest village sites appear. More than 1,000 years later, in the Ajalpán phase, earplugs for personal adornment occur. The grooved, withe-bound stone near the top is a pounder for making bark cloth.

EVOLUTION OF CORN at Tehuacán starts (*far left*) with a fragmentary cob of wild corn of 5000 B.C. date. Next (*left to right*) are an early domesticated cob of 4000 B.C., an early hybrid variety of 3000 B.C. and an early variety of modern corn of 1000 B.C. Last (*far right*) is an entirely modern cob of the time of Christ. All are shown four-fifths of natural size.

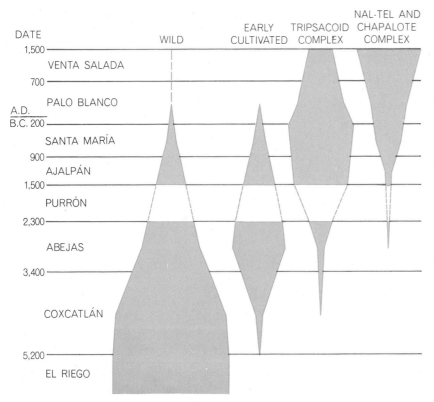

MAIN VARIETIES OF CORN changed in their relative abundance at Tehuacán between the time of initial cultivation during the Coxcatlán culture phase and the arrival of the conquistadors. Abundant at first, wild corn had become virtually extinct by the start of the Christian era, as had the early cultivated (but not hybridized) varieties. Thereafter the hybrids of the tripsacoid complex (produced by interbreeding wild corn with introduced varieties of corn-*Tripsacum* or corn-teosinte hybrids) were steadily replaced by two still extant types of corn, Nal-Tel and Chapalote. Minor varieties of late corn are not shown.

Foundation in Mexico, which is particularly interested in the origins of corn. The project eventually engaged nearly 50 experts in many specialties, not only archaeology and botany (including experts on many plants other than corn) but also zoology, geography, geology, ecology, genetics, ethnology and other disciplines.

The Coxcatlán cave, where the intensive new hunt had begun, turned out to be our richest dig. Working downward, we found that the cave had 28 separate occupation levels, the earliest of which may date to about 10,000 B.C. This remarkably long sequence has one major interruption: the period between 2300 B.C. and 900 B.C. The time from 900 B.C. to A.D. 1500, however, is represented by seven occupation levels. In combination with our findings in the Purrón cave, which contains 25 floors that date from about 7000 B.C. to 500 B.C., we have an almost continuous record (the longest interruption is less than 500 years) of nearly 12,000 years of prehistory. This is by far the longest record for any New World area.

All together we undertook major excavations at 12 sites in the valley of Tehuacán [*see bottom illustration on page 14*]. Of these only five caves—Coxcatlán, Purrón, San Marcos, Tecorral and El Riego East—contained .remains of ancient corn. But these and the other stratified sites gave us a wealth of additional information about the people who inhabited the valley over a span of 12,000 years. In four seasons of digging, from 1961 through 1964, we reaped a vast archaeological harvest. This includes nearly a million individual remains of human activity, more than 1,000 animal bones (including those of extinct antelopes and horses), 80,000 individual wild-plant remains and some 25,000 specimens of corn. The artifacts arrange themselves into significant sequences of stone tools, textiles and pottery. They provide an almost continuous picture of the rise of civilization in the valley of Tehuacán. From the valley's geology, from the shells of its land snails, from the pollen and other remains of its plants and from a variety of other relics our group of specialists has traced the changes in climate, physical environment and plant and animal life that took place during the 12,000 years. They have even been able to tell (from the kinds of plant remains in various occupation levels) at what seasons of the year many of the floors in the caves were occupied.

Outstanding among our many finds was a collection of minuscule corncobs that we tenderly extracted from the lowest of five occupation levels at the San Marcos cave. They were only about 20 millimeters long, no bigger than the filter tip of a cigarette [*see top illustration on opposite page*], but under a magnifying lens one could see that they were indeed miniature ears of corn, with sockets that had once contained kernels enclosed in pods. These cobs proved to be some 7,000 years old. Mangelsdorf is convinced that this must be wild corn—the original parent from which modern corn is descended.

Cultivated corn, of course, cannot survive without man's intervention; the dozens of seeds on each cob are enveloped by a tough, thick husk that prevents them from scattering. Mangelsdorf has concluded that corn's wild progenitor probably consisted of a single seed spike on the stalk, with a few pod-covered ovules arrayed on the spike and a pollen-bearing tassel attached to the spike's end [*see bottom illustration at right*]. The most primitive cobs we unearthed in the valley of Tehuacán fulfilled these specifications. Each had the stump of a tassel at the end, each had borne kernels of the pod-popcorn type and each had been covered with only a light husk consisting of two leaves. These characteristics would have allowed the plant to disperse its seeds at maturity; the pods would then have protected the seeds until conditions were appropriate for germination.

The people of the valley of Tehuacán lived for thousands of years as collectors of wild vegetable and animal foods before they made their first timid efforts as agriculturists. It would therefore be foolhardy to suggest that the inhabitants of this arid highland pocket of Mexico were the first or the only people in the Western Hemisphere to bring wild corn under cultivation. On the contrary, the New World's invention of agriculture will probably prove to be geographically fragmented. What can be said for the people of Tehuacán is that they are the first whose evolution from primitive food collectors to civilized agriculturists has been traced in detail. As yet we have no such complete story either for the Old World or for other parts of the New World. This story is as follows.

From a hazy beginning some 12,000 years ago until about 7000 B.C. the people of Tehuacán were few in number. They wandered the valley from season to season in search of jackrabbits, rats, birds, turtles and other small animals, as well as such plant foods as be-

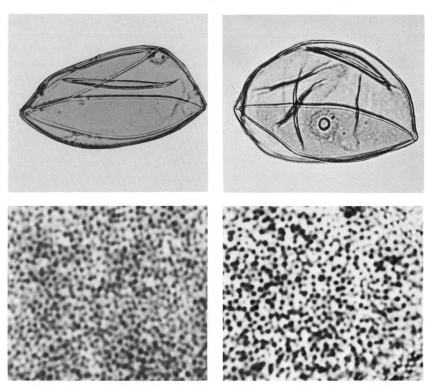

ANTIQUITY OF CORN in the New World was conclusively demonstrated when grains of pollen were found in drilling cores taken from Mexico City lake-bottom strata estimated to be 80,000 years old. Top two photographs (*magnification 435 diameters*) compare the ancient corn pollen (*left*) with modern pollen (*right*). Lower photographs (*magnification 4,500 diameters*) reveal similar ancient (*left*) and modern (*right*) pollen surface markings. The analysis and photographs are the work of Elso S. Barghoorn of Harvard University.

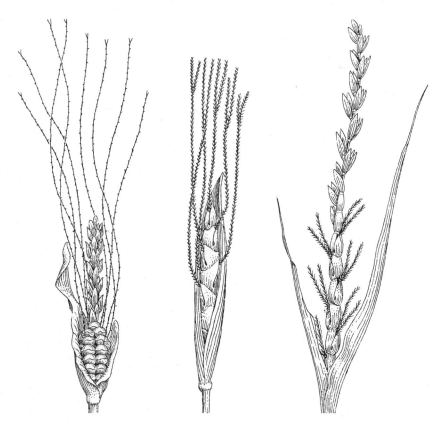

THREE NEW WORLD GRASSES are involved in the history of domesticated corn. Wild corn (*reconstruction at left*) was a pod-pop variety in which the male efflorescence grew from the end of the cob. Teosinte (*center*) and *Tripsacum* (*right*) are corn relatives that readily hybridized with wild and cultivated corn. Modern corn came from such crosses.

came available at different times of the year. Only occasionally did they manage to kill one of the now extinct species of horses and antelopes whose bones mark the lowest cave strata. These people used only a few simple implements of flaked stone: leaf-shaped projectile points, scrapers and engraving tools. We have named this earliest culture period the Ajuereado phase [*see illustration on pages 16 and 17*].

Around 6700 B.C. this simple pattern changed and a new phase—which we have named the El Riego culture from the cave where its first evidences appear—came into being. From then until about 5000 B.C. the people shifted from being predominantly trappers and hunters to being predominantly collectors of plant foods. Most of the plants they collected were wild, but they had domesticated squashes (starting with the species *Cucurbita mixta*) and avocados, and they also ate wild varieties of beans, amaranth and chili peppers. Among the flaked-stone implements, choppers appear. Entirely new kinds of stone tools—grinders, mortars, pestles and pounders of polished stone—are found in large numbers. During the growing season some families evidently gathered in temporary settlements, but these groups broke up into one-family bands during the leaner periods of the year. A number of burials dating from this culture phase hint at the possibility of part-time priests or witch doctors who directed the ceremonies involving the dead. The El Riego culture, however, had no corn.

By about 5000 B.C. a new phase, which we call the Coxcatlán culture,

had evolved. In this period only 10 percent of the valley's foodstuffs came from domestication rather than from collecting, hunting or trapping, but the list of domesticated plants is long. It includes corn, the water-bottle gourd, two species of squash, the amaranth, black and white zapotes, the tepary bean (*Phaseolus acutifolius*), the jack bean (*Canavalia ensiformis*), probably the common bean (*Phaseolus vulgaris*) and chili peppers.

Coxcatlán projectile points tend to be smaller than their predecessors; scrapers and choppers, however, remain much the same. The polished stone implements include forerunners of the classic New World roller-and-stone device for grinding grain: the mano and metate. There was evidently enough surplus energy among the people to allow the laborious hollowing out of stone water jugs and bowls.

It was in the phase following the Coxcatlán that the people of Tehuacán made the fundamental shift. By about 3400 B.C. the food provided by agriculture rose to about 30 percent of the total, domesticated animals (starting with the dog) made their appearance, and the people formed their first fixed settlements—small pit-house villages. By this stage (which we call the Abejas culture) they lived at a subsistence level that can be regarded as a foundation for the beginning of civilization. In about 2300 B.C. this gave way to the Purrón culture, marked by the cultivation of more hybridized types of corn and the manufacture of pottery.

Thereafter the pace of civilization in

the valley speeded up greatly. The descendants of the Purrón people developed a culture (called Ajalpán) that from about 1500 B.C. on involved a more complex village life, refinements of pottery and more elaborate ceremonialism, including the development of a figurine cult, perhaps representing family gods. This culture led in turn to an even more sophisticated one (which we call Santa María) that started about 850 B.C. Taking advantage of the valley's streams, the Santa María peoples of Tehuacán began to grow their hybrid corn in irrigated fields. Our surveys indicate a sharp rise in population. Temple mounds were built, and artifacts show signs of numerous contacts with cultures outside the valley. The Tehuacán culture in this period seems to have been strongly influenced by that of the Olmec people who lived to the southeast along the coast of Veracruz.

By about 200 B.C. the outside influence on Tehuacán affairs shifted from that of the Olmec of the east coast to that of Monte Alban to the south and west. The valley now had large irrigation projects and substantial hilltop ceremonial centers surrounded by villages. In this Palo Blanco phase some of the population proceeded to full-time specialization in various occupations, including the development of a salt industry. New domesticated food products appeared—the turkey, the tomato, the peanut and the guava. In the next period—Venta Salada, starting about A.D. 700—Monte Alban influences gave way to the influence of the Mixtecs. This period saw the rise of true

COXCATLÁN CAVE BURIAL, dating to about A.D. 100, contained the extended body of an adolescent American Indian, wrapped in a pair of cotton blankets with brightly colored stripes. This bundle in turn rested on sticks and the whole was wrapped in bark cloth.

cities in the valley, of an agricultural system that provided some 85 percent of the total food supply, of trade and commerce, a standing army, large-scale irrigation projects and a complex religion. Finally, just before the Spanish Conquest, the Aztecs took over from the Mixtecs.

Our archaeological study of the valley of Tehuacán, carried forward in collaboration with workers in so many other disciplines, has been gratifyingly productive. Not only have we documented one example of the origin of domesticated corn but also comparative studies of other domesticated plants have indicated that there were multiple centers of plant domestication in the Americas. At least for the moment we have at Tehuacán not only evidence of the earliest village life in the New World but also the first (and worst) pottery in Mexico and a fairly large sample of skeletons of some of the earliest Indians yet known.

Even more important is the fact that we at last have one New World example of the development of a culture from savagery to civilization. Preliminary analysis of the Tehuacán materials indicate that the traditional hypothesis about the evolution of high cultures may have to be reexamined and modified. In southern Mexico many of the characteristic elements of the Old World's Neolithic Revolution fail to appear suddenly in the form of a new culture complex or a revolutionized way of life. For example, tools of ground (rather than chipped) stone first occur at Tehuacán about 6700 B.C., and plant domestication begins at least by 5000 B.C. The other classic elements of the Old World Neolithic, however, are slow to appear. Villages are not found until around 3000 B.C., nor pottery until around 2300 B.C., and a sudden increase in population is delayed until 500 B.C. Reviewing this record, I think more in terms of Neolithic "evolution" than "revolution."

Our preliminary researches at Tehuacán suggest rich fields for further exploration. There is need not only for detailed investigations of the domestication and development of other New World food plants but also for attempts to obtain similar data for the Old World. Then—perhaps most challenging of all —there is the need for comparative studies of the similarities and differences between evolving cultures in the Old World and the New to determine the hows and whys of the rise of civilization itself.

SOPHISTICATED FIGURINE of painted pottery is one example of the artistic capacity of Tehuacán village craftsmen. This specimen, 2,900 years old, shows Olmec influences.

3

FOREST CLEARANCE IN THE STONE AGE

JOHANNES IVERSEN
March 1956

Perhaps the greatest single step forward in the history of mankind was the transition from hunting to agriculture. In the Mesolithic Age men lived by the spear, the bow and the fishing net; in the Neolithic Age they became farmers. The change came independently at different times in diverse parts of the world. Just how and when men turned to farming in Western Europe has been a subject of debate among naturalists and archaeologists for a hundred years. New methods of dating the implements of Stone Age men have recently given more factual substance to the debate. What is more, we have learned enough about the world in which they lived to test our theories about how they lived by experiment. This is a report of a set of experiments by which a group of scientists in Denmark attempted to re-enact some aspects of the hunting-to-agriculture chapter of mankind's past.

Denmark has unearthed relics of both stages—the bone and flint implements of the Mesolithic hunters and the polished stone axes of the Neolithic farmers. And in ancient lake sediments and bogs the prehistoric tools lie in recognizable strata of pollen, that marvelous dating instrument which identifies each period by its prevailing vegetation. The pollen record, as ecologists read it, tells the following story.

Toward the end of the last ice age, when vegetation was emerging and the country was still open, hunters ranged all over Denmark. Then, as forests grew dense and reduced large game, men abandoned the forested interior and retreated to the coast, where they made their living by fishing and seal-hunting. This state of affairs continued for thousands of years, until man suddenly appears in the forest, hacking out a new living. Clearings are hewed in the primeval forest. Tree pollen rapidly declines in certain regions, and we find in its place a sharp rise in pollen of herbaceous plants and the emergence of cereals and new weeds, notably plantain—the plant which the Indians of North America called "the footsteps of the white man."

Very shortly a new growth of tree species which typically follow forest clearance—willow, aspen, birch—springs up. The presence of birch strongly suggests that man used fire to help clear the forest, for on fertile soil birch succeeds

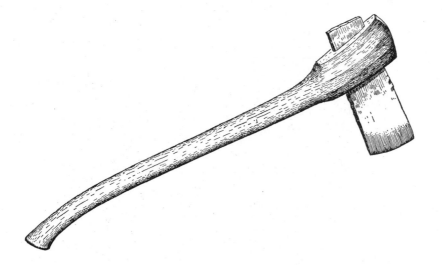

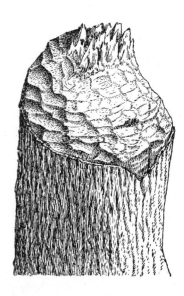

STONE AXE was reconstructed by mounting the Neolithic flint head on a copy of a Neolithic haft preserved at the bottom of a bog. It was found that the full swing of the modern woodsman often chipped or broke the head. Using short, rapid strokes, the experimenters learned to fell trees more than a foot in diameter in 30 minutes. To fell small trees they chopped all the way around the trunk.

a mixed oak forest only after burning. Meanwhile the ground flora undergoes a radical and significant change. Grasses, white clover, sheep sorrel, sheep's-bit and other pasture plants take the upper hand. We can visualize cattle grazing and browsing in grassy meadows bordered by scrub forests of birch and hazel.

Finally comes a third phase. The grasses, birch and eventually hazel decline, and a big-tree forest takes over once more. Oak now is more dominant than before; elm and linden never recover the strength they had in the primeval forest.

All this seems to mean that men cleared large areas of the original forest with axes, burned over the clearings, planted small fields of cereals and used the rest for pasturing animals. Their colonization was of short duration: when the forest grew back, they moved on to clear a suitable new area. According to the pollen record, some of their settlements can scarcely have lasted more than 50 years.

Now this is a neat, tidy theory, but there are troublesome questions. Could Neolithic man really have cleared large areas of the thick primeval forest with his crude flint axes? Could he have burned off the felled trees and shrubs in his clearings? Our team of ecologists and archaeologists decided to put these questions to the test of field experiment. We obtained the needed funds and permission to clear a two-acre area in the Draved Forest of Denmark, which is a mixed oak forest like that of Neolithic times.

Two archaeologists, Jörgen Troels-Smith and Svend Jörgensen, took charge of the axe tests. They were able to obtain a number of Neolithic flint axe blades from the National Museum in Copenhagen, and a model for the wooden haft was available in the form of the famous Sigerslev hafted axe excavated from a Danish bog. In Neolithic axes, whose hafts were of ash wood, the blade was inserted in a rectangular hole in the haft [see drawing on opposite page]. Jörgensen and Troels-Smith demonstrated that if the haft was not to be split, it must not hold the blade too tightly but must leave room for a little sidewise play of the blade when it struck.

After making a number of hafted axes fitted with Stone Age man's blades, the two archaeologists, together with two professional lumberjacks, went forth into the forest in September, 1952. When the party attacked the trees, it soon became

TREES WERE BURNED by covering them with brushwood and igniting a 30-foot strip. When the strip was almost burned out, the larger logs were used to light the next one.

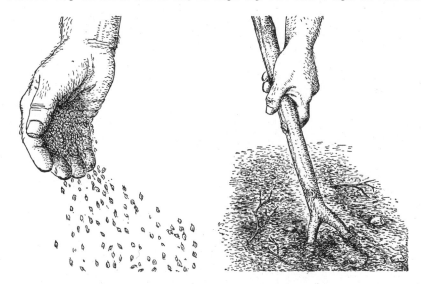

SEED WAS SOWN by hand in the still-warm ash (*left*). Then the seed bed was raked with a forked stick (*right*). The plants sown were barley and two primitive varieties of wheat.

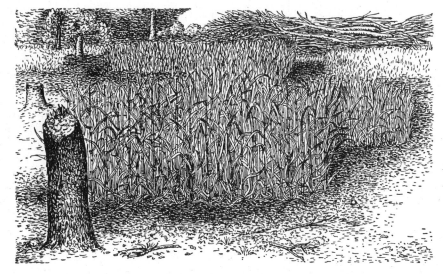

BARLEY HAD GROWN to this height six weeks after it had been sown in the ash of the burned brushwood and trees. Barley sown in plots not covered with ash grew very poorly.

apparent that the usual tree-chopping technique, in which one puts his shoulders and weight into long, powerful blows, would not do. It often shattered the edge of the delicate flint blade or broke the blade in two. The lumberjacks, unable to change their habits, damaged several axes. The archaeologists soon discovered that the proper way to use the flint axe was to chip at the tree with short, quick strokes, using mainly the elbow and wrist. Troels-Smith, working with an axe blade which had not been sharpened since the Stone Age, employed it effectively throughout the whole clearing operation without damaging it.

When the two archaeologists reached

peak form, they were able to fell oak trees more than a foot in diameter within half an hour. Small trees they dropped by cutting all around the trunk; on substantial-sized ones they used the slower method of hewing through notches on the opposite sides, in order to control the direction of fall. We realized that for clearing purposes it would be advantageous to have all the trunks lying in the north-south direction; for example, the wood would dry more quickly.

In this manner we cleared the two acres of forest, letting the largest trees stand but killing them by cutting rings through the sapwood. Troels-Smith and Jörgensen concluded that Neolithic men could have cut large clearings in the

forests with their flint axes without great difficulty.

The next problem was to learn how they might have burned off their clearings. For help in this phase of the experiment we called on Kustaa Vilkuna of the University of Helsinki, who is an expert on primitive burning techniques which were still being used quite recently by farmers in the spruce forests of Finland.

Without waiting for the wood to dry, we first tested two burning methods, one modern, the other primitive. The modern method, though effective in forests of conifers, failed completely in our deciduous forest. The primitive method, how-

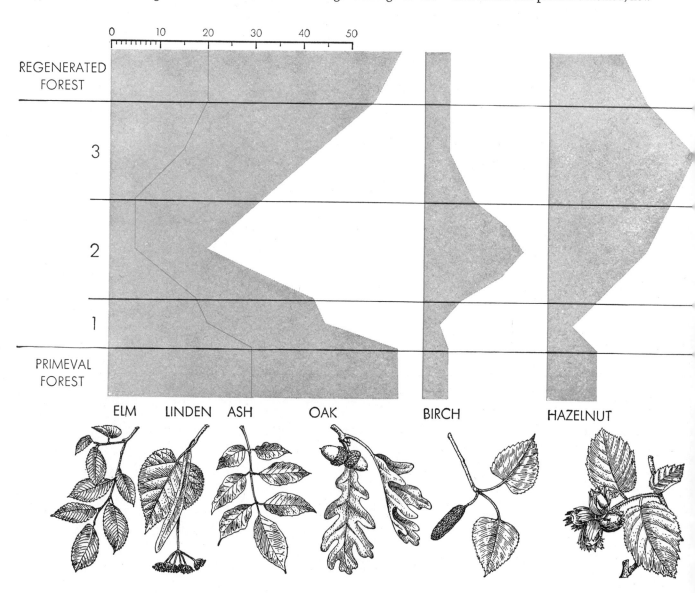

POLLEN DIAGRAM shows the effect of forest clearance on the vegetation of Denmark between about 2500 B.C. and 2300 B.C. The diagram is based on many samples of pollen taken by boring down into bogs. The width of each colored area on the diagram represents the proportion of pollen from one species in comparison to that from all others. The scale of the proportions is given at the

ever, was successful, and we proceeded to use it in the clearing in May of 1954, after the felled trees had had more than a year to dry. Brushwood and branches cut from the trees were spread over the area to be burned. Then this material was ignited along a 30-foot-wide belt by means of torches of burning birch bark attached to stakes. When the belt was well cleared, we pushed its still burning logs forward with long poles to set fire to the adjacent area. In this way we burned off the tangle of felled vegetation belt by belt. The fire was controlled carefully, day and night, to achieve an even and thorough burning of the ground. It was rather hard work, as oak wood burns slowly, but there were no serious diffi-

culties, and in three or four days the job was finished. We burned only half of the two-acre clearing, because we wished to compare the subsequent growth on burned and unburned ground.

Immediately after the burning we sowed part of the area with primitive varieties of wheat (einkorn and emmer) and naked barley. That these cereals were grown in Denmark by Neolithic man is shown by grain impressions on excavated pottery. Axel Steensberg, an expert on agricultural methods, old and new, obtained seeds of the cereals from botanical collections and directed our agriculture.

We spread the seeds on the ground, raked them in with a forked branch, and

waited for the harvest. For comparison we sowed two sets of plots—one burned and one unburned but hoed and weeded. The contrast in results was remarkable. On the unburned ground the grain scarcely grew at all. Evidently the rather acid forest soil was not suited to cereal growing. But the burned ground produced a luxuriant crop (which Steensberg harvested, in Neolithic fashion, with a flint knife and a flint sickle). The success of the cereals in this ground was due in part to sweetening of the soil by the wood-ash and the absence of competition from other vegetation, but the burning may also have created other beneficial factors, and we are now investigating this matter. In any case,

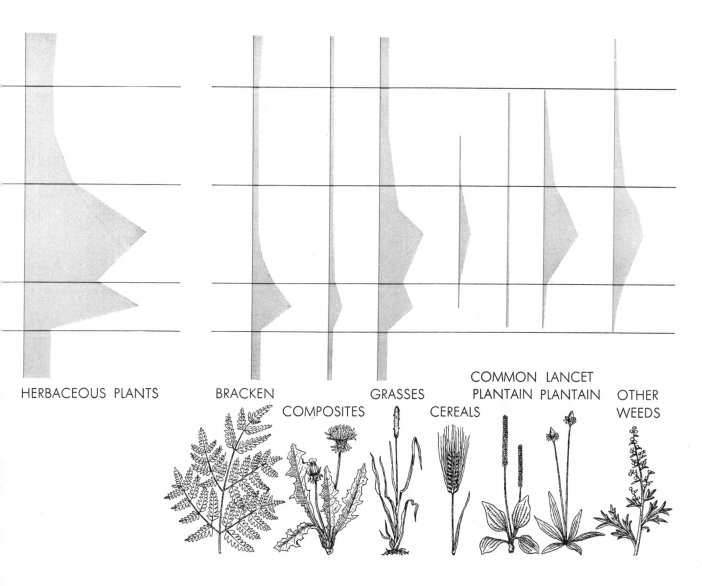

HERBACEOUS PLANTS   BRACKEN   GRASSES   COMMON LANCET PLANTAIN PLANTAIN   OTHER WEEDS
                                    COMPOSITES   CEREALS

upper left. In the primeval forest (*colored areas below the bottom horizontal line*) the distribution of pollens was 30 per cent elm, linden and ash; 30 per cent oak; 5 per cent birch; 10 per cent hazel, **and so on. During the three stages of forest clearance (*1, 2 and 3*) the distribution of pollens changed. The distribution of herb pollens is shown at the right of the break in the horizontal lines.**

NEW COMMUNITY OF WILD PLANTS grew up in the parts of the clearing that had been burned over. At the left is a species of fern called bracken. Second from the left is hazel. Both of these plants had been present in the original forest. They grew up again

whatever the factors are, they are short-lived, for the second year the burned plots yielded much smaller crops.

Now, two years after the clearing and burning, we are in the process of watching developments in the early recovery of natural plant growth. The burned and unburned areas are developing quite differently.

In the area cleared of trees but unburned, events are following an unsurprising and unexciting course. The ground vegetation consists mainly of the species that grew there before the clearing, though it is growing more luxuriously because it has more sunlight. Bracken (ferns), always abundant in this part of the world, is flourishing far more richly than when it was shaded. Grasses and sedges have increased.

The burned ground, on the other

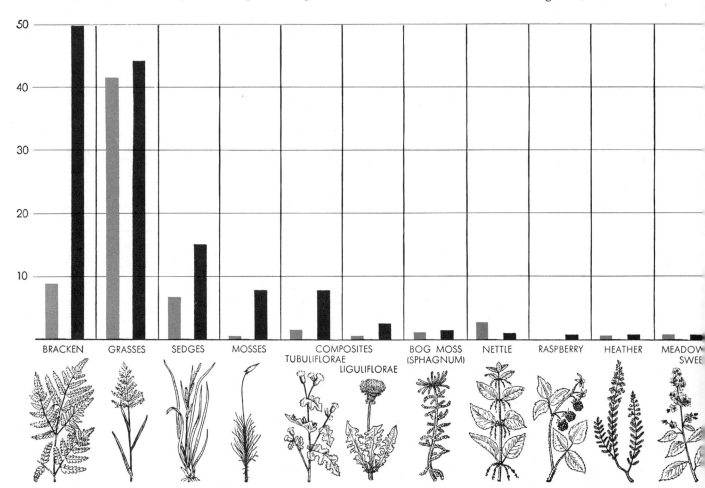

NEOLITHIC COMMUNITY OF WILD PLANTS that followed clearance and cultivation is analyzed in this pollen diagram. The colored bars indicate the amount of pollen from each plant before clearance. The black bars indicate the amount of pollen after clear-

from relatively deep roots. Third from the left are dandelions, members of a family which grows in profusion under such condi-

tions. Fourth are mosses, which had never been seen in this forest before. Their spores were blown into the clearing on the wind.

hand, is a scene of botanical revolution. Bracken is coming back here too, but most of the other old plants, having shallower roots, were killed off by the fire. In their stead we have a whole garden of new plants. Plantain has made its appearance, just as it does in the ancient

pollen record after forest clearance. There is a profusion of members of the family *Compositae*, including dandelions, daisies, sow thistle and so forth. (These plants do not bulk large in the fossil pollen record, but that is understandable because they are pollinated by insects rather than by the wind.)

A particularly interesting development is the sudden appearance of mosses and their spread over large patches of the burned area. The main species have never been seen in this forest before. Their spores have flown into the clearing on the wind, and no doubt mosses came the same way to the areas burned by Neolithic man. What makes them especially significant is that certain mosses seem to be definite indicators of fire; three species have been so identified in America, and sure enough the same three appeared in our burned clearing. Since the moss phase in a burned forest must be ephemeral, moss spores in the fossil record should enable us to pinpoint the dates of forest clearance by Neolithic man and to learn whether they burned the same clearing more than once during the existence of a continuous settlement. Unfortunately the small moss spores are difficult to recognize, and analysts of the ancient pollen deposits have not counted them hitherto. We made a small test count at the site of a Neolithic forest clearing in Denmark, analyzing the layers representing the time of the clearance and the period just before. According to our fragmentary count, there was a sharp rise in general moss growth (we made no attempt to distinguish individual species) immediately after the clearance of the area [*see chart at the left*].

Our experimental clearing in the

Danish forest is just beginning to pass into the second phase, when pioneer trees appear and the regeneration of the forest commences. Birch seedlings are starting to spring up in profusion; willow seedlings have appeared; and hazel, aspen and linden shoots are rising from roots that were not killed by the fire. We are looking forward to studying this gradual regeneration in the years to come, as well as to reliving the stage in Neolithic farming when men grazed their cattle on the re-emerging ground vegetation.

Meanwhile we can say that so far our experiment has confirmed the archaeological interpretation of the pollen record on several important counts. It has been demonstrated that the forest could indeed have been cleared by the primitive tools of Neolithic man, and that in the first stage at least the reviving vegetation follows a course very like that deduced from the ancient pollen layers.

Of course man's transition from hunting to farming may well have taken other paths besides the one we have traced in the Danish clearings. More than one type of agriculture may have existed simultaneously in Denmark. As a matter of fact, Troels-Smith has found evidences of a more primitive agriculture during the same period on the Danish coast, where the Middle Stone Age men apparently cleared no forests but practiced a little crude farming along with their hunting and fishing.

The Neolithic farming culture described in this article is so much more advanced, and begins so suddenly, that it seems to signal the arrival and invasion of a vigorous new people from another region.

RREL   PARSLEY FAMILY   COMMON PLANTAIN   LANCET PLANTAIN

ance. The scale at the left is based on grains of pollen per 1,000 grains of tree pollen.

4

THE CHINAMPAS OF MEXICO

MICHAEL D. COE
July 1964

When the Spanish conquistadores entered Mexico in 1519, they found most of the peoples in the region unwillingly paying tribute to the emperor of the Aztecs, who ruled from a shimmering island capital in a lake on the site of modern Mexico City. Less than 200 years earlier the Aztecs had been a small, poor, semibarbaric tribe that had just settled in the area after centuries of wandering in search of a home. Shortly after their arrival they fought with their neighbors and were obliged to retreat to two small islands in the lake. There they adopted a unique form of land reclamation and agriculture known as the chinampa system. This system, which had long been practiced on the margins of the lake, was one of the most intensive and productive methods of farming that has ever been devised. It provided the Aztecs with land to live on and with the first surplus of food they had ever known. Their new wealth enabled them to create a standing army that soon subjugated nearby peoples. Driven by the demands of their sun-god for sacrificial captives, and supported by chinampa agriculture (which was also practiced by some of their vassals), the Aztecs quickly expanded their empire throughout Mexico.

The Spaniards toppled the Aztecs within two years and razed their magnificent pyramid temples, but the chinampa system has persisted to the present. Now, after enduring for perhaps 2,000 years, it too appears to be facing extinction.

Chinampas are long, narrow strips of land surrounded on at least three sides by water. Properly maintained, they can produce several crops a year and will remain fertile for centuries without having to lie fallow. The important role

they have played in the long history of Mexico is probably unknown to the *chinamperos* who tend them and to the many tourists who visit the most famous chinampa center: the town of Xochimilco south of Mexico City.

In Xochimilco the guides relate the charming story that chinampas are, or once were, "floating gardens." This is a tall tale that goes back at least to 1590, when a Father Acosta included it in his *Natural and Moral History of the Indies:* "Those who have not seen the seed gardens that are constructed on the lake of Mexico, in the midst of the waters, will take what is described here as a fabulous story, or at best will believe it to be an enchantment of the devil, to whom these people paid worship. But in reality the matter is entirely feasible. Gardens that move on the water have been built by piling earth on sedges and reeds in such a manner that the water does not destroy them, and on these gardens they plant and cultivate, and plants grow and ripen, and they tow these gardens from one place to another."

Acosta may have been deceived by the rafts of water vegetation that even today are towed to the chinampas and dragged onto them as compost. The real interest of a chinampa town such as Xochimilco lies not in its fables and its tourist attractions—flower-garlanded boats plying canals, waterborne mariachi bands and floating soft-drink purveyors—but in the problem of the nature and origin of the chinampas and the relation of this form of agriculture to the rise of the pre-Columbian civilizations of central Mexico.

The chinampa zone is located in the Valley of Mexico, a landlocked basin entirely surrounded by mountains of

volcanic origin. The valley, which is a mile and a half above sea level, has an extent of some 3,000 square miles. In pre-Spanish times a sheet of water, called by the Aztecs the Lake of the Moon, covered a fourth of the valley during the rainy summer season. In the dry winter season evaporation reduced this shallow body of water to five separate lakes: Zumpango on the north, Xaltocán and Texcoco in the center and Xochimilco and Chalco on the south [*see illustration on opposite page*]. The last two were really a single lake divided by an artificial causeway. Villages were established in the valley sometime late in the second millennium B.C.; since then the valley has supported dense populations of farmers. During the first or second century A.D. the populous city of Teotihuacán, which covered at least eight square miles at the northeastern edge of the valley, came to dominate the region. Although Teotihuacán was overthrown as long ago as A.D. 600, its enormous pyramid temples still stand. The last, most powerful and best known of the civilized states of the valley before the arrival of the Spanish was the empire of the Aztecs, centered on the island of Tenochtitlán-Tlatelolco in the western part of Lake Texcoco.

Since the Spanish conquest in 1521 man has drastically changed the valley. In the colonial era the water was partly drained in the course of reclaiming land for agriculture. Far more, however, was removed by a great tunnel bored through the mountains to the north in 1900, during the rule of Porfirio Díaz. The valley has been further dried out by the tapping of springs and digging of wells to provide water for the rapid growth of Mexico City. Of the estimated six billion cubic meters of water avail-

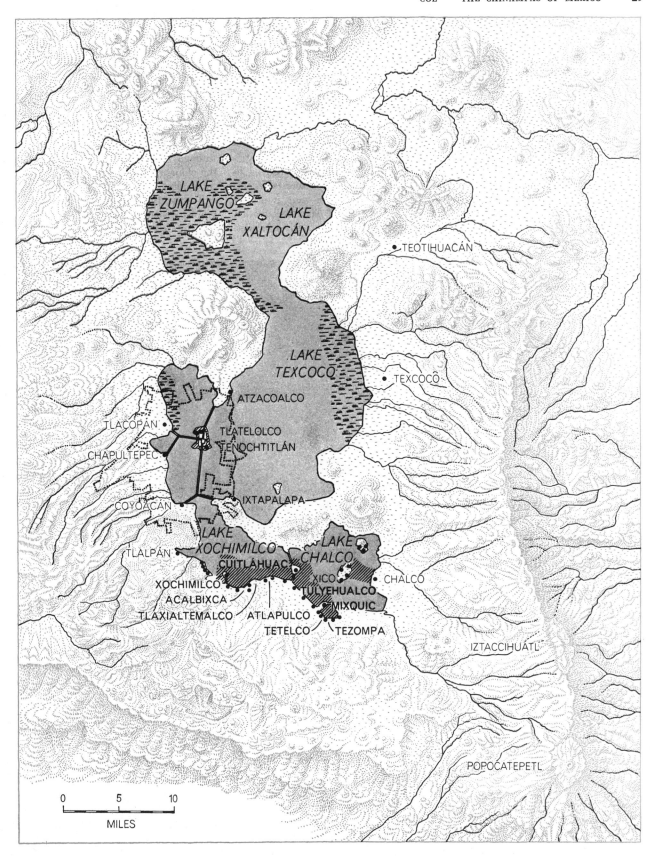

CHINAMPA AREAS (*hatched*) and the Valley of Mexico are shown as they appeared in summer at the time of the Spanish conquest in 1521. In the rainy summer season the five lakes coalesced into one large lake: the Lake of the Moon. Tenochtitlán-Tlatelolco was the Aztec capital. The dotted line marks the limits of modern Mexico City. The broken line between Atzacoalco and Ixtapalapa shows the location of the great Aztec dike that sealed off and protected the chinampas from the salty water of Lake Texcoco. Causeways and aqueducts leading to the Aztec capital are also shown. The names of the nine chinampa towns that remain today are given in heavy type. The large black dots without names are the sites of the freshwater springs that fed the chinampa zones.

able in the valley each year, 744 million cubic meters is consumed by the urban population. Most of the rest evaporates. As a result only isolated puddles of the Lake of the Moon remain, including parts of Lake Texcoco and Lake Xochimilco. Dehydration has so weakened the underlying sediments that the larger buildings of Mexico City are sinking at the rate of about a foot a year.

The removal of the water has also had a disastrous effect on the chinampas. From ancient times down to the past century or so many chinampa towns—small urban centers surrounded by the lovely canals and cultivated strips—existed on the western and southern margins of the old Lake of the Moon. Today only nine remain, and eight of them are probably doomed. Xochimilco alone may endure because of its importance as a tourist center.

In a masterly study of Xochimilco published in 1939 the German geographer Elizabeth Schilling established to the satisfaction of most interested scholars that the chinampa zone is an example of large-scale land reclamation through drainage. Recently detailed aerial photographs have confirmed her judgment. These show Xochimilco to be a network of canals of various widths laid out generally at right angles to one another to form a close approximation of a grid. This could not have been achieved by a random anchoring of "floating gardens." Departures from the pattern have probably come about through destruction and rebuilding of the chinampas, which are easily ruined by flooding and neglect.

To the trained observer the photographs reveal carefully planned canals that drained the swampy southern shore

CHINAMPA GARDENS and canals that surround each of them on at least three sides form a grid pattern in this vertical air view. The grid "tilts" about 16 degrees east of north. Many of the canals that appear to be silted up are simply covered with waterweeds. Part of the town of Xochimilco, south of Mexico City, is at lower left. First canals were dug 2,000 years ago to drain swampy areas.

of Lake Xochimilco, where water flowing in from numerous springs had been held in the spongy soil. Here the water table was higher than the surface of the open lake to the north. The canals permitted the spring water to flow freely into Lake Xochimilco and thence into Lake Texcoco, which was deeper. The peaty sediments then released much of the trapped water. Mud dug out in making the canals was piled between them, adding height to the narrow islands and peninsulas that constitute the chinampas. The sides of the garden plots were held in place by posts and by vines and branches woven between them. Later living willow trees replaced many of these wattle walls. Until a few decades ago the water flowed out of Lake Xochimilco into Lake Texcoco through the willow-bordered Canal de la Viga, which carried native women to the market of Mexico City in canoes laden with the rich produce of Xochimilco. Now abandoned, the canal is largely silted up.

In many ways this remarkable drainage project resembled land-reclamation schemes elsewhere, such as those in the fens of eastern England or the polders of the Netherlands. It was unique, however, in the kind of farm plots that resulted, in the technique of their cultivation and in their enormous productivity. Each chinampa is about 300 feet long and between 15 and 30 feet wide. The surrounding canals serve as thoroughfares for the flat-bottomed canoes of the farmers. Ideally the surface of the garden plot is no more than a few feet above the water. Before each planting the *chinamperos*, using a canvas bag on the end of a long pole, scoop rich mud from the bottom and load it into their canoes. The mud is then spread on the surface of the chinampas. In the wet season (June through October) water held in the chinampa provides enough moisture for the crops; toward the end of the dry season, when the canals are lower, the plots must be watered. After a number of years the surface of a chinampa is raised too high by the repeated application of mud and must be lowered by excavation. The surplus soil is often removed to a new or rebuilt chinampa.

New chinampas are made, naturally enough, by cutting new canals, which today is accomplished with power dredges. Older plots that have fallen into disrepair are often reconditioned. In both operations rafts of water vegetation are cut from the surface of the canals, towed to the plot and dragged

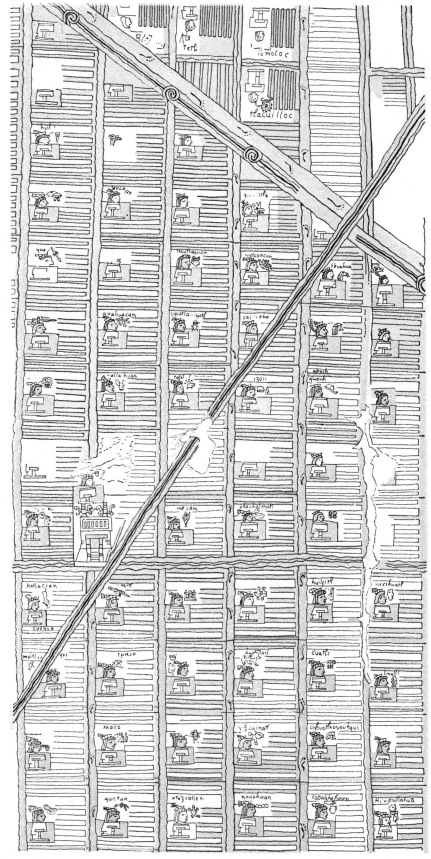

ANCIENT AZTEC MAP of a portion of Tenochtitlán-Tlatelolco shows that it was a chinampa city. Six to eight plots are associated with each house. Profile of the householder and his name in hieroglyphs and in Spanish script appear above each house. Footprints indicate a path between plots or beside a canal. This is a copy of a small part of the damaged map, which is in the National Museum of Anthropology in Mexico City.

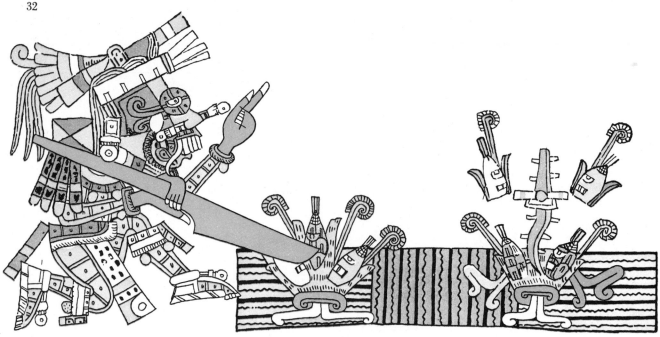

DIGGING WITH A "COA," the cultivating stick of the ancient Mexicans, the rain-god tills magic maize. The drawing is copied from a late preconquest Mexican religious work. The *coa* is considerably broader near the digging end than it is toward the handle.

into place one on top of another until they reach the desired height. After that they are covered with the usual mud. Thus each plot has its own built-in compost heap.

An essential element in chinampa farming is the technique of the seed nursery, which has been thoroughly investigated by the anthropologist Pedro Armillas of Southern Illinois University and the geographer Robert West of Louisiana State University. The nursery, at one end of the chinampa near a canal, is made by spreading a thick layer of mud over a bed of waterweeds. After several days, when the mud is hard enough, it is cut into little rectangular blocks called *chapines*. The *chinampero* makes a hole in each *chapín* with a finger, a stick or a small ball of rag, drops in the seed and covers it with manure, which now comes from cattle but in Aztec days came from humans. For protection against the occasional winter frosts the seedbed is covered with reeds or old newspapers. During dry weather the sprouting plants are watered by hand. Finally each seedling is transplanted in its own *chapín* to a place on the chinampa, which has been cultivated and leveled with a spade or hoe (the Aztecs employed a

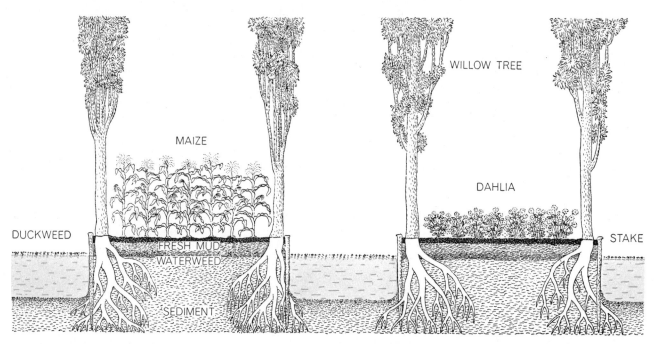

CROSS-SECTION DIAGRAM of chinampas and canals gives an idea of their construction. Fresh mud from bottom of canals and weeds for compost beneath the mud keep the chinampas fertile. Trees and stakes hold the sides of the chinampas firmly in place.

digging stick called a *coa*) and then covered with canal mud. The only crop for which the seedbed stage is not necessary is maize, which is planted directly in the chinampa.

The *chinamperos* report that they usually harvest seven different kinds of crop a year from each plot, of which two are maize. Crops raised today at Xochimilco include five varieties of maize, beans, chili peppers, tomatoes and two kinds of grain amaranth—all of which were cultivated before the Spanish arrived. Also grown are vegetables introduced from Europe, such as carrots, lettuce, cabbages, radishes, beets and onions. Xochimilco means "place of the flower gardens" in the Nahuatl language spoken by the Aztecs and still used today by the older people of the chinampa towns. The growing of flowers for sale goes back to the preconquest era, when flowers were offered on the altars of the pagan gods. Native species have imaginative Nahuatl names: *cempoaxóchitl* ("twenty flower," a marigold), *oceloxóchitl* ("jaguar flower"), *cacaloxóchitl* ("crow flower"). The gardens produce dozens of varieties of dahlia, the national flower of Mexico. European flowers include carnations, roses and lilies.

Carp and other fishes abound in the canals and are netted or speared by the *chinamperos*. Another inhabitant of the canals is the axolotl, a large salamander valued by zoologists as a laboratory animal and prized by the people of Xochimilco for its tender meat and lack of hard bones. Water birds were once caught in nets but are now scarce due to the indiscriminate use of firearms.

A basic question for the archaeologist and the historian is: How old are the chinampas? The traditional histories of the peoples of central Mexico list the Xochimilcas as one of eight tribes (the Aztecs were another) that came into the valley after a migration from a legendary home in the west. They were settled at Xochimilco by A.D. 1300 and were ruled by a succession of 17 lords. In 1352 and again in 1375 they were defeated by the Aztecs; finally, in the 15th century, they were incorporated into the Aztec state, which had absorbed the rest of the chinampa zone as well.

Some recent archaeological evidence makes it appear certain that Xochimilco, and by extension the other chinampa towns, existed long before the Xochimilcas arrived. A local newspaperman and booster of Xochimilco, José Farias

SEED NURSERY, made from small squares of rich mud, is an essential element of chinampa farming. Each square, or *chapín*, holds one seed and manure for it. When seedlings sprout, they will be transplanted in the *chapines* to places on the chinampa.

SCOOPING UP MUD from the bottom of the canal, the *chinamperos* load it into their canoe. They will spread the mud on the chinampa plot before setting out the new crop.

Galindo, has been collecting fragments of ancient pottery and clay figurines found by the *chinamperos* in the mud of the canals and in the garden plots. It is evident that such signs of human residence must postdate the initial digging of the canal system; until that had been done no one could have lived in the tangled marshes. Aztec bowls, dishes and figurines of gods and goddesses abound in Farias Galindo's collection, as might be expected from the many references to Xochimilco in Aztec documents. Of particular interest is the much older material that has been found. This includes a bowl of Coyotlatelco ware made between A.D. 600 and 900, heads broken from figurines of the Teotihuacán III culture, which flourished between A.D. 200 and 600, and Teotihuacán II figurine heads, which are as old as the beginnings of the great city of Teotihuacán in the first and second centuries A.D. Therefore it is likely that the chinampas of Xochimilco were planned and built almost 2,000 years ago.

Who was responsible? The only power in central Mexico at that time capable of such an undertaking was the growing Teotihuacán state, so that whoever built Teotihuacán also created the chinampas. Another piece of information points to the same conclusion. The grid of the Xochimilco canals is not oriented to the cardinal directions but to a point 15 to 17 degrees east of true north. So are the streets of the ruined city of Teotihuacán, and so are the grids of most of the other chinampa towns. We do not know why this is, but there were probably astrological reasons. It has been said that an urban civilization as advanced and as large as Teotihuacán must have been based on irrigation agriculture, but field archaeologists can find no trace of large-scale irrigation works. It seems far more likely that the growth of Teotihuacán was directly related to the establishment and perfection of the chinampas on the southern shore of the lake. Successive peoples and powers entered the valley and took advantage of the same system.

On the eve of the Spanish conquest Xochimilco was a flourishing island town under Aztec control, with at least 25,000 inhabitants—craftsmen as well as farmers. Cortes wrote of its "many towers of their idols, built of stone and mortar." The town, which was and still is on higher and drier ground than the chinampas, was approached from the south by a causeway crossing many canals. Its numerous wooden bridges could be raised to delay the approach of enemies. At the mainland end of the causeway was a large market; this is now the center of town. Xochimilco was divided into 18 *calpullis*, each with its own name. The Aztec institution of the *calpulli* is not well understood, but it seems to have been a local ward based on kinship. Their names survive today, and every *chinampero* knows to which ward he belongs. The *calpullis* were grouped into three larger units; the town as a whole was ruled by a native lord closely related to the Aztec emperor.

Wills, petitions and other documents filed early in the colonial period show that in Xochimilco land tenure, as well as the social system, was basically the same as that in the Aztec capital of Tenochtitlán-Tlatelolco. There were three categories of chinampa lands: (1) chinampas belonging to the *calpullis*, which could be used by a *calpulli* member to support himself and his family as long as he did not leave the land uncultivated for two years in succession; (2) office land, which belonged to the position filled by a noble official but not to him personally; (3) private land, which could be disposed of as the individual saw fit.

The island capital of the Aztecs was also surrounded by chinampas. The National Museum of Anthropology in Mexico City possesses a remarkable Aztec map on a large sheet of native paper made from the inner bark of a fig tree. This document, studied in recent years by Donald Robertson of Tulane University, shows a portion of the Aztec capital generally covering the section that is now buried under the railroad yards of Mexico City. In all likelihood it was drawn up as a tax record by Aztec scribes and used by bureaucrats into the period of Spanish domination. The similarity of the plan to that of modern Xochimilco is obvious. It shows a network of canals laid out in a grid, with the larger canals crossing the pattern diagonally. Roads

POTTERY FIGURINES of the types found in the chinampas and canals of Xochimilco were made in Aztec times, A.D. 1367 to 1521 (*top three*), during the Teotihuacán III period, A.D. 200 to 600 (*middle four*), and in the Teotihuacán II era, A.D. 1 to 200 (*bottom*). This ancient evidence of human occupation indicates that the chinampas are 2,000 years old.

and footpaths parallel the major canals, which the Spaniards said were crossed by wooden bridges.

The plan depicts some 400 houses, each with the owner's head in profile and his name in hieroglyphs. The Spanish later added Spanish transliterations of the names and also drawings of churches and other colonial structures. The property surrounding each house consists of six to eight chinampas. It was the cutting of canals and the construction of chinampas by the poor and hungry Aztecs who first came here in the 14th century A.D. that filled in the swampy land between the low rocky islands on which they had camped. The work eventually resulted in the coalescence and enlargement of the islands into the marvelous capital city that so impressed the conquistadores.

The more substantial houses of stone and mortar occupied the central sections of the capital, where the land was higher and firmer. In the very center were such large public buildings as the pyramid temples and the palaces of the emperor and his chief nobles. The bulk of the population was nonagricultural, consisting of priests, politicians, craftsmen, traders and soldiers. Nevertheless, Tenochtitlán-Tlatelolco was a chinampa city; the Spanish described it as another Venice. Thousands of canoes laden with people and produce daily plied the hundreds of canals, which were bright green with water vegetation. An Aztec poet has described the beauty of his native home:

The city is spread out in circles of jade,
Radiating flashes of light like quetzal plumes.
Beside it the lords are borne in boats:
Over them spreads a flowery mist.

The real basis of the native economy in the Valley of Mexico was the chinampa zone, which extended all the way from Tenochtitlán-Tlatelolco south to the shore of Lake Xochimilco and then east into Lake Chalco. The rest of the land in the valley, although it produced crops, was far less favorable to farming because of the arid climate. The chinampas, however, presented two difficult problems apart from those involved in their cultivation and day-to-day maintenance. One problem was to keep the water level high, the other was the prevention of floods.

The valley had no external outlet. Year after year over the millenniums nitrous salts had been swept down into

RUINS OF TEOTIHUACÁN, the large city that dominated much of Mexico from about A.D. 100 to 600, are still among the most impressive in Mexico. The rise of this great urban center may have been made possible by the development of the chinampas to the south.

the Lake of the Moon by the summer rains and had been concentrated by evaporation in the eastern part of Lake Texcoco. It was essential to keep the deadly salts away from the chinampas. For this reason the chinampas could only function properly if they were fed constantly by freshwater springs, which maintained the water level and held back the salt water. Such springs are found today in greatest abundance south of Lake Xochimilco, where chinampa towns still exist. Long ago there were adequate springs on the island of Tenochtitlán-Tlatelolco, but the rapid growth of the Aztec capital and its associated chinampas made the springs inadequate. The problem was solved by the construction of aqueducts to bring fresh water from mainland springs. It has sometimes been assumed that the sole purpose of the aqueducts was to carry drinking water to the inhabitants of the capital, but, as the ethnohistorian Angel Palerm of the Pan American Union has noted, their thirst must have been incredible.

These covered masonry watercourses were no mean structures. The first was

completed in the reign of Montezuma I (1440–1468); it brought water over a causeway from the west into the city from a large spring at the foot of Chapultepec hill. Cortes wrote that the flow was "as thick as a man's body." A second aqueduct was built by the emperor Ahuítzotl (1486–1502). For this aqueduct a spring at Coyoacán, on a point of land separating Lakes Texcoco and Xochimilco, was enlarged; the aqueduct ran along the causeway that led north to Tenochtitlán-Tlatelolco. Ahuítzotl's effort was initially crowned with disaster: the volume of water was so great that violent floods resulted. The flow of the spring diminished, it was recorded by pious Aztec chroniclers, only when the emperor sacrificed some high officials and had their hearts thrown into it, along with various valuable objects.

The second major problem of the chinampas—periodic flooding by salty water—was also finally solved by construction works. The nitrous salts, which had already made the waters of the eastern part of Lake Texcoco unsuitable for chinampas, rose and moved into the chinampa zone during the summer rains, in spite of the flow from the springs. The problem apparently became acute only in the Aztec period, when, according to the pollen chronology worked out by Paul B. Sears of Yale University, the climate of the region seems to have been wetter than at any time since the end of the last ice age. The floods nearly destroyed the entire economy of the Valley of Mexico. In the 15th century Nezahualcóyotl, the poet-king of Texcoco, supervised for his relative Montezuma I the construction of an enormous dike of stones and earth enclosed by stockades interlaced with branches. The dike, on which 20,000 men from most of the towns of the valley labored, extended 10 miles across the Lake of the Moon from Atzacoalco on the north to Ixtapalapa on the south. It sealed off the Aztec capital and the other chinampa towns from the rest of Lake Texcoco, leaving them in a freshwater lagoon. The three stone causeways connecting the capital with the mainland were pierced in several places and floodgates were installed to provide partial control of the water level in the lagoon.

The entire chinampa zone, then, represented a gigantic hydraulic scheme based on land drainage and the manipulation of water resources. The Aztecs refined and exploited it to establish a vast empire for the glory of their gods and the profit of their rulers. Defeated peoples were quickly organized as tributaries under the watchful eye of a local Aztec garrison and military governor. Twice a year they had to render a huge tribute to Tenochtitlán-Tlatelolco. The Aztec tribute list records that every year the capital received 7,000 tons of maize, 4,000 tons of beans and other foods in like quantity, as well as two million cotton cloaks and large amounts of more precious materials such as gold, amber and quetzal feathers. In fact, in supporting the dense population of the capital, variously estimated at 100,000 to 700,000 (the latter figure is highly unlikely), tribute greatly outstripped local production in importance.

It would probably be no exaggeration to say that the chinampas gave the ancient peoples of the Valley of Mexico intermittent sway over most of the country for 1,500 years before the arrival of the Spaniards. For this reason a detailed study of all aspects of this unique system as it now operates should be made before the chinampas disappear altogether in the name of progress.

II

PLANT GROWTH AND DEVELOPMENT

II

PLANT GROWTH AND DEVELOPMENT

INTRODUCTION

One of the most marvelous of all natural phenomena is the transformation of a single cell to a mature organism—a transformation that includes both *growth* (increase in cell number) and *development* (differentiation, specialization, and organization of cells). The cycle of changes that convert fertilized egg to seed and then to mature plant is repeated annually in the production of many crops, yet the true nature of these changes has only recently been discovered, largely as a result of various inquiries into the biochemistry of individual processes of growth and development.

Perhaps the most exciting of these has led to the gradual unraveling of the mystery of photosynthesis—the remarkable biochemical process by which plants, with the aid of the pigment chlorophyll, utilize the sun's energy to create carbohydrate out of water and carbon dioxide. In "The Role of Chlorophyll in Photosynthesis," Eugene I. Rabinowitch and Govindjee discuss the mechanisms involved in the capture of light energy, the first stage in the process that provides all the world with food and thus is the basis of growth and development of all life.

Differentiation in a multicellular organism is a result of differential growth within and among cells, each of which contains identical genetical material. The differentiation of such genetically identical cells is accomplished through the regulation of the genetic information stored in each cell. This information is turned on and off in accord with instructions contained in the cell itself, signals obtained from other cells, or some outside cue from the environment. In higher organisms certain cells take over the control of differentiation through the action of chemical messengers called hormones. Even the most minute concentrations of these naturally occurring organic substances can inhibit or promote growth and development: as little as one part per billion of some hormones exerts a measurable physiological effect. These plant hormones can be synthetically copied or they can be extracted from plant parts (for example, cytokinins and other growth substances are found in the liquid endosperm, or "water," of the coconut) and used experimentally to effect developmental changes. Starting from an already differentiated cell from a carrot root, it has been possible to produce a complete plant in artificial media; mature plants have even been produced from tobacco pollen grains.

In "The Control of Plant Growth," Johannes Van Overbeek points out that the hormones manufactured within plants must always be present, and in proper balance with each other, for development to proceed. For example, the sequence of development in germination is shown to be controlled by the interaction of such hormones as auxins, gibberellins, dormin, and cytokinins. It appears that these hormones affect the synthesis of nucleic acid through their regulation of genetic information.

The recognition and synthesis of growth regulators have had a profound influence on agriculture. Plant hormones are now available that influence such processes as rooting, flowering, plant size, fruit set, and fruit drop. Although their most significant use so far has been as herbicides, to kill vegetation or selectively remove weeds from crops, continuing study should enable plant breeders to control the balance of internally produced plant hormones in order to produce plants of desired growth patterns. The use of growth regulators can be expected to have an even greater impact on plant agriculture in the future, as additional growth-regulating materials are found and their effects understood.

A number of growth processes are regulated by an interaction of internal and external cues. Germination, the first stage of plant growth, is regulated by controls within the seed as well as in the environment. We take for granted that our crop seed will germinate without stopping to think that the capacity of such seed for prompt and uniform germination is a result of selection by generations of farmers, gardeners, and seedsmen. As Dov Koller points out in "Germination," few seeds exhibit readiness to grow soon after the fruit ripens. Both fruit and the seed coats usually contain growth inhibitors that prevent germination until the seed is released from the fruit and the compounds are leached from the coats. Once a seed is free of the influence of its chemical inhibitor, environmental control of germination takes over. Heat, moisture, and light must usually be available in a particular combination of intensity, amount, and duration for germination to proceed on schedule. Even fire can stimulate germination. In California, forest fires cause the waterproof coverings of sumac seeds to become permeable, thus increasing the possibility of germination.

The change from the vegetative to the flowering stage is a dramatic one in plant development. "What makes a plant suddenly turn from producing leaf buds and begin to form flowers?" asks Frank B. Salisbury in "The Flowering Process." In seeking an answer to this question plant scientists have turned to the lowly cocklebur, a plant despised by farmers. The flowering response of the easy-to-grow cocklebur is very sensitive to photoperiod—the relative length of day and night. Through a study of the cocklebur the steps associated with the flowering process have been clearly defined and the light response clarified, although the substance that causes the plant to begin forming flowers instead of leaves has not yet been identified.

Another aspect of plant development that has received much attention is reviewed by J. B. Biale in "The Ripening of Fruit." The growth of the fruit starts when pollination initiates enlargement of the ovary of the flower. In a normal fruit, growth continues to be stimulated by auxins and gibberellins from its own seed until it reaches full size. Once enlargement has been completed the ripening process begins. It was discovered that certain fruits can be stimulated to ripen by exposure to particular chemical compounds. For example, ethylene gas is routinely used to hasten the ripening of bananas that have been shipped in the immature green stage, although there is reason to believe that it functions as no more than a triggering compound. Ripening of many fruits can be delayed by storing them at low temperatures and in environments that contain less than the usual amounts of oxygen. The chemical and physiological changes associated with ripening parallel the growth, development, and aging of all living things.

Each time one of the physiological or biochemical processes of plant growth seems to be stripped of its mystery, many new and more fundamental problems become apparent. As each frontier is opened, the vanguard of plant science rapidly moves into the newly discovered region to probe the unexplored. As the process is repeated, old ideas are discarded or revised to accommodate new concepts based on new information. Because of this we are now beginning to comprehend more clearly the nature of the problems of growth and development.

5

THE ROLE OF CHLOROPHYLL IN PHOTOSYNTHESIS

EUGENE I. RABINOWITCH AND GOVINDJEE
July 1965

Any effort to understand the basis of life on this planet must always come back to photosynthesis: the process that enables plants to grow by utilizing carbon dioxide (CO_2), water (H_2O) and a tiny amount of minerals. Photosynthesis is the one large-scale process that converts simple, stable, inorganic compounds into the energy-rich combination of organic matter and oxygen and thereby makes abundant life on earth possible. Photosynthesis is the source of all living matter on earth, and of all biological energy.

The overall reaction of photosynthesis can be summarized in the following equation: $CO_2 + H_2O + light \rightarrow (CH_2O) + O_2 + 112{,}000$ calories of energy per mole. (CH_2O) stands for a carbohydrate; for example, glucose: (CH_2O)$_6$. "Mole" is short for "gram molecule": one gram

multiplied by the molecular weight of the substances in question—in this case carbohydrate and oxygen.

When one of us first summarized the state of knowledge of photosynthesis 17 years ago, the whole process was still heavily shrouded in fog [see "Photosynthesis," by Eugene I. Rabinowitch; SCIENTIFIC AMERICAN Offprint 34]. Five years later investigation had penetrated the mists sufficiently to disclose some of the main features of the process [see "Progress in Photosynthesis," by Eugene I. Rabinowitch; SCIENTIFIC AMERICAN, November, 1953]. Since then much new knowledge has been accumulated; in particular the sequence of chemical steps that convert carbon dioxide into carbohydrate is now understood in considerable detail [see "The Path of Carbon in Photosynthesis," by

J. A. Bassham; SCIENTIFIC AMERICAN Offprint 122]. The fog has also thinned out in other areas, and the day when the entire sequence of physical and chemical events in photosynthesis will be well understood seems much closer.

The photosynthetic process apparently consists of three main stages: (1) the removal of hydrogen atoms from water and the production of oxygen molecules; (2) the transfer of the hydrogen atoms from an intermediate compound in the first stage to one in the third stage, and (3) the use of the hydrogen atoms to convert carbon dioxide into a carbohydrate [see illustration on page 44].

The least understood of these three stages is the first: the removal of hydrogen atoms from water with the release of oxygen. All that is known is that it entails a series of steps probably requiring several enzymes, one of which contains manganese. The third stage—the production of carbohydrates from carbon dioxide—is the best understood, thanks largely to the work of Melvin Calvin and his co-workers at the University of California at Berkeley. The subject of our article is the second stage: the transfer of hydrogen atoms from the first stage to the third. This is the energy-storing part of photosynthesis; in it, to use the words of Robert Mayer, a discoverer of the law of the conservation of energy, "the fleeting sun rays are fixed and skillfully stored for future use."

The light energy to be converted into chemical energy by photosynthesis is first taken up by plant pigments, primarily the green pigment chlorophyll. In photosynthesis chlorophyll functions as a photocatalyst: when it is in its energized state, which results from the absorption of light, it catalyzes an energy-storing chemical reaction. This

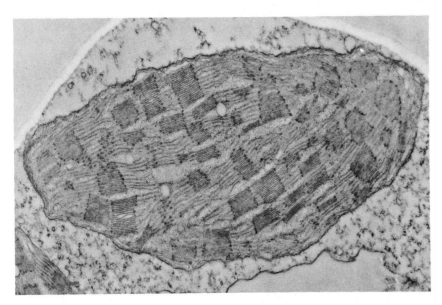

CHLOROPLAST is the organelle in a plant cell within which photosynthesis takes place. The chlorophyll is contained in the "grana," stacks of membranous sacs called lamellae, seen here in cross section. A maize-cell chloroplast is enlarged 19,000 diameters in this electron micrograph made by A. E. Vatter of the University of Colorado Medical Center.

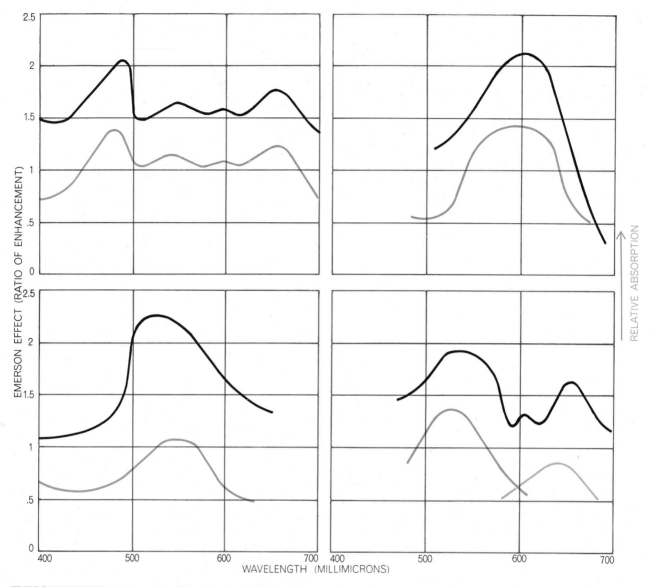

EMERSON EFFECT is shown for *Chlorella* (*top left*), the blue-green alga *Anacystis nidulans* (*top right*), *Porphyridium* (*bottom left*) and the diatom *Navicula minima* (*bottom right*). In each case the black curve shows the action spectrum of the Emerson effect, or the degree of enhancement in quantum yield as the wavelength of the supplementary illumination is varied. The curve of the action spectrum turns out to be parallel to the absorption curves (*color*) of the various accessory pigments: chlorophyll *b* in *Chlorella*, phycocyanin in *Anacystis*, phycoerythrin in *Porphyridium* and fucoxanthol (*solid color*) and chlorophyll *c* (*light color*) in *Navicula*.

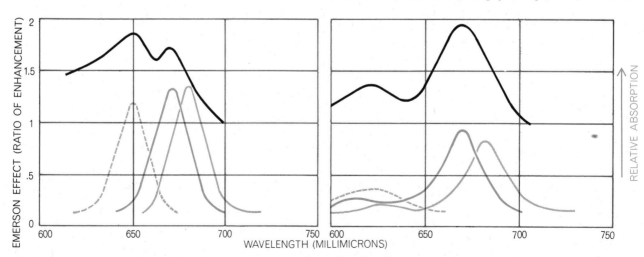

DETAILED ACTION SPECTRA of the Emerson effect reveal the presence of chlorophyll *a* 670 in *Chlorella* (*left*) and *Navicula* (*right*). The Emerson-effect peaks coincide with the absorption peaks of chlorophyll *a* 670 (*solid color*) as well as of chlorophyll *b* in *Chlorella* and chlorophyll *c* in *Navicula* (*broken curves*). The chlorophyll *a* absorption curve is also shown (*light-color curve*).

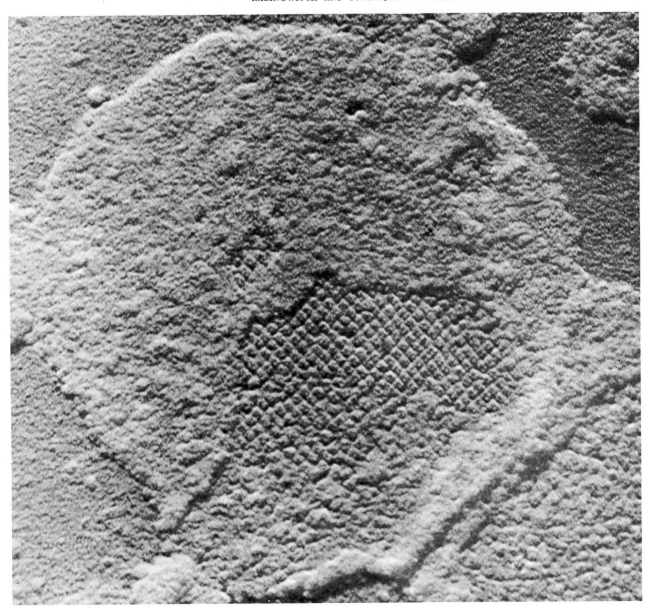

PHOTOSYNTHETIC UNITS may be the small elements, looking somewhat like cobblestones, visible in this electron micrograph made by Roderic B. Park and John Biggins of the University of California at Berkeley. In the micrograph a single lamella and a part of another one are shadowed with chromium and enlarged 175,000 diameters. Where the membrane is torn away one can see an ordered array of the units, which Park and Biggins call quantasomes and calculate could contain 230 chlorophyll molecules each.

reaction is the primary photochemical process; it is followed by a sequence of secondary "dark"—that is, nonphotochemical—reactions in which no further energy is stored.

Once it was thought that in photosynthesis the primary photochemical process is the decomposition of carbon dioxide into carbon and oxygen, followed by the combination of carbon and water. More recently it has been suggested that the energy of light serves primarily to dissociate water, presumably into hydroxyl radicals (OH) and hydrogen atoms; the hydroxyl radicals would then react to form oxygen molecules. It is better than either of these two formulations to say that the primary

photochemical process in photosynthesis is the boosting of hydrogen atoms from a stable association with oxygen in water molecules to a much less stable one with carbon in organic matter. The oxygen atoms "left behind" combine into oxygen molecules, an association also much less stable than the one between oxygen and hydrogen in water. The replacement of stable bonds (between oxygen and hydrogen) by looser bonds (between oxygen and oxygen and between hydrogen and carbon) obviously requires a supply of energy, and it explains why energy is stored in photosynthesis.

The transfer of hydrogen atoms from one molecule to another is called oxida-

tion-reduction. The hydrogen atom is transferred from a donor molecule (a "reductant") to an acceptor molecule (an "oxidant"); after the reaction the donor is said to be oxidized and the acceptor to be reduced. The transfer of an electron can often substitute for the transfer of a hydrogen atom: in an aqueous system (such as the interior of the living cell) there are always hydrogen ions (H^+), and if such an ion combines with the electron acceptor, the acquisition of an electron becomes equivalent to the acquisition of a hydrogen atom (electron + H^+ ion → H atom).

The chain of oxidation-reduction reactions in photosynthesis has some links that involve electron transfers and others that involve hydrogen-atom transfers. For the sake of simplicity we shall speak of electron transfers, with the understanding that in some cases what is actually transferred is a hydrogen atom. Indeed, the end result of the reactions undoubtedly *is* the transfer of hydrogen atoms.

In the oxidation-reduction reactions of photosynthesis the electrons must be pumped "uphill"; that is why energy must be supplied to make the reaction go. The tiny chlorophyll-containing chloroplasts of the photosynthesizing plant cell act as chemical pumps; they obtain the necessary power from the absorption of light by chlorophyll (and to some extent from absorption by other pigments in the chloroplast). It is important to realize that the energy is stored in the two products organic matter and free oxygen and not in either of them separately. To release the energy by the combustion of the organic matter (or by respiration, which is slow, enzyme-catalyzed combustion) the two products must be brought together again.

How much energy is stored in the transfer of electrons from water to carbon dioxide, converting the carbon dioxide to carbohydrate and forming a proportionate amount of oxygen?

Oxidation-reduction energy can conveniently be measured in terms of electrochemical potential. Between a given donor of electrons and a given acceptor there is a certain difference of oxidation-reduction potentials. This difference depends not only on the nature of the two reacting substances but also on the nature of the products of the reaction; it is characteristic of the two oxidation-reduction "couples." For example, when oxygen is reduced to water (H_2O) its potential is $+.81$ volt, but when it is reduced to hydrogen peroxide (H_2O_2) the potential is $+.27$ volt. The more positive the potential, the stronger is the oxidative power of the couple; the more negative the potential, the stronger is its reducing power.

When two oxidation-reduction couples are brought together, the one containing the stronger oxidant tends to oxidize the one containing the stronger reductant. In photosynthesis, however, a weak oxidant (CO_2) must oxidize a weak reductant (H_2O), producing a strong oxidant (O_2) and a strong reductant (a carbohydrate). This calls for a massive investment of energy. The specific amount needed is given by the difference between the oxidation-reduction potentials of the two couples involved in the reaction: oxygen-water and carbon-dioxide–carbohydrate. The oxygen-water potential is about $+.8$ volt; the carbon-dioxide–carbohydrate potential, about $-.4$ volt. The transfer of a single electron from water to carbon dioxide thus requires $+.8$ minus $-.4$, or 1.2, electron volts of energy. For a molecule of carbon dioxide to be reduced to CH_2O—the elementary molecular group of a carbohydrate—*four* electrons (or hydrogen atoms) must be transferred; hence the total energy needed is 4.8 electron volts. This works out to 112,000 calories per mole of carbon dioxide reduced and of oxygen liberated. In short, the pumping of electrons in the second stage of photosynthesis entails the storage of 112,000 calories of energy per mole for each set of four electrons transferred.

We know the identity of the primary electron donor in photosynthesis (water) and of the ultimate electron acceptor (carbon dioxide), but what are the intermediates involved in the transfer of electrons from the first stage to the third? This has become the focal problem in recent studies of the photosynthetic process. As a matter of fact, it is not yet definitely known what compound releases electrons from the first stage, and what compound receives them in the third; that is why these compounds are respectively labeled ZH and X in the illustration at the left. About the donor, ZH, we have almost no information; the following considerations suggest the possible nature of the primary acceptor, X.

From the study of the mechanism of respiration we are familiar with an important oxidation-reduction catalyst: nicotinamide adenine dinucleotide phosphate, or NADP (formerly known as triphosphopyridine nucleotide, or TPN). NADP has an oxidation-reduction potential of about $-.32$ volt, thus in itself it is not a strong enough reductant to provide the $-.4$-electron-volt potential needed to reduce carbon dioxide to carbohydrate. NADP can achieve this feat, however, if it is supplied with additional energy in the form of the high-energy compound adenosine triphosphate, or ATP. A molecule of ATP supplies about 10,000 calories per mole when its terminal phosphate group is split off, and this is enough to provide the needed boost to the reducing power of NADP. Furthermore, we know that NADP is reduced when cell-free preparations of chloroplasts are illuminated. Put together, these two facts led to the now widely accepted hypothesis that the sec-

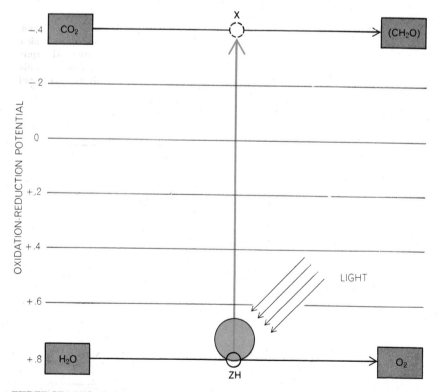

THREE STAGES of photosynthesis are the removal of hydrogen from water with the release of oxygen (*bottom arrow*), the transfer (*vertical arrow*) of the hydrogen by energy from light trapped by chlorophyll (*color*) and the use of the hydrogen to reduce carbon dioxide to carbohydrate (*top arrow*). In this scheme the oxidation-reduction potentials involved are indicated by the scale at the left, and the hypothetical "primary reductant" and "primary acceptor" intermediates are designated as ZH and X respectively.

ond stage of photosynthesis manufactures both ATP and reduced NADP and feeds them into the third stage.

At first it was assumed that NADP is identical with *X*, the primary acceptor in our scheme. Subsequent experiments by various workers—notably Anthony San Pietro at Johns Hopkins University and Daniel I. Arnon and his colleagues at the University of California at Berkeley—suggested, however, that NADP is preceded in the "bucket brigade" of electron transfer by ferredoxin, a protein that contains iron. This compound has an oxidation-reduction potential of about −.42 volt; therefore if it is reduced in light it can bring about the reduction of NADP by a "dark" reaction requiring no additional energy supply.

More recently Bessel Kok of the Research Institute for Advanced Studies in Baltimore has found evidence suggesting that compound *X* may be a still stronger reductant, with a potential of about −.6 volt. If this is so, plants have the alternatives of either applying this stronger reductant directly to the reduction of carbon dioxide or letting it reduce first ferredoxin and then NADP and using reduced NADP to reduce carbon dioxide. It seems a roundabout procedure to create a reductant sufficiently strong for the task at hand, then to sacrifice a part of its reducing power and finally to use ATP to compensate for the loss. It is not unknown, however, for nature to resort to devious ways in order to achieve its aims.

For photosynthesis to be a self-contained process the required high-energy phosphate ATP must be itself manufactured by photosynthesis. The formation of ATP has in fact been detected in illuminated fragments of bacteria by Albert W. Frenkel of the University of Minnesota and in chloroplast fragments by Arnon and his co-workers [see "The Role of Light in Photosynthesis," by Daniel I. Arnon; SCIENTIFIC AMERICAN Offprint 75]. As a matter of fact, ATP is needed not only to act as a booster in the reduction of an intermediate in the carbon cycle by reduced NADP but also for another step in the third stage of photosynthesis. According to a sequence of reactions worked out in 1951 by Andrew A. Benson and his colleagues at the University of California at Berkeley, carbon dioxide enters photosynthesis by first reacting with a "carbon dioxide acceptor," a special sugar phosphate called ribulose diphosphate. It turns out that the production of this compound from its precursor—

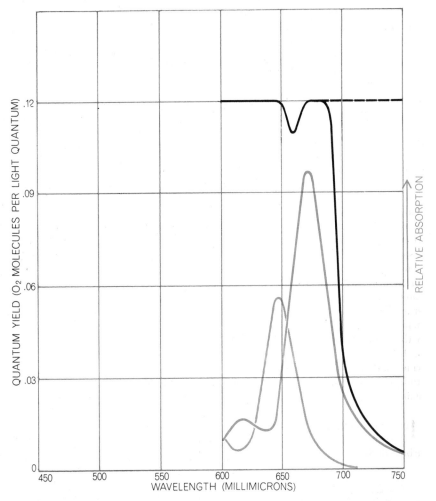

"RED DROP," the drop in quantum yield (*black curve*) of oxygen in photosynthesis under long-wave illumination, is demonstrated in the green alga *Chlorella pyrenoidosa*. Peak efficiency is restored (*broken line*) by supplementary shorter-wave illumination. Absorption curves of chlorophylls *a* (*solid color*) and *b* (*light color*) are also shown. This illustration and the next two are based on data of the late Robert Emerson of the University of Illinois.

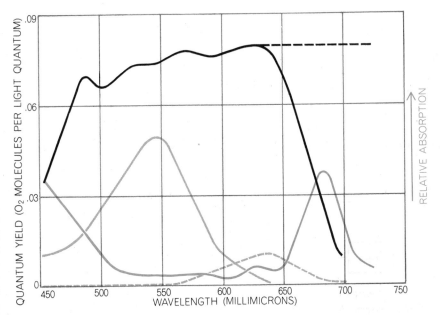

QUANTUM YIELD is similarly affected by long-wave illumination in the red alga *Porphyridium cruentum* (*black curve*). Yield drops to less than half of its maximum when absorption by chlorophyll *a* (*solid-color curve*) is at its peak. Absorption of the pigments phycoerythrin (*light-color curve*) and phycocyanin (*broken-color curve*) are also shown.

ribulose monophosphate—calls for a molecule of ATP.

ATP is produced both in chloroplasts and in mitochondria, the tiny intracellular bodies that are the site of the energy-liberating stage of respiration in animals as well as plants. The mitochondria produce ATP as their main function, exporting it as packaged energy for many life processes. The chloroplasts, on the other hand, make ATP only as an auxiliary source of energy for certain internal purposes. The energy of the light falling on the chloroplasts is stored mostly as oxidation-reduction energy by the uphill transfer of electrons. Only a relatively small fraction is diverted to the formation of ATP, and this fraction too ultimately becomes part of the oxidation-reduction energy of the final products of photosynthesis: oxygen and carbohydrate.

Let us now consider the uphill transport of electrons in greater detail. Recent investigations have yielded considerable information about this stage. Apparently the pumping of the electrons is a two-step affair, and among the most important intermediates in it are the catalysts called cytochromes.

The idea of a two-step electron-transfer process grew from a consideration of the energy economy of photosynthesis. Precise measurements, particularly those made by the late Robert Emerson and his co-workers at the University of Illinois, showed that the reduction of one molecule of carbon dioxide to carbohydrate, and the liberation of one molecule of oxygen, requires a minimum of eight quanta of light energy. The maximum quantum yield of photosynthesis, defined as the number of oxygen molecules that can be released for each quantum of light absorbed by the plant cell, is thus 1/8, or 12 percent. Since the transfer of four electrons is involved in the reduction of one carbon dioxide molecule, it was suggested that it takes two light quanta to move each electron.

Emerson and his colleagues went on to determine the quantum yield of photosynthesis in monochromatic light of different wavelengths throughout the visible spectrum. They found that the yield, although it remained constant at about 12 percent in most of the spectrum, dropped sharply near the spectrum's far-red end [see illustrations on page 45]. This decline in the quantum yield, called the "red drop," begins at a wavelength of 680 millimicrons in green plants and at 650 millimicrons in red algae.

There are two chlorophylls present in green plants: chlorophyll a and chlorophyll b. Only chlorophyll a absorbs light at wavelengths longer than 680 millimicrons; the absorption of chlorophyll b rises to a peak at 650 millimicrons and becomes negligible at about 680 millimicrons. Emerson found that the quantum yield of photosynthesis at the far-red end of the spectrum beyond 680 millimicrons can be brought to the full efficiency of 12 percent by simultaneously exposing the plant to a second beam of light with a wavelength of 650 millimicrons. In other words, when light primarily absorbed by chlorophyll a was supplemented by light primarily absorbed by chlorophyll b, both beams gave rise to oxygen at the full rate. This relative excess in photosynthesis when a plant is exposed to two beams of light simultaneously, as compared with the yield produced by the same two beams separately, is known as the Emerson effect, or enhancement.

On the basis of his discovery Emerson concluded that photosynthesis involves two photochemical processes: one using energy supplied by chlorophyll a, the other using energy supplied by chlorophyll b or some other "accessory" pigment. Experimenting with various combinations of a constant far-red beam with beams of shorter wavelength, and using four different types of algae (green, red, blue-green and brown), Emerson's group found that the strongest enhancement always occurred when the second beam was absorbed mainly by the most important accessory pigment (the green pigment chlorophyll b in green cells, the red pigment phycoerythrin in red algae, the blue pigment phycocyanin in blue-green algae and the reddish pigment fucoxanthol in brown algae). Such results suggested that these other pigments are not mere accessories of chlorophyll a but have an important function of their own in photosynthesis [see top illustration on page 42].

Certain findings concerning the behavior of pigments in living plant cells, however, seemed to make this conclusion untenable. Illuminated plant cells fluoresce; that is, pigment molecules energized by the absorption of light quanta reemit some of the absorbed energy as fluorescent light. The source of fluorescence can be identified, because each substance has its own characteristic fluorescence spectrum. The main fluorescing pigment in plants always proves to be chlorophyll a, even when the light is absorbed by another pigment. This had first been shown for brown algae in a study conducted in

1943 by H. J. Dutton, W. H. Manning and B. B. Duggar at the University of Wisconsin; later the finding was extended to other organisms by L. N. M. Duysens of the University of Leiden. Known as sensitized fluorescence, the phenomenon indicates that the initial absorber has transferred its energy of excitation to chlorophyll a; the transfer is effected by a kind of resonance process. Careful measurements have shown that certain accessory pigments—chlorophyll b, phycoerythrin, phycocyanin and fucoxanthol—pass on to chlorophyll a between 80 and 100 percent of the light quanta they absorb. For some other accessory pigments—for example carotene—the transfer is less efficient.

This puts accessory pigments back in the role of being mere adjuncts to chlorophyll a. True, they can contribute, by means of resonance transfer, light energy to photosynthesis, thereby improving the supply of energy in regions of the spectrum where chlorophyll a is a poor absorber. Chlorophyll a, however, collects all this energy before it is used in the primary photochemical process. Why, then, the enhancement effect? Why should chlorophyll a need, in order to give rise to full-rate photosynthesis, one "secondhand" quantum obtained by resonance transfer from another absorber in addition to the one quantum it had absorbed itself?

A better understanding of this paradox resulted from the discovery that there apparently exist in the cell not only chlorophyll a and chlorophyll b but also two forms of chlorophyll a. These two forms have different light-absorption characteristics, and they probably also have different photochemical functions.

In the living cell chlorophyll a absorbs light most strongly in a broad band with its peak between 670 and 680 millimicrons. In our laboratory at the University of Illinois we undertook to plot the Emerson effect more carefully than before as a function of the wavelength of the enhancing light. We found that for green and brown cells the resulting curve showed, in addition to peaks corresponding to strong absorption by the accessory pigments, a peak at 670 millimicrons that must be due to chlorophyll a itself [see bottom illustration on page 42]. It was this finding that suggested the existence of two forms of chlorophyll a. The form that absorbs light at the longer wavelengths—mainly above 680 millimicrons—seemed to belong to one pigment system, now often called System I. The form that absorbs at 670 millimicrons seemed to belong to another pigment system: System II. In the second system the form of chlorophyll *a* that absorbs at 670 millimicrons is strongly assisted by accessory pigments, probably by resonance transfer of their excitation energy. Careful analysis of the absorption band of chlorophyll *a* by C. Stacy French of the Carnegie Institution of Washington's Department of Plant Biology, and also in our laboratory, confirmed that the band is double, with one peak near 670

millimicrons (at 668 millimicrons) and another band at 683 millimicrons [*see bottom illustration on page 42*].

If chlorophyll *a* is extracted from living plants, there is only one product; we must therefore assume that in the living cell the two forms differ in the way molecules of chlorophyll *a* are clumped together, or in the way they are associated with different chemical partners (proteins, lipids or other substances). Be

this as it may, the important implication of the new finding is that photosynthesizing cells possess two light-absorbing systems, one containing a form of chlorophyll *a* absorbing around 683 millimicrons and the other a form absorbing around 670. The latter system includes chlorophyll *b* (in green-plant cells) or other accessory pigments (in brown, red and blue-green algae). Further investigation—particularly of red algae—has suggested, however, that the distribu-

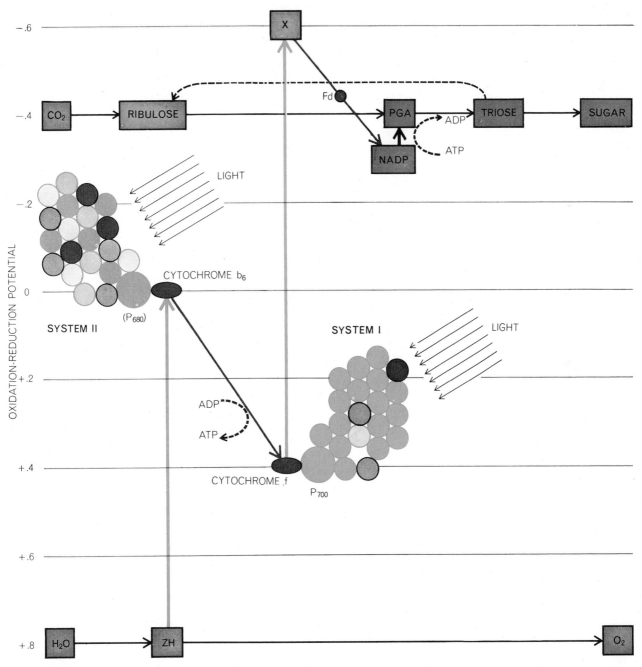

HYDROGEN TRANSFER in photosynthesis is now conceived of as a two-step process involving two pigment systems. Hydrogen atoms (or electrons) from the donor (*ZH*) are boosted to cytochrome b_6 by energy collected in System II and trapped by a hypothetical "pigment 680" (*P 680*). The pigments of System II include chlorophyll *a* 670 and such accessory pigments as chlorophyll *b* or *c*, phycoerythrin or phycocyanin, depending on the plant. The elec-

trons are passed "downhill" to cytochrome *f*, synthesizing adenosine triphosphate (*ATP*) in the process. Energy from System I (primarily chlorophyll *a*, with some accessory pigments), trapped by pigment 700 (*P 700*), boosts the electrons to a receptor (*X*), whence they move via ferredoxin (*Fd*) to nicotinamide adenine dinucleotide phosphate (*NADP*). Energy from ATP helps to move the electrons to phosphoglyceric acid (*PGA*) and into the carbon cycle.

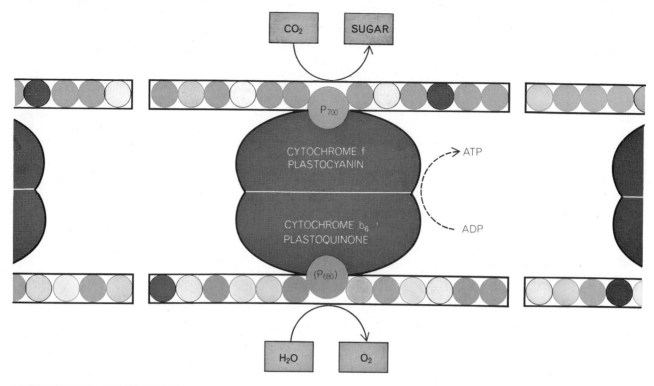

tion of these two components in the two systems may be less clear-cut. In red algae a large fraction of the chlorophyll *a* absorbing at 670 millimicrons seems to belong to System I rather than System II.

In all likelihood the two systems provide energy for two different photochemical reactions, and efficiency in photosynthesis requires that the rates of the two reactions be equal. What are these reactions? This question brings us to another significant finding, which suggested the participation of cytochromes in photosynthesis.

Cytochromes are proteins that carry an iron atom in an attached chemical group. They are found in all mitochondria, where they serve to catalyze the reactions of respiration. Robert Hill and his co-workers at the University of Cambridge first found that chloroplasts also contain cytochromes—two kinds of them. One, which they named cytochrome *f*, has a positive oxidation-reduction potential of about .4 volt. The other, which they named cytochrome b_6, has a potential of about 0 volt. In 1960 Hill, together with Fay Bendall, proposed an ingenious hypothesis as to how the two cytochromes might act as intermediate carriers of electrons and connect the two photochemical systems [see illustration on page 47]. They suggested that cytochrome b_6 receives an

electron by a photochemical reaction from the electron donor *ZH*; the electron is then passed on to cytochrome *f* by a "downhill" reaction requiring no light energy. (The oxidation-reduction potential of cytochrome *f* is much more positive than that of cytochrome b_6.) A second photochemical reaction moves the electron uphill again, from cytochrome *f* to the electron-acceptor *X* in the third stage of photosynthesis. In this sequence the photochemical reactions store energy and the reaction between the two cytochromes releases energy. Some of the released energy, however, can be salvaged by the formation of an ATP molecule; this occurs in the transfer of electrons among cytochromes in respiration. In this way ATP is obtained without spending extra light quanta on its formation, which the tight energy economy of photosynthesis does not allow.

Experiments by Duysens and his associates confirmed this hypothesis, by showing that the absorption of light by System I causes the oxidation of a cytochrome, whereas the absorption of light by System II causes its reduction. This is exactly what we would expect. The illustration on the opposite page shows that the light reaction of System II should flood the intermediates between the two photochemical reactions with electrons taken from *ZH*; the light reaction of System I should drain these

electrons away, sending them up to the acceptor *X* and into the third stage of photosynthesis.

This, then, describes in a general way the oxidation-reduction process by which the chloroplasts store the energy of light in photosynthesis. Several other investigators have contributed evidence for the two-step mechanism; notable among them are French, Kok, Arnon, Horst Witt of the Max-Vollmer Institute in Berlin and their colleagues. In detail the process probably is much more complex than our scheme suggests. Its "downhill" central part seems to include, in addition to the two cytochromes, certain compounds of the group known as quinones and also plastocyanin, a protein that contains copper.

What is known of the submicroscopic structure in which the reactions of the second stage of photosynthesis take place? There is much evidence that the photosynthetic apparatus consists of "units" within the chloroplasts, each unit containing about 300 chlorophyll molecules. This picture first emerged from experiments conducted in 1932 by Emerson and William Arnold on photosynthesis during flashes of light; it was later supported by various other observations. The pigment molecules are packed so closely in the unit that when one of them is excited by light it readily transfers its excitation to a neighbor by resonance. The energy goes on traveling through the unit, rather as the steel ball in a pinball machine bounces around among the pins and turns on one light after another. Eventually the migrating energy quantum arrives at the entrance to an enzymatic "conveyor belt," where it is trapped and utilized either to load an electron onto the belt or to unload one from it. (The steel-ball analogy should not be taken literally; the migration of energy is a quantum-mechanical phenomenon, and the quantum's location can only be defined in terms of probability; its entrapment depends on the probability of finding it at the entrance to the conveyor belt.)

How is the quantum trapped? The trap must be a pigment molecule with what is called a lower excited state; the migrating quantum can stumble into such a molecule but cannot come out of it. Kok has found evidence that System I contains a small amount of a special form of chlorophyll called pigment 700 because it absorbs light at a wavelength of 700 millimicrons; this pigment could serve as a trap for the quantum bouncing around in System I. There seems to be a proper amount of pigment 700: about one molecule per unit. Furthermore, experiments suggest that pigment 700 is oxidized by light absorbed in System I and reduced by light absorbed in System II. It has an oxidation-reduction potential of about +.4 volt. All these properties fit the role we have assigned pigment 700 in our scheme: collecting energy from a 300-molecule unit in System I, using it to transfer an electron to the acceptor X and recovering the electron from cytochrome f [see illustration on page 47].

One suspects that there should be a counterpart of pigment 700 in System II, but so far none has been convincingly demonstrated. We believe, however, that a pigment we have tentatively named pigment 680—from the anticipated position of its absorption band—does serve as an energy trap in System II. Its existence is supported by the discovery of a new fluorescent emission band of chlorophyll at 693 millimicrons, which is compatible with absorption at 680 millimicrons. This band is emitted by certain algae when they are exposed to strong light of the wavelengths absorbed by System II.

What is the spatial organization of the pigment systems in the electron-boosting mechanism of the second stage of photosynthesis? It seems that the two systems may be arranged in two monomolecular layers, with a protein layer between them containing the enzymatic conveyor belt [see illustration on page 48]. The chloroplasts are known from electron microscope studies to consist of a set of lamellae: thin alternating layers of protein and fatty material piled one atop the other. Each layer appears to consist of particles arrayed rather like cobblestones in a pavement. The particles were first observed in electron micrographs made by E. Steinmann of the Technische Hochschule in Zurich; subsequently Roderic B. Park and John Biggins of the University of California at Berkeley made clearer micrographs of the particles and named them quantasomes [see illustration on page 43]. The units comprising Systems I and II may operate independently or they may be sufficiently close together to exchange energy by resonance, when

such exchange is needed to maintain a balanced rate of operation by the two systems.

The picture of the energy-storing second stage of photosynthesis presented in this article is, of course, still only a working hypothesis. Alternative hypotheses are possible, one of which we shall briefly describe. For many years the late James Franck, who shared the Nobel prize in physics for 1925, tried to develop a plausible physicochemical mechanism of photosynthesis. In 1963 he proposed, together with Jerome L. Rosenberg of the University of Pittsburgh, a concept according to which the two consecutive photochemical steps occur in one and the same energy trap. In other words, according to Franck, the same chlorophyll molecule that takes the electron away from the initial donor ZH and transfers it to a cytochrome then supplies energy for the transfer of the electron from the cytochrome to the acceptor X. In the first transfer, Franck suggested, the chlorophyll a molecule functions in the short-lived "singlet" excited state (in which the valence electrons have opposite spins); in the second transfer it functions in the long-lived "triplet" state (in which the valence electrons have parallel spins). Franck's hypothesis avoids certain difficulties of the "two trap" theory, but new difficulties arise in their place. On balance the two-trap picture seems to us the more plausible one at present.

No doubt this picture will change as more information emerges. It is merely a first effort to penetrate the inner sanctum of photosynthesis, the photocatalytic laboratory in which the energy of sunlight is converted into the chemical energy of life.

6

THE CONTROL OF PLANT GROWTH

JOHANNES VAN OVERBEEK
July 1968

The growth of a plant basically calls for light and water. Simple though these requirements may sound, plant growth itself is of course quite complex. In the green plant cell sunlight splits the molecules of water; this is part of the process of photosynthesis. The O of the H_2O is released into the air to provide the oxygen that all living things, including man, need for the process of respiration. The H_2 reacts with carbon dioxide and with nitrate to produce the major constituents of plant cells: sugar, starch, cellulose, protein and nucleic acid. When plant tissues are consumed by humans, these substances supply the building blocks of human cells and the energy needed for human life processes.

Normal plant growth and development will not take place without exceedingly small quantities of specific, in-

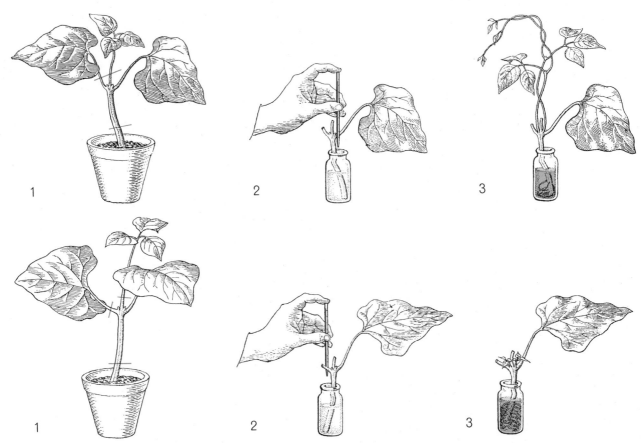

TWO TYPES OF PLANT-HORMONE ACTION were found in the author's early tests with dwarf bean plants. In both cases the original bean-plant cutting consisted of a piece of stem with one leaf and two nearly invisible buds (*step 1, top and bottom*); the fresh cuttings were then placed in small bottles of water. In one set of bottles a small amount of gibberellin, a naturally occurring plant-growth hormone, was dissolved in the water (*step 2, top*). Within a week the buds on these cuttings grew into long vinelike branches characteristic of pole beans (*step 3, top*); normally such branches would not appear at all in this particular variety of dwarf plant. In addition the normal amount of root growth in water was observed at the base of the stem. In the other set of bottles a small amount of indolebutyric acid, a synthetic auxin, was added (*step 2, bottom*). After a week bud growth was not promoted abnormally, but instead root development was greatly augmented, so that the base of the stem had the appearance of a bottle brush (*step 3, bottom*). A definite growth response was observed using as little as one part per billion of gibberellin or one part per million of indolebutyric acid.

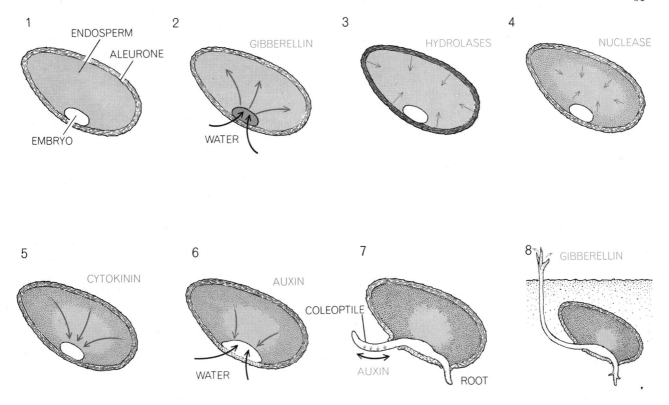

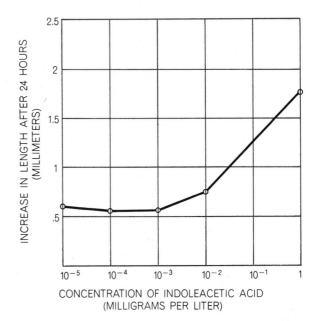

GERMINATION OF A CEREAL SEED below the surface of the soil (*1*) is regulated by a number of hormones working in sequence. First the absorption of water from the soil causes the embryo to produce a small amount of gibberellin (*2*). The gibberellin then diffuses into a layer of aleurone cells that surrounds the endosperm's food-storage cells, causing them to form enzymes (*3*) that in turn lead the endosperm cells to disintegrate and liquefy (*4*).

Cytokinins and auxins formed in this process (*5, 6*) then promote the growth of the embryo by making its cells divide and enlarge. If the shoot is pointing down into the soil, the auxins tend to migrate to the lower side of the seedling, causing it to grow faster and hence turning the growing point of the shoot upward toward the surface of the soil (*7*). Once the shoot has broken into the sunlight the plant begins to produce its own food by photosynthesis (*8*).

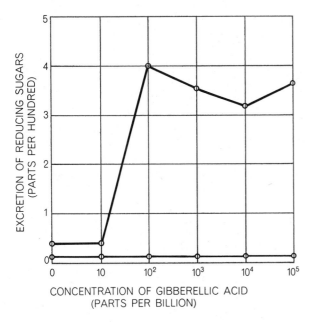

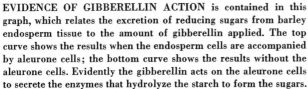

EVIDENCE OF GIBBERELLIN ACTION is contained in this graph, which relates the excretion of reducing sugars from barley endosperm tissue to the amount of gibberellin applied. The top curve shows the results when the endosperm cells are accompanied by aleurone cells; the bottom curve shows the results without the aleurone cells. Evidently the gibberellin acts on the aleurone cells to secrete the enzymes that hydrolyze the starch to form the sugars.

EVIDENCE OF AUXIN ACTION is contained in this graph, which shows the growth response of oat coleoptiles to various concentrations of the auxin indoleacetic acid (IAA). To obtain the results three-millimeter-long sections of the coleoptiles were floated on the surface of a shallow layer of hormone solution consisting of varying amounts of indoleacetic acid in distilled water. The measurements of length were made after a growing period of 24 hours.

ternally produced substances: the plant hormones. This basic rule was discovered in the 1920's by Frits W. Went, who was then working in the Netherlands. It turns out that these hormones must not only be present; they must be available in the proper balance and at the right time. Although the chemical study of the plant hormones began 40 years ago, only recently has enough knowledge been acquired to allow experiments that demonstrate the real power of the hormones in a dramatic fashion.

The basic discoveries concerning the powers of plant hormones came about in various ways. One of them grew out of an investigation of dwarf varieties of corn. In the mid-1930's at the California Institute of Technology I tried to find a physiological link between the gene deficiencies of dwarf plants and the inhibition of their growth. Could their failure to grow to normal size be traced to a hormone influence or deficiency? The only plant hormones known at the time were the auxins, typified by indole-

acetic acid. I experimented with these and found little evidence that auxins were critically involved in the dwarf plants' growth rate.

Twenty years later Bernard O. Phinney of the University of California at Los Angeles tested a new hormone on the same genetic dwarfs, and this time the outcome was different. The hormone was gibberellin, which had been discovered in Japan many years earlier but had just become available in the U.S. Phinney found that he could cause some of the corn dwarfs to grow into corn of normal size simply by placing a few drops of gibberellin solution in the center of the growing seedlings. Suddenly the old dream of being able to speed up the growth of plants at will had become an experimental reality.

The curious thing about gibberellin was that it had been discovered as the product of a fungus—a pathogenic fungus that made rice plants grow abnormally tall [see "Plant Growth Substances," by Frank B. Salisbury; SCIENTIFIC AMERICAN Offprint 110]. Strangely

enough, although the fungus apparently does not require gibberellin for its own growth, its cells produce huge quantities of the hormone. (In fact, the commercial production of gibberellin today depends on the fermentation of this fungus in vats, similar to the way penicillin is produced in vats by the cultivation of a mold.) By an odd quirk of nature the gibberellin fungus produces a chemical for which it has no use itself but which serves a vital function in higher plants. It has since been learned that all the higher plants produce gibberellin and require it for normal growth and development. We shall look into gibberellin's roles later in the article.

Another potent group of plant hormones came to light in the mid-1950's. As it happens, I was involved in the early stages of this exploration also. About 1940 the geneticist Albert F. Blakeslee asked my help on a physiological problem. In the course of solving the problem I found, with my research associate Marie E. Conklin, that coco-

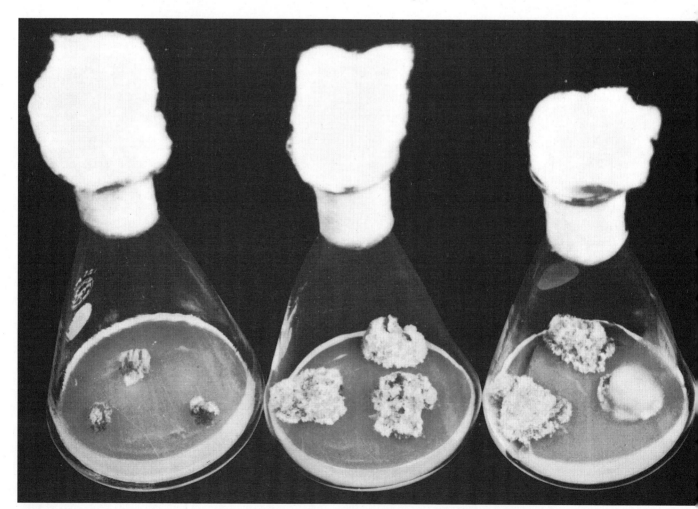

EFFECT OF CYTOKININ HORMONE on undifferentiated tissue from tobacco stems was studied by a group of investigators at the University of Wisconsin headed by Folke Skoog. By manipulating the concentration of the cytokinin in the growth medium they were

nut milk contains a new growth factor, which later turned out to be cytokinin. With the help of graduate students and chemists at Cal Tech we tried to isolate it, but after a year we had to give up. The active ingredient was always hidden in the dirtiest, stickiest residue.

In the 1950's the University of Wisconsin biologist Folke Skoog and his associates took up the pursuit of the active factor. They had been using coconut milk to grow pieces of tobacco tissue in bottles, and in order to run down the active substance they turned to other possible sources. They found that it was present in a yeast extract in a soluble form. Absorption spectra and other markers of the active fraction suggested it was a purine. Recalling that nucleic acids contain purines, one of Skoog's associates, Carlos O. Miller, searched the laboratory shelves for bottles with a nucleic acid label. He found one marked "Herring Sperm DNA," and sure enough, this material proved to be capable of causing tobacco cells to grow and divide.

I am told that when this bottle was used up, a large supply of freshly prepared DNA was ordered, but to everyone's consternation it failed to show any biological activity. The Wisconsin biological laboratories were then ransacked for nonfresh samples of DNA, and all of these proved to be active. Miller therefore returned to the freshly prepared DNA and "aged" it rapidly in an autoclave, whereupon it became active. The logical conclusion was that the growth-promoting factor must be a breakdown product of nucleic acid!

By beautiful teamwork the Wisconsin biologists led by Skoog and biochemists led by F. M. Strong then succeeded in isolating the factor. It turned out to be indeed a nucleic acid component—a derivative of adenine, one of the purine bases that make up nucleic acid. The Wisconsin group named the new hormone kinetin. They went on to synthesize several similarly active compounds, and collectively these hormones are now called the cytokinins.

Skoog and his students proceeded to

experiment with combinations of cytokinins and auxins in the culture of tobacco tissues. Starting with stem tissue, they found that simply by manipulating the relative concentrations of cytokinins and auxins in the growth medium they were able to grow roots, shoots and even flowers from the original colony of stem cells. Their results overturned the old idea that there were specific hormones for the formation of roots, leaves and stems; instead it became evident that growth and differentiation in plants are determined by the interplay of at least two growth factors.

Further information on this subject has come to light within the past two years. The breakthrough was provided by a newly discovered hormone that was first identified three years ago by Frederick T. Addicott and a team of co-workers at the University of California at Davis. They found it in extracts from cotton bolls, and they named it abscisin II, because it was believed to be responsible for the premature drop (abscission) of bolls from the plant. Meanwhile the

able to grow normal plants from the original undifferentiated cells. In the flasks shown the concentration was (*left to right*) 0, .04, .2, 1, 5 and 25 micromoles per liter. The cytokinin used was 6-(γ,γ-dimethyallylamino)purine. The growth period was six weeks.

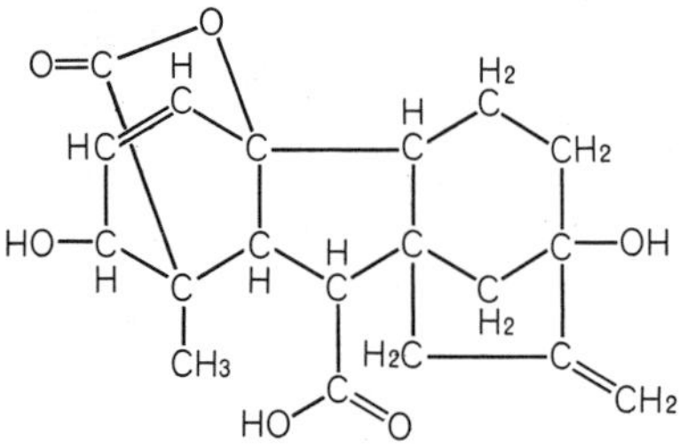

GIBBERELLIN

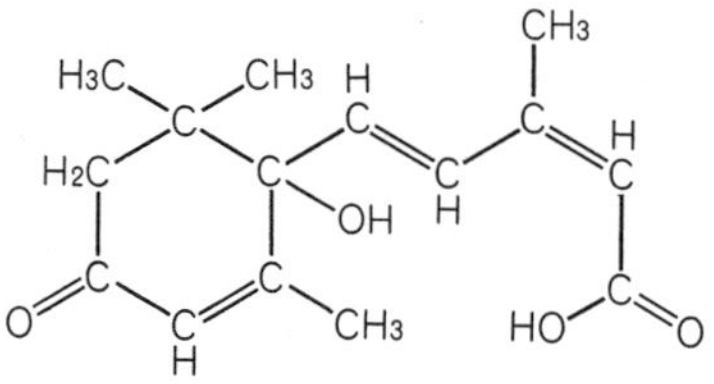

DORMIN

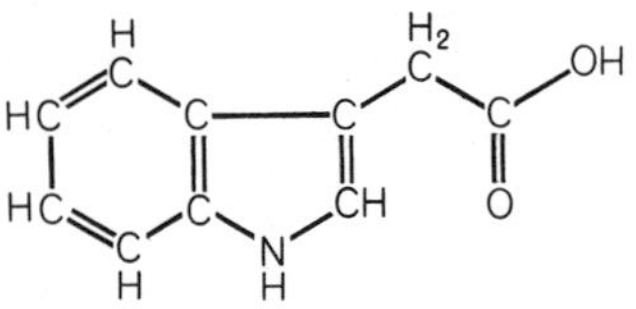

INDOLEACETIC ACID

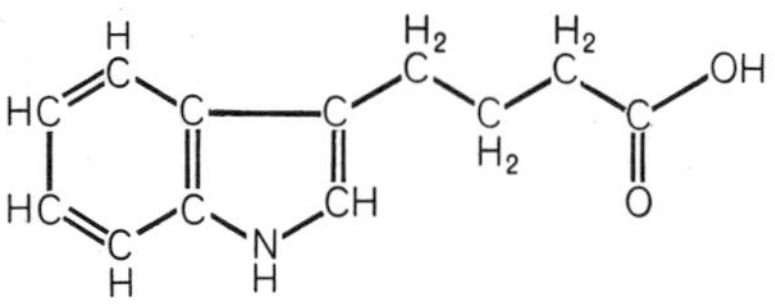

INDOLEBUTYRIC ACID

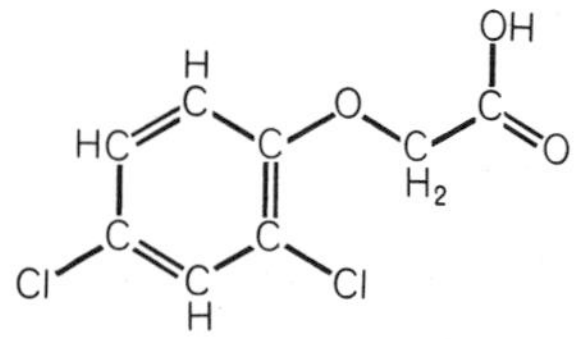

2,4-DICHLOROPHENOXYACETIC ACID

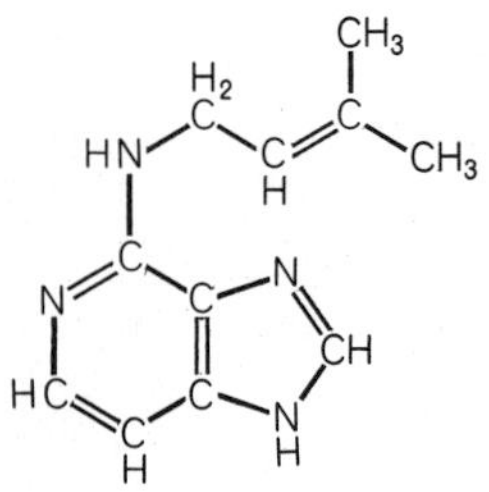

6-(γ,γ-DIMETHYLALLYLAMINO) PURINE

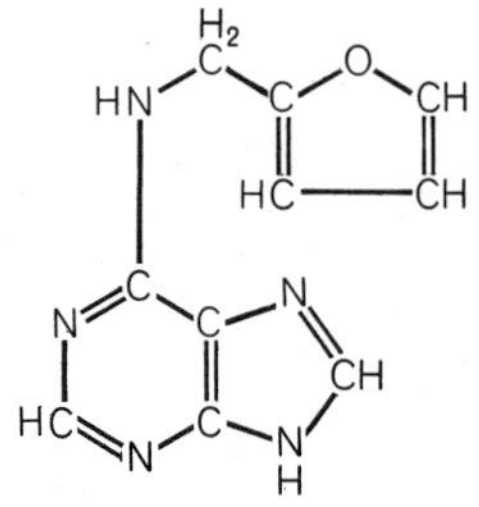

KINETIN

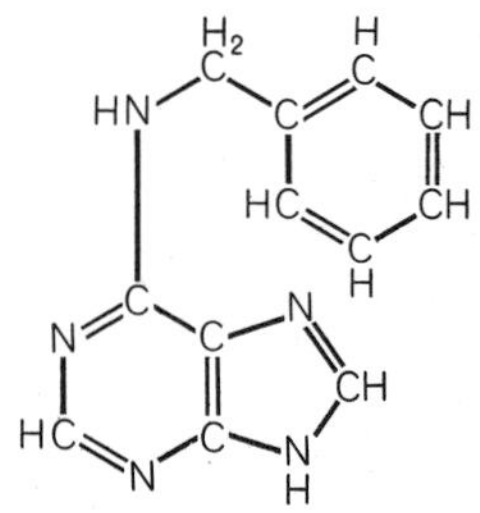

BENZYLADENINE

CHEMICAL STRUCTURES of eight of the plant-growth substances mentioned in the text are illustrated on this page. All are growth-promoting substances except for dormin, which is a growth-inhibiting hormone. Both gibberellin and dormin occur naturally; dormin has also been synthesized. Indoleacetic acid is a naturally occurring auxin. Indolebutyric acid and 2,4-dichlorophenoxyacetic acid are artificial auxins. 6-(γ,γ-dimethylallylamino)purine is a natural cytokinin. Kinetin and benzyladenine are artificial cytokinins.

same substance, found in maple leaves, was being investigated by the British chemist John W. Cornforth. It had been discovered by P. F. Wareing, an investigator of tree physiology; observing that it apparently prepared tree buds for their winter dormancy, Wareing named it dormin. Cornforth and his associates at the Milstead laboratory of Shell Research Ltd. soon succeeded in synthesizing the hormone and describing its stereochemical structure. Its full chemical name is 2-cyclohexene-1-penta-2,4-dienoic acid, 1-hydroxy-β,2,6,6-tetramethyl-4-oxo,cis-2-trans-4(d). Faced with the problem of selecting a short name for the hormone, the International Conference on Plant Growth Substances held in Ottawa in July, 1967, chose "abscisic acid." Unfortunately this name does not describe either the chemical structure or the phys-

iological activity of the substance. I shall refer to it here as dormin, because that term is descriptive of its physiological effects.

Dormin is an inhibitor of plant growth. For that reason some biologists object to calling it a hormone—a term that literally means "arousing to activity." Physiologists are now inclined, however, to classify both the promotive and the inhibitory growth regulators as hormones, because they operate in the same way (as chemical messengers) and are complementary in their actions.

In 1966, after dormin had been synthesized, Josef E. Loeffler, Iona Mason and I began detailed studies to work out its mode of action. We found that ordinary duckweed was extremely sensitive to the hormone. As little as one part per

billion of this substance in a culture solution would reduce the weed's growth rate, and one part per million was sufficient to keep this floating weed in a dormant state indefinitely. A striking feature of this action was its reversibility: as soon as the inhibiting hormone was removed, the plant tissues resumed their growth. Even after more than six weeks in the state of suspended growth, the culture could be revived simply by transferring it to a fresh medium in which the inhibiting hormone was absent.

In contrast to dormin, the cytokinins (but not auxins or gibberellin) strongly promoted the growth of duckweed cultures. We proceeded to study the mutual effects of dormin and a synthetic cytokinin (benzyladenine). Some 60 cultures were started in tubes under fluorescent light in a room at a constant temperature and were allowed to grow for a week, by which time their fresh weight had increased tenfold. Dormin, in the amount of one part per million, was then injected into some of the tubes. Within three days the growth of the cultures in these tubes slowed almost to a standstill. We now injected a very small amount of benzyladenine—one part per 10 million—in some of the tubes where growth had stopped. The cytokinin caused the cultures to resume their normal growth rate, although the dormin was still present. In short, the cytokinin overcame the dormin's inhibitory effect. We found that simply by supplying the growth medium with suitable concentrations of the opposing hormones we could apply stop-or-go control to the plant cells' growth.

It seems reasonable to conclude that in nature the growth of plants is similarly regulated by a combination of promoting and inhibiting hormones. Thus growth is controlled by an interplay of counteracting mechanisms much like that involved in driving a car. Just as we would not dream of operating a car with an accelerator and no brakes, so a plant apparently needs brakes for proper control. Dormin furnishes the brake. The accelerator, according to the particular plant and the conditions, may be cytokinin, gibberellin and/or auxin. Dormin has been detected in many plants in nature. We can speculate that the bursting forth of buds in the spring may be due in part to the decline of dormin and in part to the production of accelerating hormones such as the cytokinins.

How do the accelerating and the braking mechanisms work in the plant? By means of tracer studies with radioactive phosphate we discovered that dormin inhibits the synthesis of nucleic acids by the plant cells, and this is followed by a slowdown of growth. Conversely, the injection of cytokinin greatly accelerates the synthesis of nucleic acids. We therefore conclude that cytokinin speeds up the growth of the duckweed by increasing nucleic acid production and dormin retards it by reducing this production.

Even before our experiments Skoog's group at Wisconsin had demonstrated in the early 1950's that auxins increase the rate of nucleic acid synthesis. Our finding that dormin slows this rate has since been confirmed by Wareing, who presented the confirmation at a symposium on plant-growth regulators in London ıst winter. Obviously, since cell division and the growth of tissues require the production of nucleic acids (the cells' genetic material), this evidence of the involvement of the promoting and inhib-

itory hormones goes far toward explaining their effects on growth. It appears that the plant hormones control the fundamental biochemistry—the nucleic acid chemistry—of plant life.

We now have enough information in hand to put together a coherent picture of how the hormone system may start the growth of important crop plants such as wheat, oats, barley and rice. What happens when we sow a dry cereal seed in moist soil? The seed consists of two parts: the germ, or embryo, from which the plant will develop, and a store of food, called the endosperm, that will

nourish the developing seedling until it puts out green leaves that will enable it to produce its own food by photosynthesis. The stored food, initially in solid, undissolved form, is locked up in the endosperm cells. As the Austrian botanist Gottlieb Haberlandt showed many years ago, some action by the embryo is needed to release the food and allow it to become liquefied. What, then, is the key in the embryo that unlocks the food cellar? In 1960 Haraguro Yomo in Japan and L. G. Paleg in Australia independently discovered that the key is gibberellin.

Apparently the absorption of water

ROLE OF GIBBERELLIN in the germination process of a cereal seed is elucidated by this photograph, made by Joseph E. Varner of Michigan State University. The three barley seeds in the photograph have been cut in half and their embryos have been removed. Normally it is the embryo that produces gibberellin, which regulates the hydrolysis, or digestion, of the food-storage cells of the endosperm. The open surfaces of the three seeds have been treated with plain water (*bottom*), a solution of gibberellin in water at a concentration of one part per billion (*center*) and a solution of gibberellin in water at a concentration of 100 parts per billion (*top*). The photograph, taken 48 hours later, shows that in the gibberellin-treated seeds the digestion of the starch-filled storage tissue is already taking place.

from the soil causes the embryo to produce a small amount of gibberellin (of the order of a fraction of a part per million), which then diffuses into a layer of aleurone cells that surrounds the endosperm's food-storage cells. The gibberellin initiates a series of events that has been investigated in detail by the biochemist Joseph E. Varner at the Plant Research Laboratory of Michigan State University. Under the influence of the hormone the aleurone cells soon begin to synthesize enzymes. One (amylase) hydrolyzes the starch in the food cells into sugar; others break down those cells, disintegrating their nucleic acids and proteins. In brief, this is what happens: (1) in response to the uptake of water the embryo secretes gibberellin, (2) the gibberellin in turn causes the aleurone cells to form enzymes, (3) the enzymes go to work on the food-storage cells and cause them to disintegrate and liquefy.

Now we can reason that in the course of these events cytokinins and auxins are formed. The splitting of the nucleic acids by the newly formed enzymes can be expected to generate cytokinins, which as we have seen are derived from the breakdown of nucleic acid. Similarly, the breakdown of proteins can give rise to auxin; Skoog showed many years ago that the amino acid tryptophan is converted to indoleacetic acid in the cells of the coleoptile (the protective sheath around the seedling).

The newly generated hormones now proceed to promote the growth of the embryo. The cytokinins make its cells divide. The auxin assists by facilitating enlargement of the cells—the other requisite for cell growth. It does so by weakening the cell walls so that the cells take up water by osmosis and thus expand. The process by which auxin softens the cell walls is complex; Joe L. Key and his associates at Purdue University have found that it involves the synthesis of nucleic acids.

So finally we have our seedling growing. The shoot may, however, be pointing down into the soil. How does it make its way up to the surface and the sun? Again auxin is involved. By some geotropic process not yet understood auxin tends to migrate to the lower side of a seedling that is lying on its side. This causes the lower side to grow more rapidly than the upper, and hence the growing point of the shoot turns upward toward the soil surface. Once the shoot has broken through into the sunlight the leaves unfold, photosynthesis takes over

and the self-sustaining life of the plant begins.

As biologists we study plant hormones primarily because they can teach us a great deal about the process of growth. A fuller understanding of such life processes is the basic motivation, of course, for the work of any biologist. At the same time, plant scientists are always mindful of the practical implications of their studies. The green plant is, after all, the essential link to the sun's energy that sustains all life on the earth. It is the ultimate source of all man's food, and in itself it could supply our every food requirement. The Dutch horticulturist G. J. A. Terra has shown, for example, that calorie for calorie green leaves are as rich in essential proteins as the best of meats. A full appreciation of that fact in tropical countries could rescue those peoples from the ravages of the common protein-deficiency disease kwashiorkor. With the world population now growing very rapidly, plants have become more important than ever to man. There is no doubt that in order to cope with the increasing need for food we shall have to improve the efficiency of our agriculture. In several ways this is already being accomplished by the use of plant hormones, natural and synthetic.

One of the most useful of these, of course, is the synthetic weed killer known as 2,4-D. This substance, an auxin that retains its activity in plants much more persistently than the natural indoleacetic acid, can upset a plant's hormone balance so that growth occurs at places in the plant where it should not. It can cause roots to form on the stems in the air and at the same time slow down the normal root development underground. These abnormalities eventually lead to the death of the plant. Fortunately 2,4-D is selective in its action; it particularly attacks a number of useless weeds. As little as 500 grams per hectare (about half a pound per acre) can produce abnormal growth in a susceptible weed. Cereal plants, on the other hand, are relatively insensitive to 2,4-D; they inactivate the hormone, possibly by tying it to their proteins. Consequently the use of 2,4-D as a weed killer in grain fields has increased crop yields significantly. Synthetic auxins of the 2,4-D type have also been applied to other uses—for example to stop the premature drop of apples and pears. One might list a number of other applications of hormonal aids, for instance the use of auxins such as naphthaleneacetic acid and indolebutyric acid to propagate plants from cuttings and the use of gib-

berellins to speed up the malting process and to increase the size of grapes.

Studies of the potential uses of the newer hormones—the cytokinins and inhibitors such as dormin—are just beginning. I have obtained a patent for the utilization of cytokinins for preserving fresh vegetables. A cytokinin produced in the Shell laboratories has been found to be capable of generating viable seeds from Persian-grape plants that normally produce only male flowers; the hormone changes the developmental pattern of the flower from male to perfect hermaphrodite.

What uses dormin will have remains to be seen. Artificial growth inhibitors have been employed for some time and have produced interesting results. Materials such as CCC, B-Nine and so on have shown that flowering and fruiting can be promoted by slowing down vegetative growth. Azalea growers in the Eastern U.S. produce neat little plants that look like veritable balls of flowers by treating the plants with dwarfing agents. In the Netherlands I have seen young apple trees made to bear fruit two years before they would normally do so, simply through the use of growth inhibitors. The synthetic inhibiting chemical TIBA is widely used in the U.S. to shorten soybean plants and make them produce more branches so that they will bear more seeds.

In the long run, however, probably the most effective results will be obtained not by spraying chemicals on plants but by breeding them to produce a suitable balance of hormones of their own making. A good example of such a plant is the marvelous new variety of rice called IR-8, which has a short stem, is open in structure and gives a high yield of grain. This semidwarf variety, developed at the International Rice Research Institute near Manila, may have a genetic constitution that tips the hormone balance in favor of a vegetative-growth inhibitor. One can guess that its growth promoter may be a gibberellin and its inhibitor may be a dormin; that question remains to be determined by further research.

In the future we may measure the quality of a plant in terms of its hormone balance, just as we now define the nutritional value of a protein by its amino acid composition. Plant breeders will then be able to produce plants with specified properties by deliberately selecting genes to provide a particular ratio of growth promoters such as gibberellin or cytokinin and growth inhibitors such as dormin.

7

GERMINATION

DOV KOLLER
April 1959

In parts of the world where one season is sharply contrasted with the next, the transition from the harshest season to the mildest is heralded by a tinge of green on hill, valley and plain. Not even the humblest roadside, back yard or refuse dump escapes the gentle arrival of new plant growth. Some of this growth is represented by the sprouting buds of established plants, but by far the greater part of it is due to the shoots of newborn seedlings, sprouting from seeds that earlier in the year might have passed for inanimate crumbs of soil. It is this transformation of seed to seedling that we call germination.

To the farmer or the gardener germination seems as inevitable as the progression of the seasons. He expects every seed in his planting to sprout, or else! Meanwhile, beyond his field or garden, the seeds of wild plants also sprout in profusion. But were he to try to cultivate some of these wild plants in a similar manner, the odds are that only a pitifully small proportion of his seeds would germinate.

People who have tried and failed to grow wild plants from seed often conclude that the seeds that do not germinate are dead. Yet it would surely be paradoxical if a plant expended energy,

time and material in making dead seeds. Such a plant would be a poor evolutionary prospect indeed. Investigations spurred by this paradox have shown that practically any seed can be made to sprout under the right conditions and at the right time. We could say that when we cannot cause a seed to germinate, the fault is not the seed's but ours. A few species do produce sterile seeds, but these are exceptions.

If we take for granted the dependable germination of a commercial seed, that is a tribute to the talents and perseverance of the generations of farmers, gardeners and seedsmen who have bred it

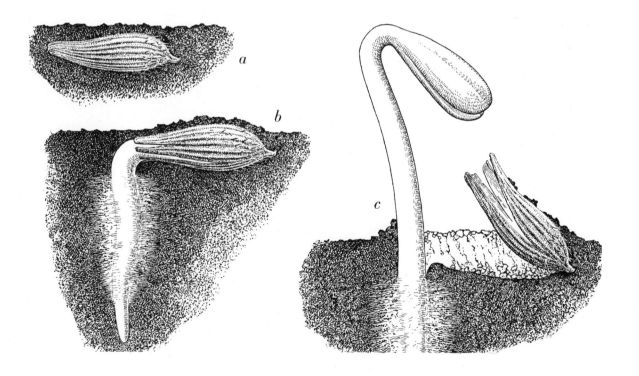

GERMINATION OF A LETTUCE SEED is depicted in these drawings. At first the seed lies "dormant" in the soil (*a*). Then, in response to environmental stimuli, a root sprouts (*b*). Finally the shoot sprouts, leaving behind the outer covering of the seed (*c*).

not only for beauty or utility, but also for full, prompt and uniform germination. For the farmer and gardener such readiness to germinate is an advantage, but for the wild plant it is a hazard. A readily germinable wild plant would literally put all its eggs in one basket; total germination, if followed by drought or disease, might lead to the total destruction of the species.

Consider what would happen in the desert if a drought followed an early shower. If all seeds germinated during the shower, all the annual species in the area would become extinct. Thus for most wild species a reluctance to germi-

nate is a condition for survival, since it ensures the maintenance of seed reserves for later germinations. The seeds of cultivated plants germinate readily, and untold numbers of them are inadvertently scattered, yet how often do we see such plants growing in the wild? The inability of domesticated plants to live under such conditions stems from their very readiness to germinate indiscriminately. If domesticated species have survived the vagaries of the environment throughout the ages, it is only because man has substituted for nature in conserving their seed.

The essential part of the seed is the

embryo it encloses. The embryo starts out as a single cell—the fertilized egg—and ends up by becoming a tiny plant consisting of a miniature root and shoot. In the usual course of development the growth of the embryo stops completely when the seed ripens. Plants that are viviparous, that is, plants in which the embryo continues to grow on the mother plant, are either genetical freaks or very specialized. The embryos of the swamp-dwelling mangrove tree, for example, grow into foot-long, javelin-like seedlings while they are still attached to the mother plant; then they plunge from the tree to embed themselves in the

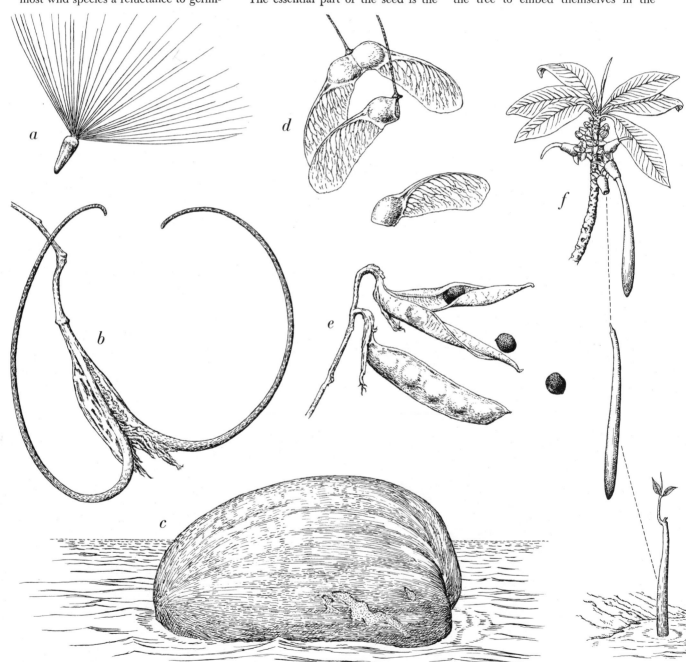

"DISPERSAL UNIT" of a plant is its seed plus other equipment, some of which is responsible for the dispersal of the seed by wind, water or other agencies. The fruit of the thistle (a) is airborne. The devil's-claw (b) and the cocklebur (g) are dispersed when they

catch in the fur of animals. The coconut (c) travels by water. The samara of the maple (d) glides on the wind. Vetch pods (e) pop when they dry, expelling their seeds. The mangrove fruit (f) is atypical in that it sprouts on the parent plant and then falls. The

bottom of the swamp. In most plants, however, the ripe seed becomes detached from the mother plant some time after the embryo has stopped growing.

The embryo thus takes the first step toward an independent existence, but it rarely takes that step unequipped. Accompanying the embryo on its journey into the unknown are several tissues and organs: a food supply, a seed coat and sometimes certain tissues of the fruit, flower, specialized leaves or other organs. The entire structure is known as the dispersal unit. Many parts of the dispersal unit serve fairly obvious functions. Thus the stored food nourishes the embryo until it becomes a self-sustaining seedling, the enveloping seed coat protects the delicate body of the embryo and its food supply, and so on. Other functions are less obvious, but a close scrutiny reveals that many things in the make-up of the dispersal unit serve to determine the fate of the embryo it accompanies, and to some extent the fate of the plant which will grow from it.

One such function is the control of the distribution in space of the progeny in relation to the parent plant. Another function inherent to the dispersal unit is the avoidance of prompt, uniform and indiscriminate germination. Both functions operate in the preservation of the species by decreasing the probability of chance annihilation.

Distribution in Space

Many dispersal units are adapted to make use of some special environmental agency for the control of dispersal. Wind is a typical agency. It can move dispersal units equipped with parachutes (*e.g.,* lettuce, thistle) or wings (*e.g.,* maple) over great distances. The same agency, but a different method, transports the various tumbleweeds (*e.g.,*

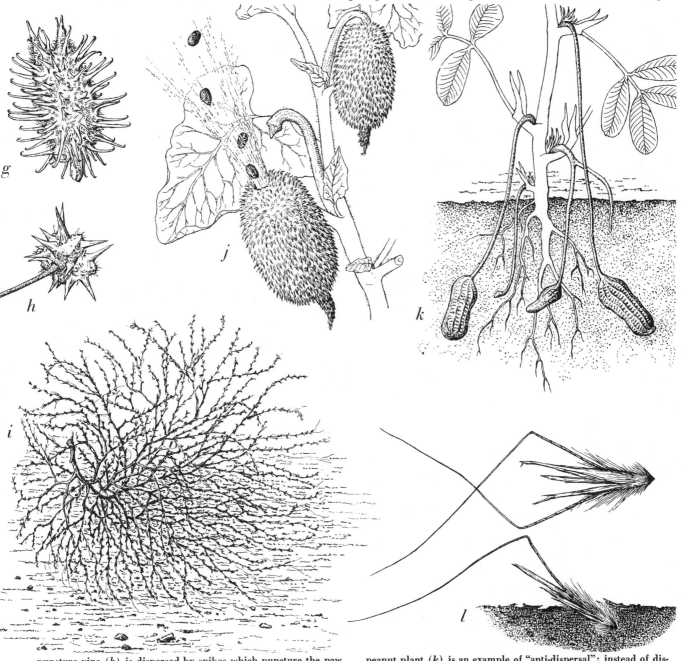

puncture vine (*h*) is dispersed by spikes which puncture the paw of an animal. The entire shoot of the tumbleweed (*i*) is a dispersal unit which scatters seed as it rolls before the wind. The squirting cucumber (*j*) ejects its seed in a sticky jet when it is touched. The peanut plant (*k*) is an example of "anti-dispersal"; instead of dispersing its seeds, it keeps them on a leash. The barbed dispersal unit of the wild oat (*l*) is propelled into the ground by the humidity-driven coiling and uncoiling of its two long "tails."

Russian thistle). Here, after seed ripening, the entire shoot breaks off at the base and rolls before the wind, scattering seed as it goes. Water is a second agency, carrying buoyant dispersal units such as coconuts to distant shores.

A third agency is aerial humidity, which operates in a variety of ways. In some species (vetch, weaver's broom, Impatiens, geranium) the fruit comprises strips of tissue joined edge to edge. As the fruit dries, tension between the strips increases until they part explosively, dispersing the seed. In other species the dispersal unit is equipped with humidity-operated devices for self-burial. The wedge-shaped dispersal units of the wild oat and stork's-bill have a long, humidity-sensitive tail that coils into a tight spring when it is dry and uncoils when it is moist. Barbs projecting from the wedge allow it to move only in the direction of its point. With daily variations in humidity the tail coils and uncoils repeatedly, driving the barbed wedge forward until it meets an obstacle or reaches a depth where humidity is constant.

Animals are a fourth agency of seed dispersal. The dispersal units of some plants (cocklebur, devil's-claw) carry hooks that catch and tangle in animal fur. Others (the puncture vine) have sharp, strong barbs that pierce horny paws. Nonpoisonous dispersal units with a tasty or nutritive fruit attract animals which propagate the seed in a variety of ways. Many small-seeded berries, such as grapes, are eaten whole; the seeds of such plants are carried off in the animal's stomach and excreted without loss of viability. Other dispersal units are

harvested as food by ants, mice or squirrels. Some fruits (e.g., bitter brush) are collected for their pulp, and the inedible seeds are left to germinate in the nest. Some seed-eating animals are too prudent, collecting more than they can consume; others cannot keep track of their numerous caches, thus leaving many seeds to germinate. The juicy, sticky fruit of the parasitic mistletoe is well adapted to dispersal by birds because it adheres to their beaks and is wiped off upon a new host branch. A remarkable dispersal unit is the fruit of the squirting cucumber, which contains a sticky fluid under great hydrostatic pressure. When it is disturbed by a passing animal, this fruit bursts, squirting its seeds in a powerful jet of fluid that glues them to the animal's skin or fur.

While many plants have evolved methods of dispersing their seed over great distances, a few species have achieved the opposite extreme: deliberate prevention of dispersal. "Anti-dispersal" of seed can be observed in the peanut and subterranean clover, the fruit of which develops at the end of a long stalk. The stalk actively grows down into the ground, thus burying the seeds in the immediate vicinity of the parent plant.

Regulated Germination

Plants can control not only their distribution in space but also their distribution in time. In many species few, if any, of the seeds will exhibit readiness to germinate soon after ripening, but over the years more and more will do so. One example is provided by some spe-

cies of mustard plant, in which only a part of each fruit opens after ripening. The seeds that are released from the fruit germinate readily, while the rest remain enclosed in the "beak" of the fruit. Their germination is thus delayed considerably. Another example is provided by many species, notably legumes, in which the embryo is denied access to soil moisture by being enclosed in a waterproof seed coat or fruit coat, thus effectively preventing its germination. The gradual relaxation of impermeability makes a small fraction of the seed population ready to germinate at any given time. Many years may thus pass before all the seeds of a given crop are germinable. In some of these "hard" seeds the impermeable coat must actually disintegrate. The more elaborate seeds of this type have ingenious valves operated by such environmental factors as atmospheric humidity.

The events that trigger germination sometimes suggest a highly specific adaptation to the environment. In California sumac proliferates after a forest fire because the fire causes a waterproof layer of the dispersal unit to become permeable. The localization of certain plant species in pastures is traceable to the fact that their hard seeds are made water-permeable by bacterial action as they pass through the digestive tract of ruminants. Far from damaging the seeds, this process enhances their prospects for germination, and moreover deposits them in a moist, manured environment. The open range is thus kept well stocked with pasture plants. This arrangement sometimes backfires, however, allowing undesirable plants such as mesquite to overrun the range and oust more desirable species.

Today botanists are exploring a new realm of mechanisms that regulate germination more specifically than by mere dispersal in time. These mechanisms help to determine the timing and locality of germination by restraining it in environments and seasons that do not afford a reasonable chance for the plant to complete a life cycle "from seed to seed." Typical of these newly discovered mechanisms is that of chemical regulation, now under investigation in the Earhart Plant Research Laboratory of the California Institute of Technology and in the Department of Botany of the Hebrew University in Israel.

Clearly there is no worse place for a tomato or melon seed to germinate than inside the growing parent fruit; such vivipary would be highly disadvantageous. The warm, moist flesh of the fruit provides just the sort of environment in

SEEDS OF MUSTARD PLANT are scattered when pod bursts. However, some seeds are retained in the "beak" of the pod (*upper right*), which disperses them later when it bursts.

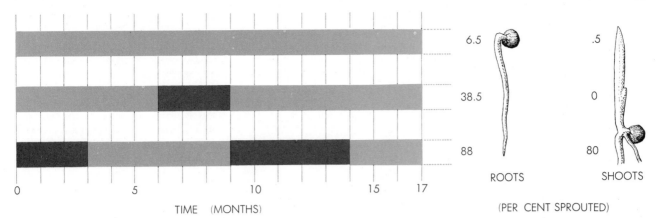

ROOTS   SHOOTS

TIME (MONTHS)   (PER CENT SPROUTED)

6.5   .5

38.5   0

88   80

TEMPERATURE triggers germination in the snow trillium, as shown in an experiment by Lela V. Barton. In this plant, root growth follows one winter; shoot growth, the second winter. In the experiment few plants germinated at a steady warm temperature (*top bar*). With one "winter" plants grew only roots (*middle bar*). Two winters produced roots and shoots (*bottom bar*).

which the seeds might be expected to sprout, yet they rarely do. How is germination delayed until the fruit has fallen and decayed? The prevention of vivipary in most fleshy fruits is due to the presence in them of substances that specifically inhibit germination. Only when the seeds are free of the pulp and juice will they germinate.

More dramatic are the "chemical rain gauges" found in many dispersal units. These are inhibitory substances that are water-soluble, and are therefore readily leached out by rainfall. The amount of inhibitor in the dispersal unit is apparently adjusted so that the amount of rainfall needed to leach it out sufficiently to permit germination will at the same time moisten the soil sufficiently to ensure the plant's subsequent growth. In the dispersal units of wild smilograss (*Oryzopsis miliacea*), local varieties are "gauged" to the rainfall pattern of their habitat. The importance of such rainfall-dependent germination control for the survival of plants in arid or semiarid zones, where rainfall is limited and erratic, will be self-evident.

Another regulatory mechanism found in many dispersal units is the "temperature gauge." In its simplest form the temperature gauge restricts germination of a species to a specific temperature range that is often very narrow and precise. This then distinguishes plants that start their lives in cool climates and seasons from those that do so in warm ones. More highly developed regulation by temperature is found in plants that will germinate only when they are submitted to a specific change in temperature. Most common are the "cold-requiring" seeds, the subject of extensive research at the Boyce Thompson Institute for Plant Research in Yonkers, N. Y. In order to germinate, these seeds require either one or two prolonged exposures

(each of several weeks) to near-freezing temperatures, alternating with one or two exposures to higher temperatures. The apple, the peach and other plants that exhibit such mechanisms are invariably denizens of temperate climates; their ability to avoid germination before prolonged exposure to cold is of high survival value, since it minimizes the danger that seeds may germinate before the hazard has passed.

Moreover, like most other plants, temperate-zone plants are habitat-specific; their entire developmental pattern (growth, flowering and fruiting) is synchronized with the climatic cycle to such an extent that they could not grow normally elsewhere. Their requirement for such "seasonal thermoperiodicity" for germination is thus an important factor in assuring them a start in life in a suitable environment, that is, one that includes a cold season. Unfortunately we have no more than fragmentary knowledge of the physiological nature of this mechanism. But its complexity can be judged from the case of the snow trillium, the root growth of which is induced by a cold period, is carried forward in a following warm period and is not followed by shoot growth unless a second cold period intervenes.

Low Temperature and Light

Especially mystifying are the workings of the temperature gauge close to the freezing point. As yet we have been unable to perceive any of the low-temperature processes at the time they take place. Our only indication that something has indeed been going on is the subsequent growth at higher temperatures. We have as yet no means of distinguishing between a cold-treated seed and an untreated one, except by germination. Under certain conditions the

naked embryos of cold-requiring seeds can be coaxed to grow without cold, but these embryos invariably give rise to plants with dwarfed, unextended shoots. The dwarfism is maintained as long as cold is denied. When the seedling is exposed to cold, it starts to grow normally. Recently it was shown that the plant-growth substance gibberellin will substitute for the cold treatment of dwarfs. Gibberellin also substitutes for cold in the so-called rosette plants, such as endive, where cold treatment causes elongation of stem and flowering. It is interesting to note that the same substance may "cure" hereditary dwarfism in the pea, in corn and in other plants [see "Plant Growth Substances," by Frank B. Salisbury; SCIENTIFIC AMERICAN Offprint 110]. Here however, the gibberellin is not substituting for cold. The nature of the relationship between gibberellin and the cold processes has been the subject of intensive research, but it is as yet unclear.

Another common, though even less understood, response to temperature variation is "diurnal thermoperiodicity," a characteristic of plants that germinate far better under daily alternations of warmth and cold than they do at any constant temperature. Ecologically, such a mechanism can prevent germination in climates, seasons or soil depths where proper temperature alternations do not occur. Physiologically, we have almost no clue to the operation of the mechanism. Rhythmical (or cyclical) phenomena have been observed in many forms of living things: plants, mammals, birds, insects and microorganisms. Many of these phenomena follow a 24-hour periodicity quite independent of the environmental, or astronomical, 24-hour cycle, but capable of being synchronized with it. It is quite likely that the study of this general phenomenon will lead to

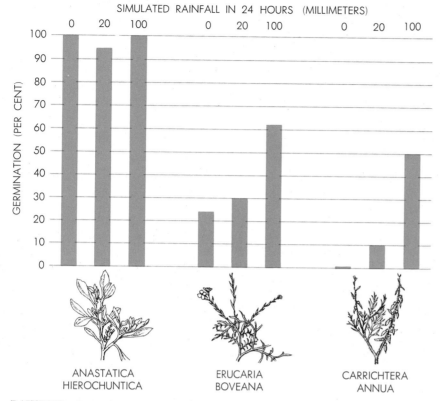

SIMULATED RAINFALL IN 24 HOURS (MILLIMETERS)

ANASTATICA HIEROCHUNTICA

ERUCARIA BOVEANA

CARRICHTERA ANNUA

RAINFALL triggers the germination of many desert plants. This chart illustrates the results of experiments by A. Soriano, who subjected desert plants to artificial rainfall. The seeds of all three species shown here were kept in moist soil. Only *Anastatica hierochuntica* did not require considerable extra rainfall to leach out a chemical that inhibited germination.

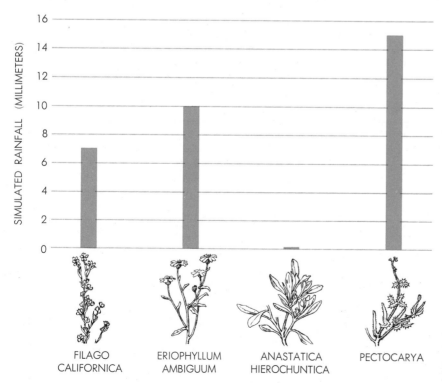

FILAGO CALIFORNICA

ERIOPHYLLUM AMBIGUUM

ANASTATICA HIEROCHUNTICA

PECTOCARYA

OPTIMAL RAINFALL for the germination of four desert plants was also determined by Soriano at the California Institute of Technology's Earhart Plant Research Laboratory.

an understanding of diurnal thermo-periodicity in germination.

Dispersal units incorporate not only rainfall and temperature gauges, but also sensitive mechanisms that respond to light. Such a mechanism in the humble lettuce plant is the subject of research at three research institutions (the U. S. Department of Agriculture laboratories at Beltsville, Md., the Hebrew University in Israel and the University of California at Los Angeles). In darkness lettuce seeds germinate tolerably well only within a narrow temperature range. Given light they germinate promptly and uniformly over a very wide range, and under a variety of conditions that would absolutely inhibit germination in the dark. Dry lettuce seed is insensitive to light, but a few minutes after the seed is moistened it becomes light-sensitive, so sensitive that exposure for a few seconds to light with an intensity of a few foot-candles suffices to produce the full effect. The obvious analogy to photographic exposure extends further: If a soaked seed is exposed to light and then dried, it will retain the "message" it received and, when it is subsequently remoistened, it will germinate in darkness.

A search of the light spectrum for the most effective wavelengths has shown that only the red portion of the visible spectrum stimulates germination. At the same time it was found that far red light (on the boundary between the visible red and the infrared) is capable of reversing the stimulation by red light, thereby inhibiting germination. A flash of red stimulates germination. A flash of far red, following closely, completely reverses the stimulation. This reversal is itself reversed when followed closely by another flash of red, and so on repeatedly. It is always the color of the final light-flash that is decisive. Our understanding of this mechanism is fragmentary. As in the case of the near-freezing of seeds, the results of the process are not immediately visible. We only perceive their end products, namely subsequent germination or nongermination in darkness.

Sensitivity to light implies the presence of a pigment that absorbs the light. The effects of the red and far red indicate some properties of this pigment, but it has yet to be extracted, purified, identified and studied—a process that may take some time, since the pigment doubtless occurs in minute amounts. Luckily for the investigator, light-sensitive mechanisms of this kind are not restricted to seeds; an identical mechanism has been observed in many developmental processes of plants, and may also occur in animals. Thus etiolated plants (plants grown in darkness) will be pale, tall and spindly and will bear unexpanded leaves; upon exposure to light they begin to grow normally. Similarly the study of the relationship between flowering and the relative length of day and night has shown that, in order for the dark period to stimulate flowering in short-day plants or inhibit it in long-day plants, the darkness must not be interrupted by light. If the plant is exposed even briefly to low-intensity light near the middle of the dark period, the effect of the darkness on flowering may be wiped out. It turns out that in both etiolation and flowering the sensitivity to light responds to the same red and far red stimuli as germination.

It may be significant that gibberellin and another plant-growth substance, kinetin, simulate the red-light stimulus that triggers germination, but to complicate matters several substances (*e.g.*, potassium nitrate and thiourea) which are not known as plant-growth regulators, also do so. Another complication is the fact that germination apparently loses its sensitivity to light when the embryos are decoated. It remains to be seen whether light acts on the embryo, somehow making it grow with vigor sufficient to overcome the resistance of the coat, or whether it works on an extra-embryonic entity, perhaps by activating an inhibitor in the coat.

Like a photographic plate, seeds can be over- and under-exposed. The brief flash of light that stimulates germination in lettuce and tobacco plants would be insufficient for the rush *Juncus maritimus*; on the other hand, although continuous illumination works as well as a flash in the case of lettuce, it would inhibit wild smilograss or the Negev saltbush, plants fully stimulated by a brief exposure. The finding that some seeds are as sensitive as fully mature plants to relative length of day and night does not, therefore, come as a surprise, in view of the fact that both have the same responses to the red and far red.

We can deduce some implications of this mechanism for the ecology of plants. Sensitivity to the period of light and dark may determine the season of germination just as it determines flower initiation and the onset or end of dormancy in the buds of trees and shrubs. Inhibition by overexposure to light may be of value in preventing germination from occurring on an exposed soil surface, where treacherous conditions such as rapid drying or high temperature are common. This may be why the germination of many desert plants is inhibited by overexposure. Conversely, inhibition by underexposure may be of value in preventing germination from taking place in poorly illuminated or overpopulated localities. This may explain why many aquatic and marsh plants require light for germination.

It should be said that this account has described only a few of the better known germination-regulating mechanisms. Moreover, it must be understood that several mechanisms are often found in a single seed. A well-known example of interdependent mechanisms occurs in the ordinary garden cress, which germinates only in response to a combination of light and temperature stimuli. *Trigonella arabica*, on the other hand, is a desert annual that has a dispersal unit equipped with at least four independently operating controls: a water-soluble inhibitor, a "hard" seed coat and sensitivity to both temperature and light.

Germination and Evolution

The life of every plant includes critical phases at which it is more than usually susceptible to the vicissitudes of the environment. Apparently it is at these phases—the change from seed to seedling, from the active to the dormant bud and vice versa, from the vegetative to the flowering plant—that regulatory mechanisms, operated by environmental signals, are particularly important. Taken together, they serve to maintain the harmony of the plant and its environment. The regulation of germination does more than this: it also acts to preserve the species by conserving embryonic material and by helping to select a favorable environment for further development of the offspring.

In evolutionary terms the origin and spread of these regulatory mechanisms may be easily imagined. Once created, whether by mutation or the reassortment of genes, the higher survival value that they imparted to their bearers provided the latter with distinct advantages over their kin that lacked these advantages. Selection and breeding by man have in many cases reversed this process, producing plants that germinate at man's will rather than in response to natural signals. These tame plants have minimal germination control. On the other hand, nature's own selection and hybridization have been, and apparently still are, tending toward more efficient control and regulation of germination, as a means for the preservation of plant species.

8

THE FLOWERING PROCESS

FRANK B. SALISBURY
April 1958

What makes a plant suddenly turn from producing leaf buds and begin to form flowers instead? There are at least two good reasons for exploring this mysterious process of nature. Firstly, the question goes to the heart of the origin of form, one of the most interesting and fundamental problems in modern biology. Secondly, if we learned how to control flowering, we could probably find radical and wonderful new ways of growing crops to our desires—in short, we could revolutionize agriculture.

A great deal of thought and experiment has been devoted to the attempt to understand the flowering process, and several years ago the findings up to that time were reviewed in an article in this magazine [see "The Control of Flowering," by Aubrey W. Naylor; SCIENTIFIC AMERICAN Offprint 113]. Since then there have been some rather exciting further developments.

As a preface to this sequel we must recall some of the basic facts that had been established at the time of Naylor's article. Experiments had shown that a plant begins to flower only after it has reached a certain maturity, known as "ripeness-to-flower"; that the main environmental factors which trigger or control flowering are temperature and light, particularly the length of day; that some plants need a long day to start to flower while others require a short day (*i.e.*, a certain extended period of darkness per day); and that the change-over which causes a plant's stem tips to begin forming flowers instead of leaves is controlled by some substance produced in the plant's leaves and transported to the growing tips. This substance has not yet been identified. But there is conclusive evidence that it exists. One of the clinching pieces of evidence is that a plant

incapable of flowering (because it has been deprived of the necessary dark period) can be made to flower by grafting onto it a plant already in flower; clearly the graft must produce this effect by supplying some hormone.

Following up these findings, a number of investigators have tried to analyze the flowering process by breaking it down to a series of steps in the transformation of the plant. This effort is making headway, and I would like to tell about some of the experiments and describe the present picture as I see it.

A favorite experimental plant of plant physiologists working on this problem is the cocklebur. This common weed is beautifully convenient for studying flowering. It is easy to cultivate; it grows well even when defoliated to a single leaf for experimental purposes; and it is highly sensitive to light conditions. It will flower only when its period of night is more than eight and a half hours long. A single exposure to a continuous dark period of the critical length will trigger it to come into bloom. Its growing points develop into clusters of little green flowers and finally into burs.

Apparently the first step toward flowering is accumulation of a sufficient food reserve by the plant. Karl C. Hamner at the University of Chicago discovered this by an elegantly simple experiment. A plant, of course, builds its material by photosynthesis, for which it needs light—the more intense, the better. Hamner starved cocklebur plants by greatly reducing their rations of light, allowing them only flashes at two-hour or three-hour intervals. Then he gave them the long night of more than eight and a half hours, which causes normal plants to start forming flowers. But these specimens refused to bloom.

However, three hours' exposure to sunlight was enough to restore them so that they were able to flower after the necessary dark period. Later James Liverman and James Bonner at the California Institute of Technology showed that the plants' deficiency was indeed one of food substance. They found that they could endow light-starved cockleburs with the ability to flower simply by supplying sugar to the plants.

The first step, then, is the build-up of material in the plant. The second step toward flowering must take place in the required eight-and-one-half-hour dark period. What processes go on during this time? And how does a cocklebur measure eight and one-half hours?

Workers at the U. S. Department of Agriculture research station in Beltsville, Md., succeeded in tracking down one clear clue. It was known that the dark period must be continuous: if it is interrupted by even a single flash of ordinary light, the cocklebur will not flower. The Beltsville workers proceeded to experiment with light of various wavelengths (colors). They discovered that light at the far-red end of the spectrum did not inhibit flowering: on the contrary it counteracted the effect of inhibiting wavelengths. Red light at the shorter wavelengths (*i.e.*, orange-red) was most effective in preventing flowering. But if, after interrupting the dark period with a flash of this light, the experimenters then followed with a flash of the longer-wavelength red light, the cocklebur would flower. In other words, the far-red light reversed the effect of the orange-red light. The plant's behavior depended on which wavelength it was exposed to last: if orange-red, the plant could not flower; if far-red, it did flower.

The Beltsville group developed the

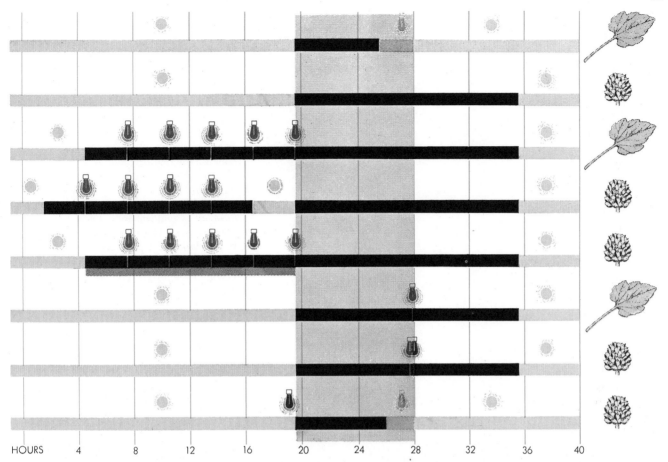

HOURS 4   8   12   16   20   24   28   32   36   40

FLOWERING OF THE COCKLEBUR requires darkness (*black*) preceded and followed by intense light (*sun or yellow bulb*). Too short a night (*top bar*), or light interrupting it (*sixth bar*), results in no flowers (*leaf symbol*). The inhibiting effect of red light is reversed (*fourth*) by daylight. Plant starved of light (*fifth*) flowers if fed sucrose (*brown*) before the critical dark period (*blue band*). Far-red light (*dark red*) reverses the red-light effect (*seventh*). An intense dose (*eighth*) can shorten the dark period required.

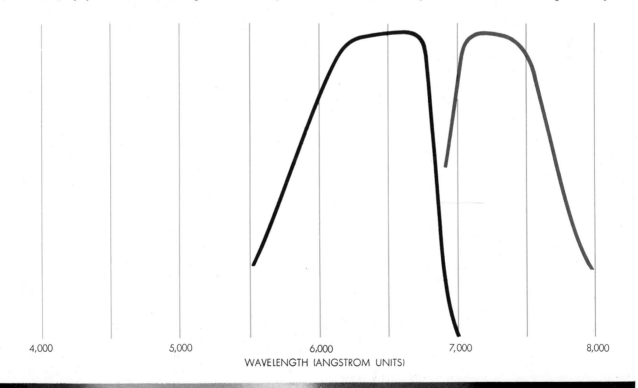

4,000   5,000   6,000   7,000   8,000
WAVELENGTH (ANGSTROM UNITS)

LIGHT interrupting the critical night inhibits flowering most strongly if it is in the orange-red range (*dark curve*). Inhibition is reversed by light of longer wavelengths (*red curve*). Far-red light, near the end of the visible range, has the strongest effect.

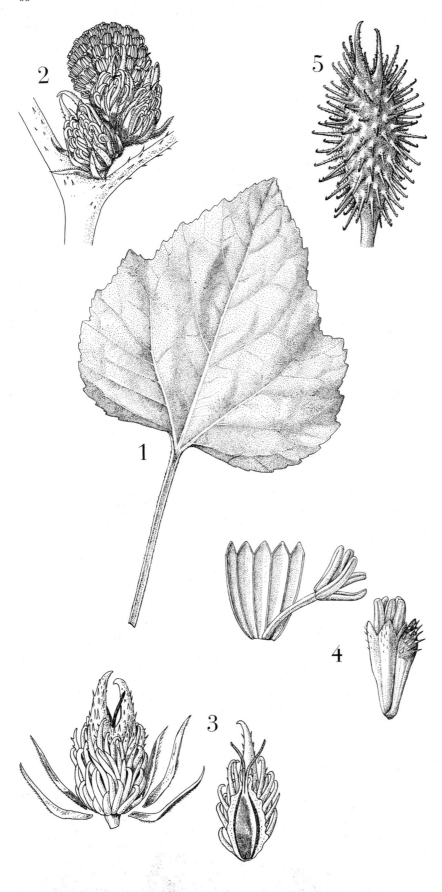

PARTS OF THE COCKLEBUR are shown in these drawings. In the center is the leaf (1). The flower head (2) with both male and female flowers is drawn about seven times natural size. A single female flower is shown whole and in cut section (3). A single male flower is shown whole and cut open (4). The seed pod or bur (5) develops from the female flower.

following theory. Light can produce chemical effects only through some absorbing pigment. Therefore a pigment must be involved in the chemical reactions that lead to the plant's flowering. Apparently this pigment, upon absorbing orange-red light, is converted to a different form which is sensitive to far-red light, and *vice versa*. The form produced by exposure to orange-red light must be the one that inhibits flowering. Since sunlight contains more orange-red than far-red light, when a plant is removed from ordinary light to darkness most of its pigment must be in the inhibitory form. This means that one function of darkness in preparing the plant to flower must be to allow the pigment to change spontaneously to the noninhibitory form.

The next question was: How long does the spontaneous conversion of the pigment take? Does it account for the full critical period; is this the clock that tells the cocklebur when eight and a half hours have elapsed? The Beltsville group investigated by exposing cockleburs to intense far-red light before retiring the plants to darkness, the object being to convert all of the pigment to the noninhibitory form. They found that at most this treatment shortened the critical dark period by only two hours—that is, the plant still needed six and a half hours of darkness to flower. Thus the experiments indicated that complete conversion of the pigment takes about two hours, and this conclusion has been supported by other evidence.

This leaves six and a half hours to account for. What are the further preparations for flowering that go on during that time? Unfortunately we do not yet have any idea: this is still a dark chapter (figuratively as well as literally) in the story of the flowering process. For want of any definite information this unknown step is merely labeled the "preparatory reaction."

The next step is somewhat clearer. The cocklebur needs *more* than eight and a half hours of darkness to flower normally. If it is exposed to light shortly after the end of that critical period, it will flower eventually, but the flower buds develop very slowly. If the dark period is extended further, the buds then develop more rapidly; in fact, the rate at which they will develop after exposure to light depends on the length of the dark period beyond eight and a half hours. This was shown by systematic experiments in which groups of plants were submitted to dark periods of various lengths and then examined under a microscope nine days later. The progress of each group's flower-bud development was directly proportional to the excess of its dark period over eight and a half hours, up to a certain limit.

What does this mean? The simplest and most plausible explanation is that the plant synthesizes the flowering hormone in the late stages of the dark period, and the more time it has for synthesizing the hormone, the faster the flower bud will develop. However, diminishing returns set in after about 12 hours of darkness; in fact, it appears that when the dark period is prolonged to 20 hours or so, the hormone begins to be destroyed.

This brings us to the somewhat controversial next step. The cocklebur will often flower even if it is left in the dark for nine days. But James Lockhart and

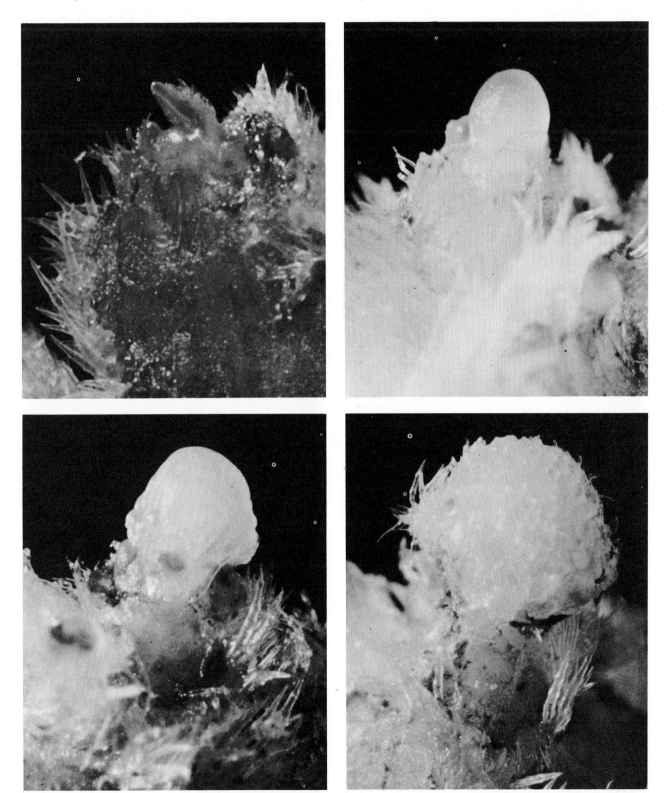

DEVELOPMENT OF THE MALE FLOWER of the cocklebur is divided into eight stages. At upper left is the apical meristem from which the bud emerges, as seen in the third-stage bud at upper right. Small bumps which begin to appear at the fourth stage (*lower left*) develop into flowerets seen in the seventh stage (*lower right*). Floral stages are used to measure rate of flowering.

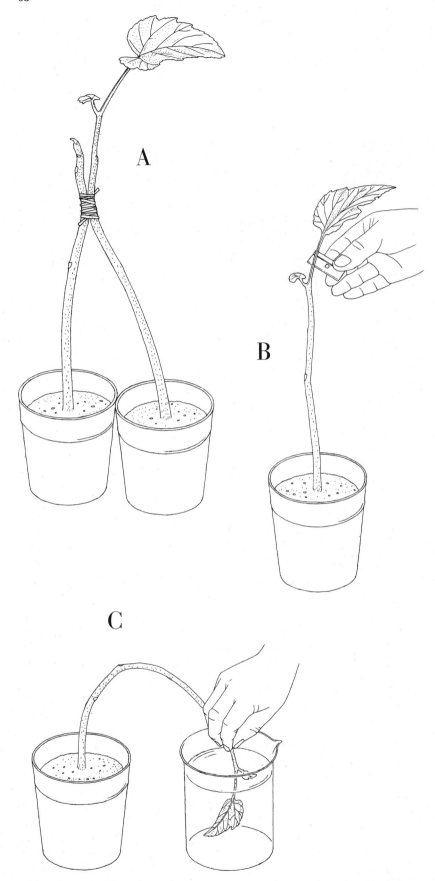

A

B

C

EXPERIMENTAL PROCEDURES used in flowering studies are depicted. A rubber band holds two grafted plants together (A) so that hormone can pass from one to the other. The one leaf on an experimental cocklebur plant is cut off with a razor blade in order to stop translocation—the passage of hormone from the leaf to the bud (B). Cocklebur plants are treated with chemicals by dipping a leaf into a solution of the chemical (C).

Hamner, working at the University of California at Los Angeles, have found that for full flowering development the plant seems to need a re-exposure to high-intensity light, e.g., sunlight, after the dark period. Consequently it may be that this exposure represents another chemical step in the normal flowering process: the light may be needed to "stabilize" the flowering hormone or to stop some process in the leaves that destroys the hormone.

A very simple experiment has disclosed another distinct stage in the requirements for flowering: namely, transport of the flowering hormone from the leaves to the growing stem tips where the flowers form. If a cocklebur's leaves are removed at the end of the critical dark period, the plant will not flower. This must mean that the flowering hormone is still in the leaves and none has yet reached the stem tips.

Finally, there are obviously two more steps in the flowering process: the flowering hormone acts upon the cells in the growing tips in some way to transform, or differentiate, them from the vegetative to the flowering form of growth, and the bud then develops into the specific structures that make up a flower.

To sum up, the flowering of a plant has been resolved so far into about eight steps: (1) the build-up of needed substances in the plant by photosynthesis, (2) conversion of a pigment in the leaves to the noninhibitory form in darkness, (3) another preparatory reaction in the darkness, (4) synthesis of the flowering hormone, also in darkness, (5) a possible further chemical reaction requiring exposure to intense light, (6) transportation of the flowering hormone from the leaves to the growing stem tips, (7) alteration of the vegetative cells there to the flowering mode of growth, (8) development of the flower bud.

All this gives us only a tantalizing general view of the process, as if we were watching the building of a house and could see the foundation, floors, walls and roof go up but were too far away to see any of the details (wiring, plumbing, etc.) that make it a living home. We long for a more intimate view, which means a closer and more direct look at the chemical reactions involved in generating a flower.

At Colorado State University we have recently been doing some experiments which are opening some enticing new avenues of approach. Mainly these experiments take the form of testing the effects of various chemicals applied to the plant before, during or after the dark period.

Some chemicals, known to be general inhibitors of plant growth (*e.g.*, maleic hydrazide), have proved to inhibit flowering in our cockleburs regardless of when they were applied. Others seem to exert a more specific action. For example, the growth hormones (auxins) interfere with flowering only if they are applied before the flowering hormone has traveled from the leaves to the growing tips. Our experiments suggest that they may play some part in destruction of the hormone in the leaves. Another chemical, the growth inhibitor dinitrophenol, prevents flowering only if it is applied to the cocklebur during the dark period. We think that it probably blocks synthesis of the flowering hormone. Dinitrophenol is known to interfere with the production of energy-rich phosphate bonds. If such bonds are the source of energy for the synthesis of the flowering hormone, dinitrophenol may exert its inhibiting effect by cutting off the power supply, so to speak.

One of our most useful tools for dissecting the flowering process has been the ion of cobalt. This chemical inhibits flowering only if applied during the early hours of the dark period. Thus we can narrow down its action to interference with one of two processes: the conversion of the pigment involved in flowering or the "preparatory reaction." Furthermore, among all the chemicals tested so far the cobalt ion is the only one that affects the length of the dark period needed for flowering. In some experiments plants treated with cobaltous chloride did not flower unless their dark period was at least 11 hours long. In other words, cobalt slows down the clock

that ticks off the critical night for the cocklebur.

With refined probing tools such as cobalt and other specific growth regulators we have hopes of eventually being able to pinpoint the substances and reactions that bring the cocklebur to flower. But already the experiments have opened wider horizons. The specific wavelengths of orange-red and far-red light that so powerfully influence the

flowering of cockleburs seem to control many other plant processes—not only the flowering of many plants but also the coloring of tomato skins, the germination of lettuce seeds, the growth of seedlings in the dark, the winter dormancy of tree buds, and so on. Perhaps we are on the track of a key pigment which plays a large role in the plant kingdom, and possibly even among some species of the animal kingdom.

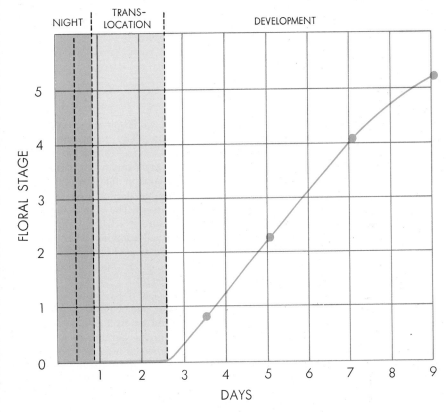

DEVELOPMENT OF THE FLOWER begins after hormone reaches the bud, and continues at a fairly constant rate. In a typical experiment the cocklebur bud reaches the first floral stage in about four days, fourth stage in seven days and fifth stage in nine days.

9

THE RIPENING OF FRUIT

J. B. BIALE
May 1954

Fruits have inspired literature and history ever since the first lady of the race reached out for the apple of the tree of knowledge. The wisdom of the sages of India was nourished by the banana, known as *Musa sapientum*—the fruit of the wise man. It is common folklore that an apple inspired a 17th-century physicist to formulate the laws of gravitation. In the pre-gunpowder age fruits served as a fairly consequential weapon. We are told that the ancient Hebrews revolted against Herod with a barrage of citrons—a fruit like the lemon but many times larger. And of course fruits have fed man through all his history, since long before he knew language or agriculture.

Today biologists are interested in fruit not only as a food but as a fascinating material for the study of life processes. The career of a fruit, from its early embryonic stage to senescence, from the cradle to the grave, spans only a short time. In the brief months of its lifetime it goes through essentially the same chemical and physiological transformations as other living things, and its development can be followed under controlled laboratory conditions.

What is a fruit and how is it formed? A layman and a botanist may have divergent ideas on this question. We all agree that the apple, pear, plum, orange, banana and the more exotic mango and cherimoya are fruits. But how about the avocado, tomato, cucumber, squash, eggplant? A layman would call these vegetables, but a botanist has no hesitation about identifying them as fruits. Any time you bite into the expanded ovary wall of a plant's flower, you eat a fruit. Botanically a fruit is composed of structures responsible for the reproductive function of the plant. The general form of these structures is shown in the dia-

gram of a typical flower at the top of the opposite page. The parts that play the main roles in fruit formation are the stamen—the male organ that bears the pollen or sperms of the plant—and the ovary, which has one or more ovules—the potential seeds. The ovule, a tiny egg-shaped structure attached by a placenta to the ovary wall, contains an embryo sac in which at the time of fertilization there are eight nuclei [*see middle illustration on opposite page*].

In most cases one of these nuclei must be fertilized if a fruit is to be formed. The pollen grains germinate; a pollen tube grows down through the stigma and style in the pistil and through the ovary wall; the tube delivers pollen nuclei within the sac; one of these sperm nuclei fuses with an ovule nucleus to form an embryo, or seed; and the ovary wall then expands, forming the fruit. But as everyone knows, some fruits develop without forming seeds; for instance, the banana, the pineapple, the seedless orange. In some cases there is pollination but no fertilization. The pollen tube stops growing before it has penetrated into the ovule. How can mere pollination stimulate the ovary to grow? The first clue came when it was discovered that dead pollen or extracts from pollen could cause a plant to set fruit. A search for the active substance narrowed it down to an extract which had properties like those of auxins, or plant hormones. By using synthetic auxins investigators succeeded in producing seedless tomatoes and a number of other fruits.

So the present idea is that auxin is required for the growth of the ovary into a fruit. One auxin commonly found in plant tissues is the growth hormone indole acetic acid. Experiments have shown that this hormone can induce the forma-

tion of seedless fruit. Further, it has been demonstrated that indole acetic acid can be made by the oxidation of tryptophane, an amino acid known to be present in pollen. It seems likely that pollen contains an enzyme which causes the conversion of tryptophane or other substances into auxin.

In the case of fruits with seeds we know that a major role in the growth of the fruit is played by auxin in the seeds. This was brought to light clearly and beautifully by the experiments of Jean Nitsch of Harvard University on strawberries. The strawberry develops not from an ovary but from a tissue to which the ovary is attached, known as the receptacle. Its "seeds" (more correctly, its fruits) are those tiny tufts that appear all over the surface of the strawberry. Nitsch found that if all these seeds are removed during the fruit's development, the growth of the strawberry stops completely. When he left a narrow ring of seeds around the strawberry, it grew into a flattened shape. Then he discovered that if a denuded strawberry is covered with naphthoxyacetic or indole butyric acid in a lanolin paste, the fruit develops normally. Chemical analysis has shown that the seeds are rich in the auxin required for growth. Each seed feeds auxin to a limited region.

In fruits with larger seeds a single seed may supply sufficient auxin for the whole fruit. Recently it has been found that an even richer source of auxin is the endosperm, the tissue in which the seed is embedded. Apparently the seed and the fruit compete for the auxin produced by the endosperm; while the seed is growing rapidly, the growth of the fruit is rather slow. Fruit can be forced to a bigger size if synthetic auxins are applied during this period.

Every fruit has four stages of develop-

71

ment. In the first stage, immediately after fertilization, it grows by division of cells. The apple, for example, forms 500,000 to one million cells by successive divisions in the first three or four weeks. Then the cells stop dividing and begin to grow by expansion. During this second stage, lasting up to two or three months, the cytoplasm of each cell moves out toward the edge and the interior is occupied by sap-filled vacuoles, which may take up 80 per cent of the cell volume. Sugars accumulate in the vacuoles; the cytoplasm, which up to now has consisted chiefly of proteins, adds starch. When the fruit has reached its full growth, there follows a third stage in which substances responsible for its aroma and flavor are formed. At the end of this stage, which lasts about as long as the second stage, the fruit is ripe, or at least ready to be picked for ripening. In the fourth stage—senescence—the fruit undergoes conspicuous chemical changes.

Most fruits will eventually ripen to the edible stage on the tree, but some (e.g., the pear and the banana) are best picked green and allowed to ripen under controlled conditions, while a few (e.g., the avocado) will never ripen if left on the tree. The avocado will hang on the tree for months without dropping or reaching the soft, edible state (except when its stem is injured), but if picked when mature it will ripen in one or two weeks at room temperature. The contrasting behavior of the avocado on and off the tree is an intriguing problem. One theory is that as long as it is attached to the tree the fruit is continuously supplied with a substance that inhibits ripening.

The ripening process of fruits has been studied intensively under well-controlled conditions in the laboratory. We have learned a great deal from these studies. Part of what we have learned has to do with the chemical composition of fruits, which is important from the point of view of nutrition as well as the ripening process. The table on page 6 shows the principal chemical components of a number of fruits. In our society, with its wide range of available foods, the fruits are important mostly as a source of vitamins. Nearly all of them are high in vitamin C; the citrus fruits, the strawberry, mango, papaya, black currant and guava especially so. Fruits are not the best source for the B vitamins, but these vitamins are crucial to the fruit. In combination with specific proteins the B vitamins make up respiratory enzymes. And respiration plays the dominant role in the development of fruit. For fruits, as for

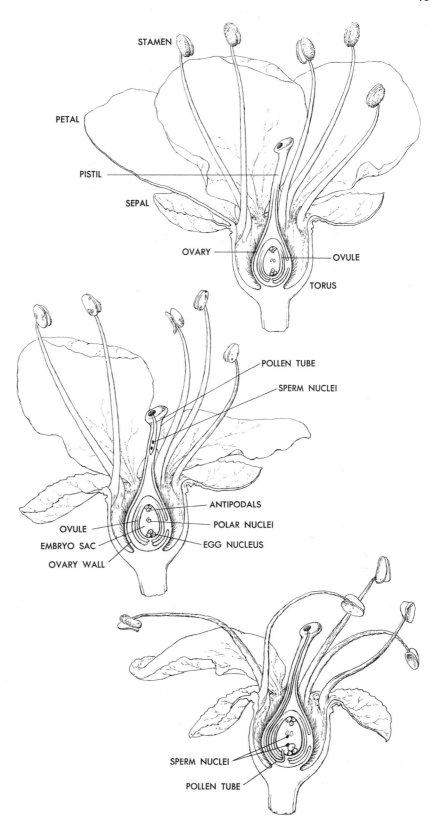

REPRODUCTION OF FRUIT is illustrated by these partially diagrammatic drawings of the apricot flower. The top drawing shows the fresh blossom. One pollen grain from an open stamen has fallen on the pistil. In the middle drawing the pollen tube, bearing the sperm nuclei, has begun to grow. Fertilization takes place when the pollen tube enters the ovule and a sperm nucleus unites with an egg nucleus to produce a new seed, or fruit. In the bottom drawing this begins to grow. By then some petals have fallen and the stamens withered.

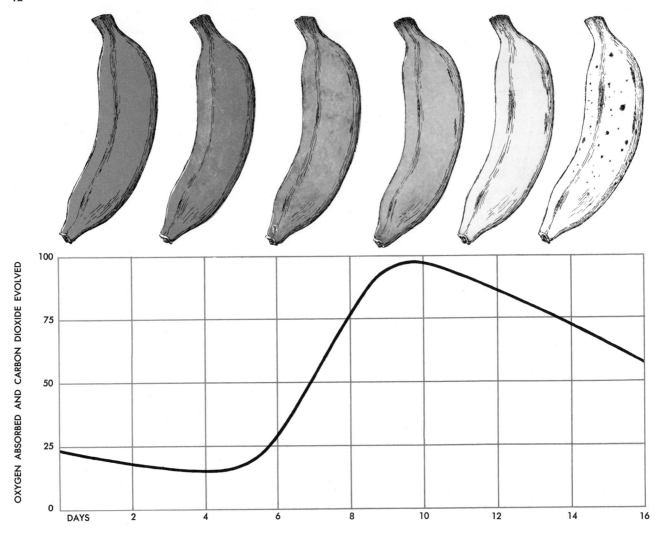

OXYGEN ABSORBED AND CARBON DIOXIDE EVOLVED

RIPENING BANANA undergoes a progressive change of color from deep green to full yellow after ripening is begun. The respiration curve shows that oxygen absorption equals carbon dioxide evolution. The fruit is ripe just after the high point or "climacteric."

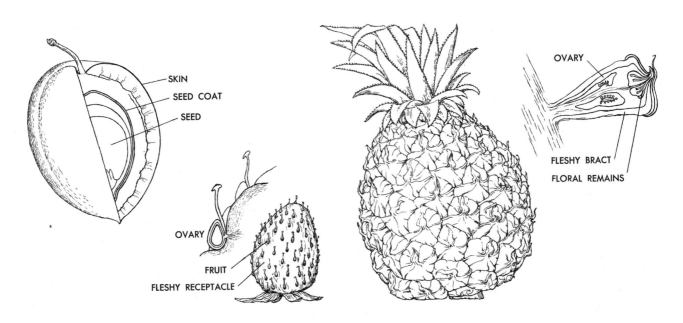

TYPES OF FRUIT are represented by the plum, strawberry and pineapple. The first is a simple fruit, so described because it has one seed. The second is an aggregate fruit with many seeds or fruits in one tissue. The third is a multiple fruit from many flowers.

leaves, roots, bacteria, animals and every living thing, respiration is the very cornerstone of life.

In general terms respiration involves the reaction of one molecule of sugar with six molecules of oxygen to yield six molecules of carbon dioxide, six molecules of water and 680 kilocalories of energy. We find that the respiration of fruit, measured by the cells' uptake of oxygen, is high during the stage of cell division and gradually decreases during enlargement and maturation of the cells. After the cells mature, there comes a sharp rise in respiration rate, followed by a decline. This phase of respiration has been named the "climacteric." Before the climacteric the fruit apparently is resistant to disease. After it, senescence begins to set in, and the fruit becomes susceptible to physiological diseases and invasions by fungi. It is during or immediately after the climacteric that fruits ripen: apples, pears and bananas turn from green to yellow, some avocados from green to dark brown.

Of the chemical changes associated with ripening, perhaps the most characteristic involves the pectins—gel-forming substances. Pectins consist of long chains of galacturonic acid with methyl groups ($-CH_3$) attached to the carboxyls (COOH). When all the methyl groups are split off, the remaining galacturonic acid chain is referred to as pectic acid. As fruit ripening sets in, a water-insoluble protopectin decreases sharply and the pectin content increases. After the fruit is fully ripe, the total amount of pectic substances declines. Meanwhile in some fruits (such as the banana) starch is converted to sugar; in others (e.g., the persimmon) tannins disappear; in many the green pigment chlorophyll disappears. As the masking effect of chlorophyll is removed, other pigments become visible and the fruit assumes its characteristic "ripe" color. In some cases the color is deceiving. A green orange may have as high a sugar content as a golden one, for the changes in the rind of the fruit do not always reflect the condition of the flesh.

Temperature has a marked effect on the course of respiration during the climacteric stage. Within certain limits, the higher the temperature, the sharper the rise and the higher the peak. Low temperatures tend to suppress or completely obliterate the climacteric. If the fruit is kept for a prolonged period at rather low temperatures, the ripening process may be affected adversely. Some fruits are more sensitive than others to low-tem-

perature injuries. Bananas, for example, should not be kept below 53 degrees Fahrenheit. For best results avocados should first be ripened at room temperature and afterwards placed in the refrigerator, rather than *vice versa*. Lemons should be stored at about 55 degrees and grapefruit at 50 degrees.

Fruit physiologists have attempted to slow down the ripening process by storing fruits in atmospheres with less than the usual amount of oxygen (21 per cent in dry air). A low oxygen level retards the onset of the climacteric, but if it is too low, it promotes fermentation and causes accumulation of toxic substances. For many fruits storage in an atmosphere with 5 to 10 per cent oxygen seems to be best: it delays the climacteric, cuts down respiration, prolongs storage life and improves the quality of the fruit.

In the early years of this century many fruit growers made a practice of ripening citrus fruits by forced "curing" in a room with a kerosene stove. They supposed it was the heat that turned the fruit from green to yellow, but investigations showed that incomplete combustion products of the kerosene were responsible. The most active gas was eventually identified as ethylene. As little as one part per million of ethylene in the air will speed the onset of the climacteric rise in respiration. In some fruits, such as citrus, it merely brings about an accelerated disappearance of chlorophyll from the peel. In the banana, on the other hand, it causes an earlier transformation of starch to sugar.

Since ethylene was observed to be effective in minute quantities, it became essential to find a sensitive method for detecting this gas. Pea seedlings were found to be highly sensitive indicators. By the use of this test it was discovered that a number of fruits not only were affected by ethylene but actually produced this gas during the climacteric stage. In a typical experiment air is passed at a constant rate through a jar of fruit and then into a container with pea seedlings kept in the dark. If ethylene is given off by the fruit, the seedlings grow thick, short and curly instead of tall and spindly [*see photograph on next page*]. This test has shown that the ethylene emanations occur only during the fruit's climacteric.

Whether ethylene is the cause or the effect of the rise in the respiration rate is difficult to tell from the pea test. With more precise chemical methods we have recently found evidence that in many fruits the production of ethylene definitely follows the rise in respiration. Moreover, the mango exhibits a typical climacteric but no detectable ethylene. We are inclined to favor the view that while ethylene affects the ripening process, it is not the master reaction. The search for the trigger mechanism that sets off the master reaction is now in progress.

This takes us into a new and fascinating field: the metabolic processes of particles in the cytoplasm of the cell. In order to study the chemical role of these particles in respiration, they must be separated from the other components of the cell. This is accomplished by a series

PROTOPECTIN

COO·CH₃   COO·CH₃   COO·CH₃   COO·CH₃

PECTIN

COOH   COOH   COOH   COOH

PECTIC ACID

CHEMICAL CHANGE characteristic of ripening is the appearance of pectic acid. This derives from pectin, which consists of galacturonic-acid chains with methyl groups attached.

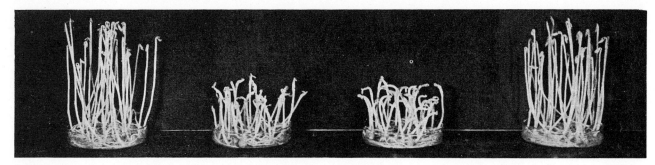

PEA-SEEDLING INDICATORS show the ethylene production of ripening avocados. At right are control seedlings grown without exposure to ethylene. Seedlings at left were grown in air from jar of avocados kept at 5 degrees C.; they show the fruit produced no ethylene. Next two sets of seedlings show effects of ethylene in air from avocados ripening at temperatures of 15 and 25 degrees C.

of centrifugations at different speeds. First the fruit cells are broken up in a Waring blender and centrifuged at low speed (500 times gravity). This removes from the fluid the cell nuclei, cell fragments and some whole cells. The remaining fluid is then subjected to a high-speed centrifugation (17,000 times gravity) which separates the tiny cytoplasmic particles, ranging in size from about half a micron to three microns.

These particles contain highly organized enzymes responsible for metabolic processes common to all living things. The enzymes break down pyruvic acid (a three-carbon compound derived from sugars) into carbon dioxide and water, and they also oxidize certain acids which play a vital role in the oxidation of pyruvic acid. Some of these acids have been found in fruits in fairly large quantities. In addition to their oxidative power, the particles in the cytoplasm of fruit cells are able to utilize the energy of oxidation to form key substances in the energy cycle of cells. They take up inorganic phosphate and add it to adenosine monophosphate (AMP) to form the diphosphate (ADP) and triphosphate (ATP)—a process known as oxidative phosphorylation. The phosphate acceptors or donors supply energy for the cell's chemical factory and for work performed by the organism.

The ability of cytoplasmic particles from avocados to carry on oxidative phosphorylation has been utilized to study the biochemistry of the ripening process during the climacteric rise. Particle suspensions were prepared from fruits at various stages in the climacteric cycle. Throughout the period of the ripening process the particles showed a high efficiency of phosphorylation: that is, a high ratio of phosphorus uptake per atom of oxygen used in respiration. But the addition of a small amount of dinitrophenol prior to the climacteric rise markedly reduces this efficiency. The oxygen uptake is not affected, but much less phosphorus is incorporated into the organic acceptors. Dinitrophenol is a substance which dissociates oxidation from phosphorylation; thus respiration can proceed at a fast rate without the formation of high-energy phosphate compounds. It was found, however, that during the climacteric cycle treatment with this reagent does not slow down phosphorylation. Evidently the relationship between these two major biochemical reactions—oxidation and phosphorylation—undergoes a change in the course of the climacteric.

We are inclined to conclude that fruit cells exhibit essentially the same laws that govern the growth, development and aging of every living thing—whether a plant, a microorganism or liver tissue.

	WATER	PROTEIN	FAT	CARBO-HYDRATES	TOTAL ACID	MINERALS			VITAMINS		
						CALCIUM	IRON	PHOSPHORUS	C	B₁	B₂
TEMPERATE ZONE											
APPLE	84	.3	.4	12	.6	7	.4	12	6	.03	.01
APRICOT	85	1	.1	10.4	**1.6**	13	.6	24			
CHERRY	82	**1.2**	.5	10.5	.5	19	.4	30	9	.05	.02
GRAPE	82	.8	.4	**14.9**	.5	11	.3	10	4	.05	.03
PEACH	87	.5	.1	8.8	.6	10	.3	19	8	.05	.05
PEAR	83	.7	.4	8.9	.4	15	.3	18	4	.06	.07
PLUM	86	.7	.2	8.3	.9	20	.6	27	6	.13	.04
STRAWBERRY	90	.6	.6	5	**1.3**	**34**	.7	28	**55**	.03	.05
SUBTROPICAL											
AVOCADO	68	**1.7**	**20**	5		10	.6	38	16	.06	.13
DATE	20	**2.2**	.6	**75**		**72**	2.1	**60**	0	.09	.10
FIG	78	**1.4**	.4	**20**	.4	**54**	.6	32	2	.06	.05
GRAPEFRUIT	89	.5	.2	10	**1.7**	22	.2	18	**40**	.04	.02
LEMON	89	.9	.6	8	**5.5**	**40**	.6	22	**50**	.04	
ORANGE	87	.9	.2	11	1	33	.4	23	**50**	.08	.03
TROPICAL											
BANANA	75	**1.2**	.2	**23**	.4	8	.6	28	10	.04	.05
MANGO	81	.7	.2	**17**	.5	9	.2	13	**41**	.06	.06
PAPAYA	89	.6	.1	10	.2	20	.4	23	**55**	.03	.04
PINEAPPLE	85	.4	.2	**14**	1	16	.3	16	25	.08	.02

CONSTITUENTS OF FRUITS grown in temperate, subtropical and tropical climates are listed. The numbers in the first five columns refer to grams per 100 grams; those in remaining columns, to milligrams per 100 grams. Abundant elements are in bold type.

III

PLANT ENVIRONMENT

III

PLANT ENVIRONMENT

INTRODUCTION

The growth of plants depends on an interaction between genetic endowment and environment. The environmental milieu includes the radiant energy from the sun, which supplies heat and light; the soil, which provides nutrients and water; and the air, which contains carbon dioxide and oxygen. Agricultural productivity is based largely on the amelioration and control of the environment of plants. The problem of supplying more and superior agricultural products to a hungry and expanding world population requires a knowledge of what plants require for maximum productivity and how these requirements can be satisfied.

Light influences a great number of physiological reactions in plants. The trapping of light energy and its subsequent transformation into useful chemical energy by photosynthesis have been reviewed in the previous section. At energy values much lower than those required for photosynthesis, light affects such processes as flowering, dormancy, tuberization, and seed-stalk formation in some species. Plants respond to the length of light and dark periods (photoperiodism), detecting changes in day length through a light-sensitive pigment common to all plants. In "Light and Plant Development," W. L. Butler and Robert J. Downs review the detective work that resulted in the discovery of phytochrome and its role in the mediation of the light reactions of plants. (See also Salisbury's "The Flowering Process," page 64.) Phytochrome has now been extracted from dark-grown oat seedlings using protein chromatographic methods. The molecular weight has been determined to be on the order of 150,000. The chromophore has been split from the protein and classified as a cyclic tetrapyrrole like a bile pigment. Absorption spectra and thin layer chromatograms of the phytochrome chromophore indicate a remarkable similarity to allophycocyanin—the structure predicted from physiological evidence by Hendricks and Borthwick prior to 1950.

The soil provides the habitat for almost half the bulk of most plants. Soil, in the agricultural sense, is a loose collection of weathered rocks and organic detritus that provides moisture, minerals, and the mechanical stability for the growth of roots. As Charles E. Kellogg points out in "Soil," five factors shape its development: parent rock, land relief, organic matter, climate, and time. Because all of these factors have a large amplitude of variation, and are nowhere to be found in precisely the same combination, classification of soils is sometimes difficult. Yet it is possible to recognize a number of large soil groups, and Kellogg discusses seven of them in some detail. The prescriptions for managing soil for agricultural purposes must be based on an intimate knowledge of what soil can provide in relation to crop needs.

Enough rain and snow fall on the earth each year to cover its 197 million square miles with almost three feet of water. In spite of this huge amount of precipitation, its distribution in time and space are such that vast areas of potentially productive land are deserts. In "Water," Roger Revelle discusses some of the problems encountered in the management of this resource, the principal consumers of which are man's crops. The basic technique of conserving water in arid regions is to concentrate the runoff from a large area and divert it to the point of use. Cultural practices can help prevent unnecessary losses of collected water, and crops with low water requirements can be chosen. In the future we can expect to see desalinized water used for irrigation in arid areas close to oceans. But water is not free, and to meet the cost of new irrigation systems, breeding programs, and cultural techniques, farmers must produce more food per acre-foot of water than in the past.

The weather proverbially engenders much talk and little action. Although recent attempts to control weather have had limited success (cloud-seeding, for example), it must be said that man has yet to succeed in changing the weather. Nevertheless, he has learned to adjust agricultural practices to fit the climate and to match crops with sites for maximum efficiency.

Investigations designed to obtain a detailed knowledge of the environmental requirements of plants are the topic of Frits Went's article "Climate and Agriculture." These investigations have been made possible by the development of highly specialized greenhouses called phytotrons, in which most environmental factors—humidity, temperature, water, light intensity and duration—can be precisely controlled. By using the wide range of artifically duplicated environmental conditions that can be created in the phytotron, it is possible to grow a small crop of any annual plant at any time of the year and determine whether it is climatically adapted to a particular region. With detailed information on the effects of climatic factors on plants, ways can be found to modify climate to suit crops, through such practices as irrigation, mulching, supplemental lighting, shading, and light interruption, and ways can be found to adjust crops to climate through selection and breeding for adaptation. To develop an increasingly productive agricultural system, the total environment of plants must be considered a resource to be wisely managed.

10

LIGHT AND PLANT DEVELOPMENT

W. L. BUTLER AND ROBERT J. DOWNS
December 1966

Various kinds of plant germinate, grow, flower and fruit at different times in the year, each in its own season. Thus some plants flower in the spring, others in the summer and still others in the autumn. And in the autumn, trees and shrubs stop growing in apparent anticipation of winter, usually well before the weather turns cold. What is the nature of the clock or calendar that regulates these cycles in the diverse life histories of plants? Some 40 years ago it was discovered that the regulator is the seasonal variation in the length of the day and night. Since this is the one factor in the environment that changes at a constant rate with the change of the seasons, in retrospect the discovery does not seem so surprising. But how do plants detect the change in the ratio of daylight to darkness? The answer to this question is just now becoming clear. It appears that a single light-sensitive pigment, common to all plants, triggers one or another of the crises in plant growth, from the sprouting of the seed to the onset of dormancy, depending upon the plant species. This discovery is a major breakthrough toward a more complete understanding of the life processes of plants, and it places within reach a means for the artificial regulation of these processes.

The pigment has been called phytochrome by the investigators who discovered it at the Plant Industry Station of the U. S. Department of Agriculture in Beltsville, Md. It has been partially isolated, and it has been made to perform in the test tube what seems to be its critical photosensitive reaction: changing back and forth from one of its two forms to the other upon exposure to one or the other of two wavelengths of light that differ by 75 millimicrons. (A millimicron is a ten thousandth of a centimeter.) Phytochrome appears to be chemically active

in one of its forms and inactive in the other. In the tissues of the plant it functions as an enzyme and probably catalyzes a biochemical reaction that is crucial to many metabolic processes.

The first step toward the discovery of phytochrome was taken in the 1920's, when W. W. Garner and H. A. Allard of the Department of Agriculture recognized "photoperiodism" [see "The Control of Flowering," by Aubrey W. Naylor; SCIENTIFIC AMERICAN Offprint 113]. They showed that many plants will not flower unless the days are of the right length—some species flowering when the days are short, some when the days are long. Indeed, some plants seem to react not simply to seasonal changes but to changes in the length of the day from one week to the next. Photoperiodism explained why plants of one type, even though planted at different times, always flower together, and why some plants do not fruit or flower in certain latitudes.

Other investigators soon reasoned that if a plant needs a certain length of day to flower, then keeping it in the dark for part of the day would inhibit its flowering. They tried to demonstrate such an effect in the laboratory. Nothing happened: the plants always bloomed in the proper season. The riddle was solved when the reverse experiment was tried, that is, when the night was interrupted with a brief interval of light. Chrysanthemums, poinsettias, soybeans, cockleburs and other plants that flower during the short days and long nights of autumn and early winter remained vegetative (*i.e.*, nonflowering). Moreover, they could be made to bloom out of season, when the night was lengthened by keeping them in the dark at the beginning or the end of a long summer day.

Conversely, a brief interval of light interrupting the long winter night induced flowering in petunias, barley, spinach and other plants that normally bloom in the short-night summer season. Artificial lengthening of the short summer night kept these plants vegetative. Interrupting or prolonging the nighttime darkness correspondingly affected stem growth and other processes as well as flowering in many plants.

It was evident that light must act upon a photoreceptive compound, or compounds, to set some mechanism that runs to completion in darkness. As a first step toward elucidating the chemistry of the process H. A. Borthwick, Marion W. Parker and Sterling B. Hendricks of the Department of Agriculture in 1944 set out to determine the wavelength or color of light that is most effective in inhibiting flowering in long-night plants. They exposed each of a series of Biloxi soybean plants, from which they had stripped all but one leaf, to different wavelengths of light from a large spectrograph [*as in middle illustration on opposite page*]. Several days after the treatment the plants were examined for the effect of this exposure upon the formation of buds. Red light with a wavelength of 660 millimicrons proved to be by far the most effective inhibitor of flowering. The cocklebur and other long-night plants gave the same response. By plotting on a graph the energy of light required to inhibit flowering at various wavelengths, the investigators obtained the "action spectrum" of the mechanism that inhibits flowering. This showed that to interfere with flowering, much more light energy is required at, for example, 520 or 700 millimicrons than at (or very near) 660 millimicrons [*see illustration at bottom of page 84*]. The wavelength at which the unknown substance absorbs light

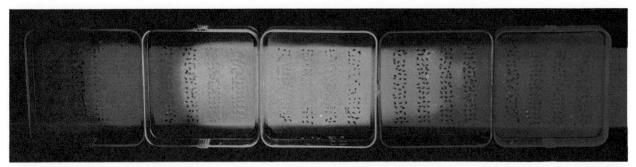

LIGHT FROM SPECTROGRAPH exposes lettuce seeds to different wavelengths. Only 2 per cent of those in far-red light at far left will sprout, while 90 per cent of those in red will germinate. In this photograph only a few of seeds in red region are visible.

CATALPA TREE SEEDLINGS, kept on short days and long nights, are exposed to this spectrum in the middle of the night. Those seen in red will grow. Seedlings in far-red light at far left and all of the others will stop growing and become dormant.

CHRYSANTHEMUM PLANTS at left have had long nights interrupted by period of red light. This divides night into two short dark periods, and plants will not bloom. Chrysanthemums at right also had nights interrupted by red light, but they were irradiated with far-red, as seen here. Far-red light reversed effects of red light, and the plants bloomed just as if nights had been uninterrupted.

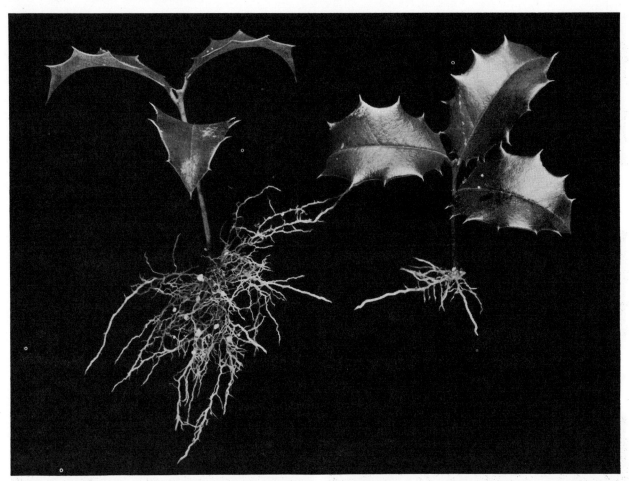

ROOTS OF AMERICAN HOLLY show response to light. Both cuttings were rooted during winter. That at right was kept on natu-ral days and nights. Long nights were interrupted with 30 foot-candles of light for plant at left; its roots grew prodigiously.

MATURE APPLES do not turn red if they are kept in the dark while ripening (right). Ethyl alcohol collects instead of red pig-ment. The apples can manufacture anthocyanin, the red pigment, only if they are exposed to light when they are mature (left).

most efficiently was thus shown to be 660 millimicrons.

Borthwick and his associates then turned to short-night plants. The same spectrographic experiments yielded exactly the same action spectrum. But in this case the effect of the exposure was to promote—not inhibit—flowering! Since the same wavelength of light caused the greatest response in both cases, the investigators could only conclude that a single photoreceptive substance was involved in these two diametrically opposed responses.

Subsequent experiments implicated the same compound in the control of stem elongation and leaf growth. Recent work has shown that light at 660 millimicrons also acts upon mature apples to turn them red by enabling them to make the pigment anthocyanin. The same red light controls the production of anthocyanin in a number of seedling plants.

With the collaboration of a research group headed by Eben H. Toole, Borthwick and his associates next set out to determine which wavelengths of light trigger germination in those seeds that must be exposed to light in order to grow. Many weed and crop seeds are of this type. The action spectrum for the promotion of seed germination turned out to be essentially the same as that established for other plant responses. Whereas about 20 per cent of the seeds germinated in the dark or when they were exposed to green, blue and other shorter-wavelength colors, more than 90 per cent sprouted after irradiation by red light at 660 millimicrons.

This finding led the two groups of investigators to the study of an entirely different effect of light upon germination. It had been observed in the late 1930's that germination was inhibited when seeds were exposed to the longer wavelengths of far-red light which are invisible to the human eye. That observation was speedily confirmed. In fact, seeds that had been pushed to maximum germinative capacity by exposure to red light failed to germinate when they were subsequently irradiated with far-red light. The plotting of the action spectrum for this effect showed that far-red light at 735 millimicrons wavelength is the most potent in inhibiting the germination of seeds.

Still more interesting was the discovery that the diametrically opposed effects of red and far-red light upon germination are fully reversible. After a series of alternate exposures to light of 660 and 735 millimicrons, the seeds re-

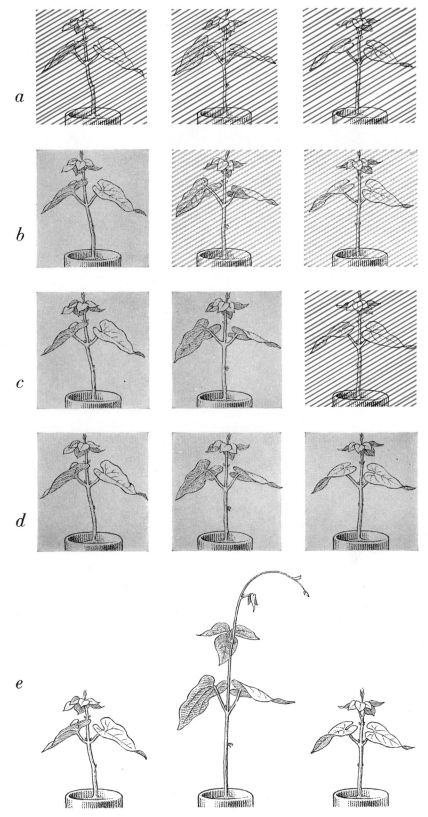

STEM ELONGATION in pinto-bean seedlings is promoted by exposure to far-red light on four successive evenings. All three plants are in red-irradiated condition at end of each day (*a*). Second and third plants are exposed briefly to far-red light (*b*); third plant is given dose of red light, which reverses effects of far red (*c*). Then all three have normal nights (*d*). Some days later first and third plants are still short, but center plant is tall (*e*).

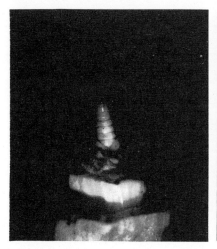

GROWING TIPS OF BARLEY, photographed through dissecting microscope at magnification of 20 diameters, show effects of long and short nights. Tip at left, dissected out from leaves, displays only leaf buds and little growth after being kept on long nights for two weeks. Plant at right had short nights; its tip has grown long and produced many tiny flower buds.

sponded to the light by which they were last irradiated. If it was red, they germinated; if far-red, they remained dormant [*see illustrations at top of pages 84 and 85*]. Now all of the phenomena of growth and flowering had to be re-examined for the effect of far-red as well as of red light. In each case the experiments demonstrated that irradiation by far-red light reversed the effects obtained by irradiation with red light.

The reversibility of these reactions and the clear definition of their action spectra strongly suggested that a single light-sensitive substance is at work in every case, and that this substance exists in two forms. One form, which was designated phytochrome 660, or P_{660}, absorbs red light in the region of 660 millimicrons. When P_{660} is irradiated at this wavelength, it is transformed into phytochrome 735 (P_{735}), which absorbs far-red light at 735 millimicrons. When P_{735} is irradiated with light at 735 millimicrons, it reverts in turn to the P_{660} form.

In order to substantiate these deductions phytochrome had to be extracted from the plant. This required a method for detecting the presence of the compound other than the responses of a living plant. Since those responses occur at sharply defined wavelengths, there was reason to expect that phytochrome itself would prove to be more opaque, or "dense," to light at the wavelengths of 660 and 735 millimicrons when examined in a spectrophotometer—an instrument that measures the intensity of transmitted light at discrete wavelengths. The transformation of phytochrome from one form to the other would also show up well.

Measuring the very small amounts of phytochrome present in plants was not a simple task. K. H. Norris and one of the authors (Butler), respectively an engineer and a biophysicist in the Department of Agriculture, had been studying the pigment composition of intact plant-tissue, and had developed some sensitive spectrophotometers that measured the absorption of light by leaves and other plant parts. With Hendricks, a chemist, and H. W. Siegelman, a plant physiologist who had been investigating the chemistry of phytochrome, they formed a research group to look for the reversible pigment. Initially the absorption of light by plant parts failed to reveal the presence of phytochrome. The plant tissue, however, contained large amounts of chlorophyll, which absorbs strongly at 675 millimicrons. Apparently the absorption of light by chlorophyll—at a wavelength so near the phytochrome absorption peak of 660 millimicrons—was masking the absorption by phytochrome.

It was no great problem, however, to get around this obstacle. Seedlings can be sprouted in the dark and can grow for a while on the food energy stored in the seed; they do not begin to synthesize chlorophyll until they are exposed to the light. Corn seedlings were accordingly grown for several days in complete darkness. They were then chopped up and exposed to red light to put the phytochrome into the P_{735} form in which it absorbs far-red light. In the spectrophotometer a weak beam of light at 735 millimicrons was projected through the sample, weak light being used so that the phytochrome would not change

form. The absorption spectrum showed that the phytochrome was indeed absorbing light at 735 millimicrons; the same sample passed light at 660 millimicrons. On the other hand, after exposure to relatively bright far-red light the chopped seedlings were found to absorb more light at 660 millimicrons and less light at 735 millimicrons. Repeated demonstration of this reversibility fully confirmed all that had been predicted from the responses of whole, growing plants. No doubt remained that a single compound was responsible for the reversible changes in growing plants.

The spectrophotometer measures the changes in "optical density" with such high sensitivity that it can be used to assay the amount of phytochrome in plant tissue. Thus with the help of this instrument a search was instituted for a plant that would supply phytochrome in sufficient abundance for chemical separation. Certain plant tissues, such as the flesh and seed of the avocado and the head of the cauliflower, showed a relatively high concentration of phytochrome. The cotyledons (the first leaves,

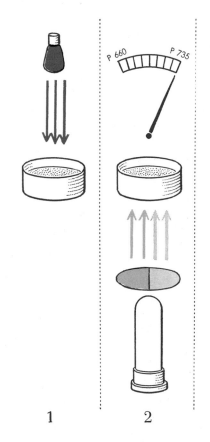

1   **2**

SPECTROPHOTOMETER TESTS, diagrammed here, show effects of red and far-red light on chopped corn seedlings in glass-bottomed dish. Dish is exposed to bright

which feed the seedling) of most legumes synthesize phytochrome, and the concentration reaches its maximum about five days after the seeds start soaking up water. The growing shoot, or hypocotyl, as well as the first leaves of the legumes, contain less phytochrome than the cotyledons. In cabbage and turnip seedlings that have been sprouted in the dark the cotyledons are also a good source of phytochrome. However, five-day-old, dark-grown corn seedlings proved to be the best source, because they develop a high phytochrome content, have large stems and are easy to grow and harvest.

The preliminary and partial chemical isolation of phytochrome was easily accomplished. Corn shoots were ground up in a blender along with water and a mild alkaline buffer, and a clear solution was separated from the solid material by filtration. This extract exhibits exactly the same reversible optical-density changes at 660 and 735 millimicrons as the chopped seedlings themselves. Thorough study of the partially purified material has developed no evidence that any compound other than phytochrome participates in the photoreaction.

Though phytochrome has not yet been isolated in pure form, the outlook is favorable. The compound shows all the properties of a relatively stable soluble protein, and Hendricks and Siegelman are now using the techniques of protein chemistry to purify it. They have subjected it to dialysis (diffusion through a porous membrane), and it has retained its photochemical activity. Oxidizing and reducing agents do not affect it. Moreover, the photoconversion occurs at zero degrees centigrade as readily as at 35 degrees, as would be expected of a strictly photochemical reaction. Higher temperatures denature the protein and destroy its photochemical activity.

Measurements of the amount of red and far-red light necessary to bring about the conversion show that P_{660} consumes only one third as much energy in being transformed to P_{735} as P_{735} consumes in being changed into P_{660}. In both cases, however, the energy consumption is relatively small, indicating that the phytochrome absorbs light efficiently. The pigment should turn out to be a blue or blue-green, the colors complementary to red, but the concentration achieved so far has been too low to make the color visible.

Experiments with growing plants had indicated that P_{735} slowly changes back into P_{660} in darkness, whereas red-absorbing P_{660} is stable. Direct measurement of the changes in the form of phytochrome in intact corn seedlings have confirmed these indications. In seedlings that have never been exposed to light, phytochrome occurs entirely in the red-absorbing, or P_{660}, form. These seedlings are exposed briefly to red light to convert the P_{660} to P_{735}. They are then returned to the darkroom and are examined with the spectrophotometer at intervals thereafter. Such measurements show that it takes about four hours at room temperature for the P_{735} to change back into P_{660}.

This conversion in the absence of light is apparently mediated by enzymes. It is markedly retarded by lowering the temperature, and it does not occur at

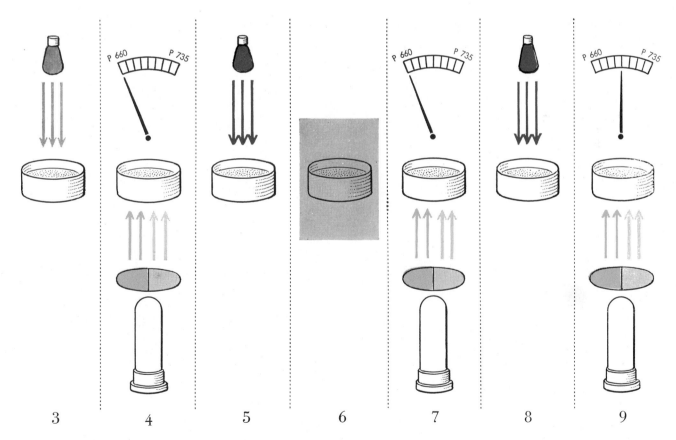

3   4   5   6   7   8   9

red light (1). Filters are used in spectrophotometer to project weak red and far-red light through dish (2); dial shows absorption at wavelength of 735 millimicrons. Now dish is exposed to bright far-red light (3). Next test (4) indicates phytochrome has changed to form that absorbs light at 660 millimicrons. After second exposure to bright red light (5) seedlings are placed in dark for four hours (6), and P_{735} again changes to P_{660} (7). Exposure to red light (8) shows that phytochrome lost half its activity in dark (9).

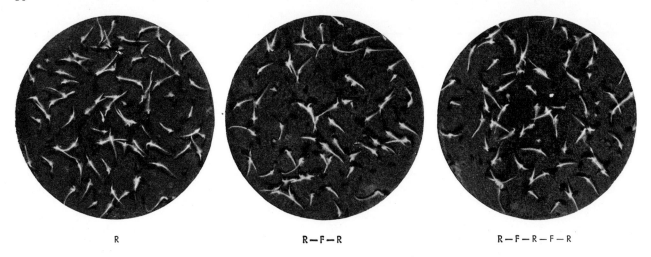

R   R—F—R   R—F—R—F—R

GERMINATION OF LETTUCE SEEDS placed on moist disks of blotting paper is promoted by exposure to red light. "R" indicates an exposure to red light; "F," to far-red light. The last type of light given in the series determines whether the seeds sprout.

all in the absence of oxygen. In the partially purified clear liquid extracts of seedlings, however, P_{735} is stable in the dark, indicating that the dark-conversion enzyme system has been removed. Half the phytochrome activity is lost in intact seedlings during the dark-conversion of P_{735} to P_{660}. After a second illumination with red light, total phytochrome activity declines still more. In continuous light, phytochrome activity is quite low, but is still detectable. It was fortunate that the presence of chlorophyll made it necessary to grow seedlings in darkness for the early experiments. If the seedlings had received even small amounts of light, the unstable P_{735} would have formed and would have soon lost its activity to such an extent that phytochrome might never have been detected.

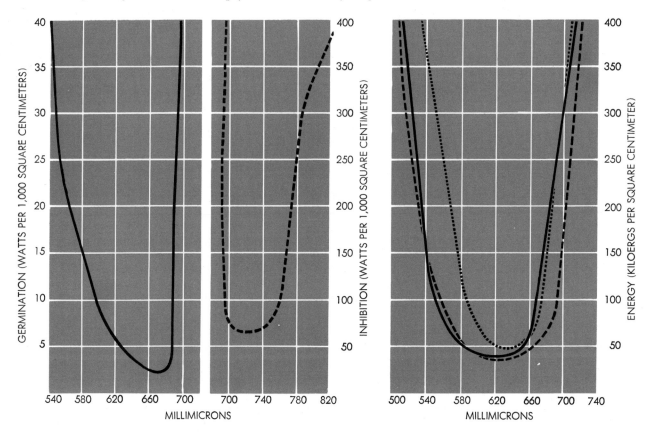

ACTION SPECTRA for the promotion (*left*) and inhibition (*middle*) of germination in lettuce seeds show energy of light required (*vertical scale*) at each wavelength (*horizontal scale*) to produce the desired effect in 50 per cent of the seeds. Curves at right are spectra for the promotion of flowering in barley (*solid line*), and for the inhibition of flowering in soybeans (*long dashes*) and cockleburs (*short dashes*). Barley flowers during short nights, and the other two flower when the nights are long.

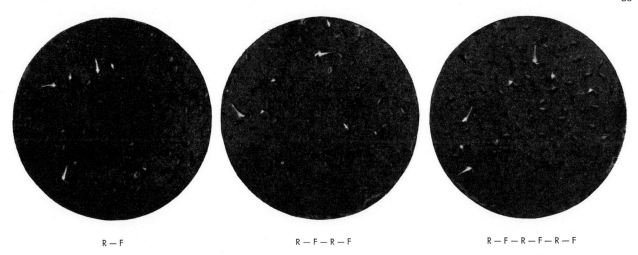

R — F   R — F — R — F   R — F — R — F — R — F

GERMINATION IS INHIBITED by irradiation with far-red light. While the great majority of seeds germinated after final exposure to red light (*opposite page*), practically none of the lettuce seeds irradiated last with far-red light will ever be able to germinate.

How phytochrome exerts its manifold influences on plant growth is still unknown. P_{735} seems to be the active form, while P_{660} appears to be a quiescent form in which the plant can store the potentially active compound. At the close of any period of exposure to light, phytochrome is predominantly in the far-red-absorbing form. The rate at which P_{735} is then carried through the dark conversion back to P_{660} provides the plant with a "clock" for measuring the duration of the dark period. The rate of conversion probably varies from one plant to another, and must depend in part upon such factors as temperature. The effective dark period might be the time required for the complete conversion, or it might be the time in darkness after the conversion is finished. In either case a brief interval of light in the middle of the dark period would cause a plant to respond as though the dark period were short.

Phytochrome is undoubtedly an enzyme—a biological catalyst. Its ability to control so many kinds of plant response in so many different tissues suggests that it catalyzes a critical reaction that is common to many metabolic pathways. Several reactions of this kind are known. One is the reaction that forms the so-called acetyl coenzyme-A compounds. These compounds are essential intermediates in fat utilization and fat synthesis, in cellular respiration and in the synthesis of anthocyanin and sterol compounds. The regulation of the supply of acetyl coenzyme-A compounds would provide an ideal control for growth processes. More than three fourths of all the carbon in a plant is incorporated in this coenzyme at some stage or other.

The extraction and partial purification of phytochrome is the starting point of a major forward movement in the understanding of plant physiology. Further research on this remarkable protein, ubiquitous in the plant world, should answer many questions concerning germination, growth, flowering, dormancy and coloring. It should also provide a means to control all of these plant processes to the great benefit of agriculture.

ABSORPTION

600   650   700   750   800
MILLIMICRONS

ABSORPTION SPECTRA for the two forms of phytochrome are shown here. The form known as P_{660} (*solid line*) absorbs the most light at a wavelength of 660 millimicrons, while P_{735} (*broken line*) is far more absorbent, or opaque, to light at a wavelength of 735 millimicrons. The reactions of plants to these wavelengths indicate that P_{735} is the active form.

11

SOIL

CHARLES E. KELLOGG

July 1950

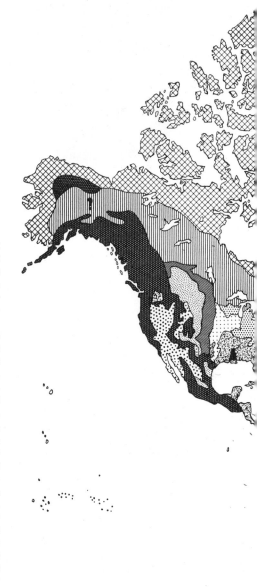

WHAT is soil? A farmer would answer simply: "Something to grow crops in." An agricultural chemist would probably answer: "A foothold from which plants get their water and most of their nutrients." A pedologist (soil scientist), on the other hand, would have quite a different reaction to the question. From his point of view the query itself is awkwardly vague—like asking what is rock, or mineral, or gas. His reply would be that there are thousands of kinds of soils, each as individual in character and properties as a species of rock, of plant or of animal. Every soil has its own unique history; like a living organism it is the creature of a dynamic process of evolution, and a revealing subject for scientific study.

All three of these points of view have contributed to the development of modern soil science. Over the centuries farmers have learned by trial and error how to use their soil more efficiently. During the past 150 years biologists and chemists have determined rather accurately, through thousands of experiments in laboratories, greenhouses and field plots, what nutrients plants need. We now know that plants require some 15 elements for growth. From the air and water they get carbon, hydrogen and oxygen. From the soil they need phosphorus, potassium, calcium, magnesium, nitrogen, iron, sulfur and trace amounts of boron, manganese, copper, zinc and molybdenum. The result of this knowledge has been the development of more effective methods of fertilization, crop rotation and so on. But this is far from the whole story. The physical texture, structure and chemical relations of a soil play a vital part in its ability to store nutrients and supply them to plants. Hence for the scientific management of soil it is necessary to study the history and fundamental characteristics of the particular soil concerned.

Broadly speaking, soil is that part of the outer mantle of the earth that ex-tends from the surface down to the limit of biological forces, *i.e.*, the depth to which living organisms penetrate. Five factors shape the evolution of a soil. They are (1) the parent rock, or hereditary material; (2) relief, or the configuration of the land where the soil lies; (3) climate and (4) living matter, acting on the parent materials, and (5) time. The history of their combined effects is etched in the profile of the soil. Climate, plants and soil are so intimately related that they form predictable patterns: thus black soil, subhumid climate and tall grasses usually go together; light-colored leached soil, cool, moist climate and evergreen forest form another combination; deep-red leached soil, hot, moist climate and rain forest form still another.

Every type of soil in the world has a unique profile. When the profile is laid bare, as in a fresh road-cut or in a deep pit, it can be seen that the soil consists of a series of differing layers, which the soil scientist calls "horizons." The profile is divided into three general levels called A, B and C, corresponding loosely to the surface soil, the subsoil and the substratum, or weathered material beneath the soil proper. Each of these main horizons may have several subdivisions. Soils vary greatly in depth, of course; they tend to become deeper toward the equator and thinner toward the poles. In the U. S. farm soils range from about two to five feet deep, taking the A and B horizons together.

Components of the Soil

Most soil horizons are mixtures of sand, silt and clay, along with some organic matter. Of these constituents, clay has the greatest influence on the chemical activity of the soil. This is because many of the chemical reactions occur at the surfaces of soil particles, and the clay particles, being very small (less than .002 millimeter), are most numerous and offer the most surface. The clay parti-

PRAIRIE SOILS, DE-GRADED CHER-NOZEM

CHER-NOZEMS AND REDDISH CHESTNUT SOILS

DARK GRAY AND BLACK SOILS OF TROPICAL SAVANNAS

SOILS OF THE WORLD are located in a general way on this map. Although each pattern on the map is

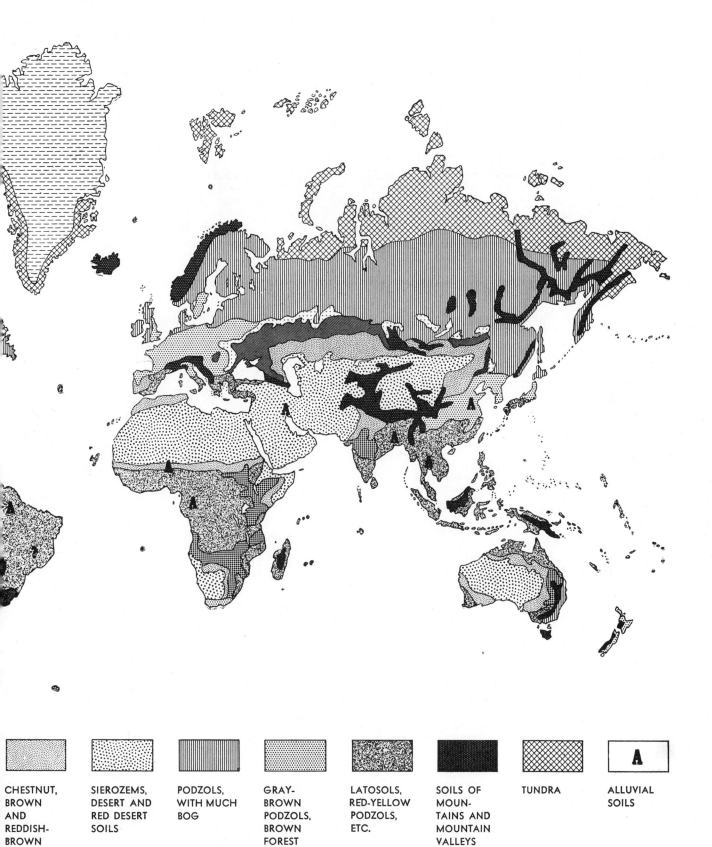

CHESTNUT, BROWN AND REDDISH-BROWN SOILS	SIEROZEMS, DESERT AND RED DESERT SOILS	PODZOLS, WITH MUCH BOG	GRAY-BROWN PODZOLS, BROWN FOREST SOILS	LATOSOLS, RED-YELLOW PODZOLS, ETC.	SOILS OF MOUN-TAINS AND MOUNTAIN VALLEYS	TUNDRA	ALLUVIAL SOILS

a rough approximation of the soil type of the region, many of the patterns include thousands of soil types. The symbol A, which stands for alluvial soils, denotes only a few of the most important alluvial areas. Many small but important alluvial areas are not shown. The pattern on the interior of Greenland is its ice cap.

88

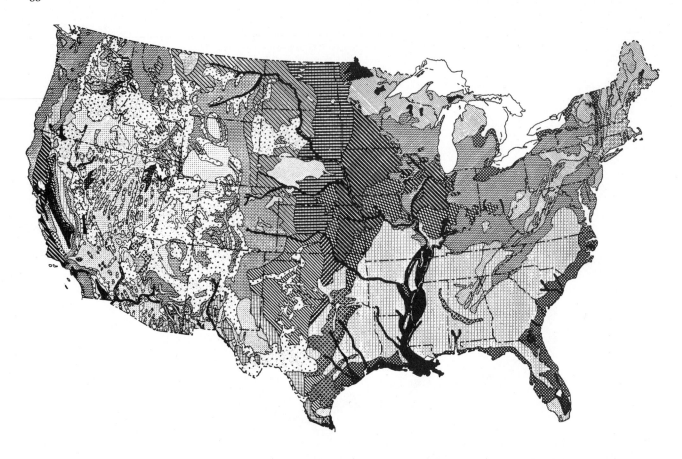

ZONAL

Great groups of soils with well-developed characteristics, reflecting the dominating influence of climate and vegetation.

PODZOL SOILS
Light-colored leached soils of cool, humid forested regions.

BROWN PODZOLIC SOILS
Brown leached soils of cool-temperate, humid forested regions.

GRAY-BROWN PODZOLIC SOILS
Grayish-brown leached soils of temperate, humid forested regions.

RED AND YELLOW PODZOLIC SOILS
Red or yellow leached soils of warm-temperate, humid forested regions.

PRAIRIE SOILS
Very dark brown soils of cool and temperate, relatively humid grasslands.

REDDISH PRAIRIE SOILS
Dark reddish-brown soils of warm-temperate, relatively humid grasslands.

CHERNOZEM SOILS
Dark-brown to nearly black soils of cool and temperate, subhumid grasslands.

CHESTNUT SOILS
Dark-brown soils of cool and temperate, subhumid to semiarid grasslands.

REDDISH CHESTNUT SOILS
Dark reddish-brown soils of warm-temperate, semiarid regions under mixed shrub and grass vegetation.

REDDISH BROWN SOILS
Reddish-brown soils of warm-temperate to hot, semiarid to arid regions, under mixed shrub and grass vegetation.

BROWN SOILS
Brown soils of cool and temperate, semiarid grasslands.

NONCALCIC BROWN SOILS
Brown or light reddish-brown soils of warm-temperate, semiarid regions, under mixed forest, shrub, and grass vegetation.

SIEROZEM OR GRAY DESERT SOILS
Gray soils of cool to temperate, arid regions, under shrub and grass vegetation.

RED DESERT SOILS
Light reddish-brown soils of warm-temperate to hot, arid regions, under shrub vegetation.

INTRAZONAL

Great groups of soils with more or less well-developed characteristics reflecting the dominating influence of some local factor of relief, parent material or age over the normal effect of climate and vegetation.

PLANOSOLS
Soils with strongly leached surface horizons over claypans on nearly flat land in cool to warm, humid to subhumid regions, under grass or forest vegetation.

RENDZINA SOILS
Dark grayish-brown to black soils developed from soft limy materials in cool to warm, humid to subhumid regions, mostly under grass vegetation.

SOLONCHAK (1) SOLONETZ (2) SOILS
(1) Light-colored soils with high concentration of soluble salts, in subhumid to arid regions, under salt-loving plants.

(2) Dark-colored soils with hard prismatic subsoils, usually strongly alkaline, in subhumid or semiarid regions under grass or shrub vegetation.

BOG SOILS
Poorly drained dark peat or muck soils underlain by peat, mostly in humid regions, under swamp or marsh types of vegetation.

WIESENBODEN (1), GROUND WATER PODZOL (2), AND HALF-BOG SOILS (3)
(1) Dark-brown to black soils developed with poor drainage under grasses in humid and subhumid regions.

(2) Gray sandy soils with brown cemented sandy subsoils developed under forests from nearly level imperfectly drained sand in humid regions.

(3) Poorly drained, shallow, dark peaty or mucky soils underlain by gray mineral soil, in humid regions, under swamp-forests.

AZONAL

Soils without well-developed soil characteristics.

LITHOSOLS AND SHALLOW SOILS (ARID-SUBHUMID)
Shallow soils consisting largely of an imperfectly weathered mass of rock fragments, largely but not exclusively on steep slopes. (HUMID)

SANDS (DRY)
Very sandy soils.

ALLUVIAL SOILS
Soils developing from recently deposited alluvium that have had little or no modification by processes of soil formation.

SOILS OF THE U. S. are located here in greater detail than on the map on pages 86 and 87. With the key to each pattern on this map is the name and a brief description of each soil type and the climate and vegetation with which it is associated. Within the area of each pattern are areas of other soils too small to be shown.

cles hold on their surface various positive ions, notably those of hydrogen (H^+), sodium (Na^+) and calcium (Ca^{++}). The acidity or alkalinity of a soil depends on which of these ions predominates: a soil dominated by hydrogen is acid, and one in which sodium prevails is highly alkaline. The type of ions held by a clay affects its transportability through the soil. Clays dominated by sodium or by hydrogen are more sticky and more easily suspended in water than those bearing calcium. Hence if the fine clay becomes acid or highly alkaline, in the absence of excess salts, it can be carried in suspension to lower horizons or even be leached completely out of the soil by percolating waters. Calcium clay, on the other hand, tends to stay where it is.

The proportions of sand, silt and clay in a soil horizon, which may range from pure sand through the various loams to pure clay, determine its texture. Obviously a soil's texture affects its permeability to water and root penetration. Even more important is the soil's structure. This depends on how the primary particles in it are clumped. There are four kinds of structure: granular, blocky (lumpy), columnar (vertically elongated blocks) and platy. For most plants a granular structure or mixed granular and blocky structure is best. A soil's texture stays practically constant, but its structure can be changed drastically, for better or worse. In soil management the maintenance of proper soil structure is as important as the maintenance of plant nutrients.

Also vital for a good growing soil is organic matter: it promotes granular structure, aids in the retention of plant nutrients, supplies food for microorganisms and small animals that help enrich the soil, and, when decomposed, itself furnishes a balanced, slowly available supply of plant nutrients. The organic matter of the soil is a complex of things that includes living roots, microorganisms, a fairly stable brown or black decomposition product called humus and a host of intermediate products. The organic material ultimately breaks down, of course, to water, ash, carbon dioxide and a small amount of other gases.

It is the processes of solution, decomposition, addition and leaching, acting on the original mixture of minerals, that account for most of the great differences in chemical composition among soil horizons and types. These processes cause losses of some compounds from the parent material and increases of others.

Chemical Activity

In most soils only a very small part of the material present is soluble in water at any one time. If this part is leached out, a bit more comes into solution. This mechanism for maintaining a small but replenishable supply of soluble material is essential to plant life. Most crop plants require soils that have no more than .5 per cent readily soluble material, and many have a much lower tolerance. Only a very few wild plants will live in soils that contain as much as 10 per cent of such material.

There is also a self-balancing mechanism that maintains the content of free ions in a soil. The clay and organic matter absorb ions from the soil solution when the ion content is high and release them when it is low. This buffering effect causes a soil to resist any permanent change in its degree of acidity or alkalinity. For example, in an acid soil the free hydrogen ions can be quickly neutralized with lime, but then other hydrogen ions come into solution from the insoluble clays and humus. Thus to reduce the acidity permanently one must keep adding lime until the reserve of hydrogen ions has been used up; or, we could also say, until the calcium ions have replaced nearly all the replaceable hydrogen ions in the humus and clay. Since it is the clay and organic matter that hold the reserve of ions, much more lime is needed to change a heavy clay soil than a sandy soil low in organic matter. We say that the clayey soil is well buffered. Speaking loosely, similar mechanisms maintain low but stable concentrations of potassium, phosphate, magnesium and other ions in the soil.

Every gardener knows that the acidity of a soil is measured on the so-called pH scale. A neutral soil has a pH of 7; a lower pH means that the soil is acid and a higher one that it is alkaline. Natural soil horizons range in pH from about 3 (extremely acid) to above 10 (very strongly alkaline). For most crop plants the most productive pH value is between 6 and 7. A few plants, notably blueberries and cranberries, need an acid soil with a pH of about 4.5.

Over a long period the cumulative effects of the cycle of plant feeding, growth, death and decay greatly influence the character of the soil in which the plants grow. The different types of plants, and even individual species, vary greatly in the kind and amount of nutrients they take from soil, air and water and deposit in the soil when they decompose. Under desert shrubs the amount of nutrients deposited in the soil is very low; under tropical rain forests it is very high. Even under adjacent desert shrubs the soils often differ, ranging from slightly alkaline to very strongly alkaline and from very low in soluble salt to very high, because of the different feeding habits of unlike species and the different salts returned to the surface. In New Zealand, soils under kauri gum trees have a very thick, exceedingly acid, white horizon leached of clay and humus; under adjacent broad-leafed trees the soil is brown, well charged with organic matter and only moderately leached. In the northern part of the U. S., hardwood trees deposit roughly twice as much calcium, magnesium, potassium and phosphorus on the soil as do pines and spruces.

In subhumid, temperate regions, near the boundary between trees and grass, the grasses feed heavily on mineral nutrients, and the soils become dark-colored. Under the nearby forest more organic acids are produced, and the resulting soils are lower in nutrients and organic matter. If, however, we count the nutrients embodied in the plants and animals as well as those in the soil, the forest area has more nutrient material, acre for acre, than the adjacent prairie.

In the humid tropics, near the boundary between rain forest and savanna (grassy plain), the soils present another picture. There during the hot rainy season everything soluble leaches out of the soil; the only nutrients retained are those held by plants in their roots, stems and leaves. In the forest the trees hold a large supply of nutrients; the grasses of the savanna, on the other hand, hold very little, and of this little some is lost because the grass usually is burned off during the dry season and much of the ash washes away with the first heavy rains of the wet season, especially through cracks in the earth.

Thus the fertility of soils varies with climate in a paradoxical fashion: in a subhumid region such as the Minnesota-Dakota boundary more productive soils develop under wild grasses than under forest, whereas near the tropical Congo the reverse is true.

This continual cycle of nutrients out of the soil into plants and back to the soil is perhaps the most important fact of soil dynamics. If the cycle is broken by the removal of nutrients through the harvesting of crops, the soil may deteriorate rapidly. On the other hand, farmers and gardeners can build up the soil to a much higher level of nutrients than its natural condition by adding chemical fertilizers or organic matter "from over the hedge."

Other cycles of change influence the soil. Rains alternately aerate the soil and expel air from it, so that gases are continually exchanged between soil and air. Dust from deserts and distant volcanoes also brings important additions to soils: some highly productive soils, as in Java, exist in the very shadow of volcanoes. The periodic overflow of great rivers and small streams leaves a film of fresh new minerals on the flooded soil. Perhaps nearly one third of the world's population gets its food supply from such alluvial soils, continually being renewed.

Erosion

What does erosion do to the soil? Contrary to a common impression, erosion

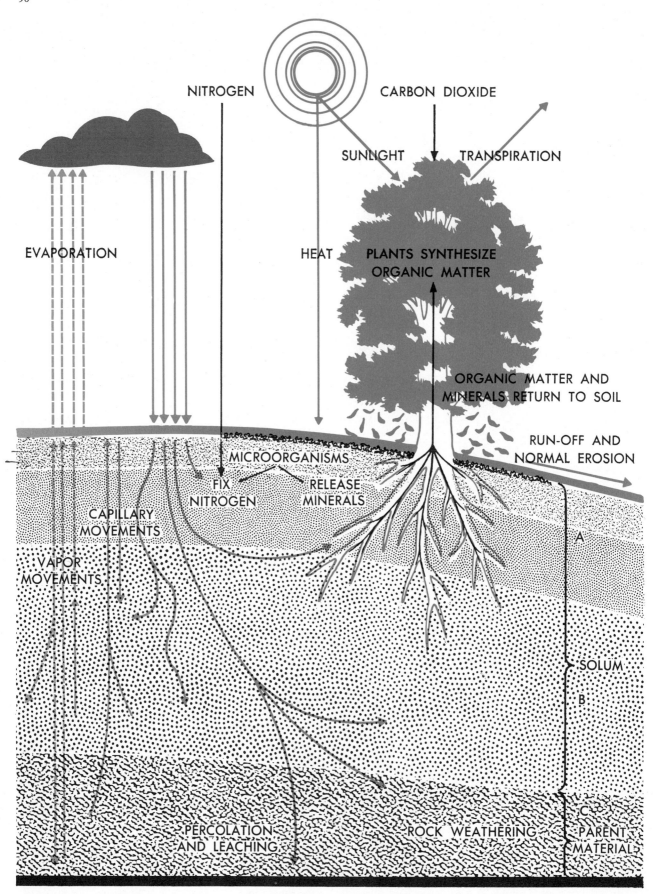

NITROGEN   CARBON DIOXIDE

SUNLIGHT   TRANSPIRATION

EVAPORATION

HEAT   PLANTS SYNTHESIZE ORGANIC MATTER

ORGANIC MATTER AND MINERALS RETURN TO SOIL

RUN-OFF AND NORMAL EROSION

MICROORGANISMS

FIX NITROGEN   RELEASE MINERALS

CAPILLARY MOVEMENTS

A

VAPOR MOVEMENTS

SOLUM

B

PERCOLATION AND LEACHING   ROCK WEATHERING

C PARENT MATERIAL

PHYSICAL AND BIOLOGICAL processes participate in the life of the soil. In the biological process plants remove water and various nutrients from the soil and combine them with carbon from the air. The various plant structures eventually return to the soil and decompose, releasing nutrients for a new cycle of plant growth.

has beneficial aspects as well as harmful ones. It is one of the great natural processes forming the landscape and soil materials. As a result of erosion old lands are continually being removed and new ones built. Slow, normal soil erosion is an essential process by which upland soils maintain their productivity. As the surface of the upper horizon of a plant-covered soil is gradually eroded away, at the rate of perhaps one inch in 100 to 1,000 years, the upper part of the horizon beneath is gradually changed and becomes part of the top horizon, and so on in turn down through the soil profile, until finally new, fresh minerals from the underlying rock come within reach of plant roots and are incorporated into the bottom horizon of the productive soil.

The damage comes when natural erosion is greatly accelerated, as a result of excessive burning, grazing, forest-cutting or tillage, so that the horizons are removed faster than they form. The most serious erosion injuries to soils are not chemical, not the loss of plant nutrients, but the changes in structure arising from the loss of granular surface horizons and the exposure of massive clay, hardpans or even rock.

We cannot stop all rapid erosion. In regions of active uplift and near the margins of spreading deserts, active erosion is inevitable, even under wild vegetation. But we can and should control it where practicable. The basic method of control is to maintain a protective cover of plants. This sounds simple enough, but in practice it is complicated by the fact that the cover must be productive as well as protective.

These, then, are some of the general principles governing the formation of soils. They should make clear why many thousands of soil types exist in the world. Now let us look at some actual soils in the field.

The Gray-Brown Podzolic

Suppose we drive out into Marion County, Iowa, some 30 miles southeast of Des Moines. Here we see a gray-brown soil such as can be found in almost any county east to the Atlantic Ocean. This type of soil is known as Gray-Brown Podzolic, from the Russian word *podzol,* meaning ashes. Until 150 years ago the Gray-Brown Podzolic soils supported the heartland of Western civilization. They have developed under a humid temperate climate and are generally associated with hardwood forests (here in Marion County the trees are mainly oaks and hickories).

The profile of this soil is not particularly dramatic. At first glance it looks like just a plain, brown-colored garden soil. Generally the upper six inches of the A horizon is very pale brown or light brownish-gray silt loam, crumbly and mellow. This surface horizon receives the decomposition products of fallen leaves, and in it the small animals, surface roots and microorganisms are most active. Directly underneath it the soil is more leached. This A_2 horizon, some five to 10 inches thick, is a yellowish-gray, floury silt loam. The particles are oriented into thin plates, easy to crumble. Next comes a transitional zone in which soil of the B horizon is slowly being transformed to A. Under this thin transitional layer lies the main B, some eight to 12 inches thick. It is much richer in clay, coarser in structure and a darker brown than the A above. Its silty clay loam has a blocky structure; the blocks are the size of filberts or walnuts and often sharply angular at the corners. Its coarse, open structure is maintained by the penetration of growing tree roots and by the slight movements of the roots as the wind sways the trees. If it is cleared, plowed and exposed directly to the sun and rain, this horizon is likely to become massive—sticky when wet and hard when dry. Beneath the main B is another B subdivision some four to eight inches thick; this is a dark yellowish-brown, tough clay, packed into larger blocky aggregates with sharper angles.

Thus the A and B horizons of this soil together are some 30 to 36 inches thick. The C horizon, or parent material from which the soil is made, is a blocky, light yellowish-brown and pale olive clay.

Gray-Brown Podzolic soils may develop from limestone, shale, glacial drift or other rocks. However derived, all of them have similar profiles, though they may differ in some details; some are more sandy, a few richer in clay, some reddish in color because of the red color of the parent rocks. The Gray-Brown Podzolic soils are normally acid—too acid for the best growth of crops. So liming is essential. Because these soils need occasional rotations of alfalfa and other good legume hays to maintain organic matter and good structure, the best farming systems usually keep some livestock on them.

Perhaps more is known about this group of soils than about any other. The soil of the famous Rothamsted Experimental Station in England—the oldest agricultural experiment station in the world—is quite a bit like this. After 106 years of continuous culture for a single crop—wheat—without fertilizers, manure or legumes, the soil of this station still yields 10 to 12 bushels of wheat per acre, a little over the world average! But of course this is no way for a farmer to handle such soil. By using lime to correct acidity, phosphate and potash fertilizers, farm manure from the livestock and rotation plantings of clover, a farmer can get 40 to 45 bushels of wheat per acre from this soil.

The outstanding feature of this type of soil is the wide range of cereals, grasses and legumes, root crops, fruits and vegetables that grow well in it. If well managed, it is dependable. But some farmers are neglecting lime, minerals and legumes, and even allowing the soil to erode. With poor farming, this soil deteriorates rapidly to low levels of productivity. Thus on what was originally the same kind of soil one can find both very poor fields and excellent fields that have been made even more productive than the natural soil.

The Clarion

Suppose now we drive on west to another spot near Des Moines. Here the original native vegetation was tall grass rather than forest. The soils are nearly black. We shall cut our profile pit in a gently sloping place where the last great glaciers left a limy mixture of sand and boulder clay, along with some stones. We find that the surface A horizon, some eight to 14 inches thick, is a very dark brown to brownish-black fine granular loam. Below this the soil fades gradually into the other horizons; the boundaries are not so sharp and the differences are less prominent than in the Gray-Brown Podzolic. There is a transitional horizon some eight to 10 inches thick of a brownish-gray to dark yellowish-brown loam, with some streaks of the darker soil from above. These two upper layers are only slightly acid, while those beneath are neutral or calcareous. The top B layer under the transitional horizon is a yellowish-brown loam seven to 11 inches thick, and below this is another B of light yellowish-brown heavy loam or light clay loam about one foot thick. The A and B horizons together have a depth of some three feet. Below them lies a deep, pale yellow, mixed limy glacial till.

What we have been looking at is the Clarion soil—a Prairie soil representative of a group of nearly black soils that are dominant in the Corn Belt. Curiously, nowhere else in the world can one find another large area of soils exactly like these. The Clarion soil is in a transition zone between the forests of the more humid East and the grasses of the drier West. It has high fertility and nearly as wide a range of crops as the Gray-Brown Podzolic. Severe droughts are rare. The grasses of the region have been heavy feeders on minerals from the whole soil, and have concentrated them, along with abundant amounts of humus, in the A horizon—a great storehouse of mineral nutrients and nitrogen. The humus content of this soil can be maintained only under continuous grass. Any tillage and exposure of the soil to the sun is certain to reduce its humus, and no reasonable additions of manure or compost can maintain it at the original level. But a rotation of corn, small grains and mixed grasses and legumes can keep the humus at about 70 per cent of the natural level.

FOUR IOWA SOIL TYPES are viewed in profile. At the left is the profile of the Weller silt loam, a Gray-Brown Podzolic soil. Second from the left is the profile of the Clarion loam, a dark brown Prairie soil. Third from the left is the profile of the Edina silt loam, a Planosol char- acterized by a gray, leached horizon over a dense clay- pan. Fourth from the left is the profile of the Webster silty clay loam, a Wiesenböden soil characteristic of regions with poor drainage. All four of these photo- graphs show the soil to a depth of about four feet.

The Prairie soils usually need some phosphate fertilizers. Most of them also respond to extra nitrogen, especially for corn. Soil erosion should be controlled, for the loss of the surface horizons of this soil means serious depletion of organic matter, of nitrogen and of phosphorus. Generally, however, erosion of a Prairie soil does not expose an intractable B horizon, so the soil can be built back rather easily even after a great amount of erosion, short of deep gullying.

The Chernozem

A bit farther west and north—say west of Fargo, N. D.—we enter a region of subhumid grassland with another type of nearly black soil called Chernozem, from the Russian for "black earth." This soil is found over large areas of central Eurasia and North America. These great areas in the interior of continents were little cultivated before the age of modern science; not until railroads crossed the continents did the Chernozem soils become important in world commerce. Now a large proportion of the bread grains of the world come from them.

In contrast to the Gray-Brown Podzolic, the Chernozem soils have a limited number of adapted crops—mostly cereals. Lately breeders have developed some varieties of corn that also do fairly well in them. Unlike the Gray-Brown Podzolic soils, which need to be built up in productivity, the Chernozem soils have their maximum fertility when first plowed. Because there is relatively little natural leaching, these soils require little fertilizer unless the land is artificially irrigated; then both phosphorus and nitrogen are valuable. It has been found that under normal conditions soluble phosphates give worth-while returns on many of the Chernozems. These soils need no lime; one of their characteristics is a layer of accumulated lime just beneath the B horizon.

In the next drier belt beyond the Chernozem toward the Western desert are the Chestnut soils, somewhat like the Chernozem soils but lower in organic matter and even less leached of lime. The climate is still more uncertain; almost every year is drier or wetter than the "average." The uncertainty of climate and the rather limited range of crops that can be grown in the soil make its farming more hazardous. This has had an effect on the region's politics. In the areas of the Gray-Brown Podzolic soils, which are dependable and support a very wide range of crops, farmers developed a preference for individual effort and *laissez faire*. Farmers on the Chernozem-like soils, on the other hand, are forced to think more in terms of cooperation. They think more about controlled railroad rates, crop insurance, cooperative credit and the like. Many of the so-called radical ideas of government—that is, radical

to believers in *laissez faire*—that have developed in the U. S. during the past 100 years originated in this area of grassland soils.

The Podzol

North of the Gray-Brown Podzolic region in the U. S., and over large areas in the northern part of the U.S.S.R. and northern Europe, is another type of soil known simply as Podzol. It is named for the ashlike, nearly white leached horizon that lies just under its organic mat. This soil is found in humid cool-temperate regions of coniferous forest or mixed conifers and broad-leafed trees. The soil is normally quite acid. In summers it is commonly fairly dry, for a while at least. But during part of the time leaching is severe—more so than in the Prairie or Chernozem.

In the woods the Podzol has a striking profile. Decomposition is slow in this cool climate, so a mat up to one foot thick of partly decomposed and matted leaves and twigs lies over the mineral soil. Beneath the organic mat, the A_1 horizon is less than half an inch thick. Directly under this is the nearly white A_2 horizon of maximum leaching; it is a loamy soil structured in thin fragile plates. Most of the roots are in the upper part of this horizon or just under the organic mat, although a few penetrate deeper. When the Podzol has been severely burned by forest fires, however, the vegetation often changes from the rather shallower-rooted pines to the more deeply rooted oaks until a new layer of organic matter has formed again. Ordinarily the white A_2 horizon is only two or three inches thick, but well-developed ones average more than a foot, and the Kauri Podzols of New Zealand have A_2 horizons up to 10 feet thick!

Below the white A_2, sometimes separated from it by a very thin transitional zone, lies a dark brown layer, the main B horizon. In this layer are accumulated iron and organic and mineral colloids leached down from above. When the soil is largely sandy, the B is cemented into a continuous but irregular hardpan. Many Podzols in Britain have a very thin, solidly cemented hardpan or "iron stone" just at the top of the B horizon. But the hardpan of the Podzol is not very impermeable to water, and strong roots also can get through it. Good management, liming, use of fertilizers and the growing of legume-grass meadows in time disintegrates the hardpan. The whole soil changes to something like a Prairie soil, only brown instead of brownish black, gradually lightening in color down to the C at some 15 to 30 inches.

Most of the Podzols are in geologically young landscapes. Many are stony. Usually swampy soils are mixed with them. In the Northern Lake States, for

example, perhaps half of the land is swampy, not counting lakes. Only rarely do we find large areas of stone-free Podzols suitable for farming. The usual thing is to find relatively small areas of arable soils intricately mixed with others.

Nevertheless, Podzols are responsive to management and can be made economically productive. They can grow vegetables, especially potatoes and root crops, small grains, and good pasture feed and hays for dairy cattle. After clearing the land, the first step in the improvement of the soil is liming, so that legume-grass mixtures may be grown. Recent experience in Wisconsin suggests that liming also helps to release the native sources of phosphorus. It is necessary, however, to add phosphate fertilizers, especially at first, and potash as well; the native sources of potash in Podzols are only very slowly released.

Because a number of efforts to cultivate the very sandy Podzols in the Northern Lake States have failed, some say that this region "should all go back to forest." Some of it should. But it is equally important that the potentially arable Podzols in the region be used for farming in order to develop a good balance among forestry, mining, recreation and agriculture. With the recent improvement in the agricultural arts, farming on Podzols has a much better outlook than formerly.

The Tundra

Now let us look at some of the soils of the "frozen North." North of the Podzols, beyond the Brooks Range of Alaska, we come to a great treeless plain of Tundra soils. From a plane on a sunny day, the Tundra looks like a great grassy pasture. Most of the land is well covered with low plants—dwarf willow trees some three to six inches high, little arctic shrubs and flowering plants, grasslike sedges and reindeer moss. Little swamps and ponds dot the plain until near the Arctic Ocean there is nearly as much water as land.

Though the annual rainfall is light—no greater than that of the semi-arid West—the climate is humid, because of low temperatures and the consequent low evaporation. North of an average annual temperature of some three to five degrees Fahrenheit the subsoil is permanently frozen, except where the substratum is gravelly and drainage is very good. The soil lying on this permafrost has poor drainage in summer. As winter approaches again, the surface freezes, leaving a wet, squashy mass of soil between the two frozen layers. Pressures develop that force the viscous soil into mounds and blisters of many sizes, shapes and forms. Coarse sedges or other plants growing on the higher spots add to the mound-building with thick mats of roots.

If you dig into the Tundra soil near

Barrow in early July, you will find a surface layer some three to four inches thick of acid, brown, loamy soil, filled with fine roots and rich in humus. Beneath this is a grayer, mushy soil, also rich in humus. At nine to 12 inches you will strike the permafrost, here a rock-like, brown and grayish-brown loam, streaked and mottled with nearly white ice.

So far little beyond uncertain gardening and extensive grazing has been tried on the Tundra. No crop plants with the ability to grow at low temperatures such as the native plants can tolerate have been developed. Some day they may be.

Before we leave Alaska, we might take a quick look at the main productive soil of the Matanuska Valley. It is south of the permafrost line and developed under forest. During its formation strong winds from the northeast blew silt out of the bed of the Matanuska River—a glacier-fed stream. This gradual accumulation of fresh, fine mineral, still going on, gradually covers the soil, burying the roots, leaves and twigs. The result is a dark brown, silty soil, fairly rich in organic matter and some two to six feet deep. The soil is covered with a mat of moss, roots and other coarse material which, when plowed into the surface, adds to its stock of organic matter. This soil has good promise for vegetables and meadows, especially with improvements in plant varieties and culture practices.

Tropical Soils

It is in the tropics that we find the greatest variety of soils. Because of severe weathering there and a great diversity of seasonal combinations of rainfall, temperature and humidity, probably more soil types exist within the tropics than in all the land outside.

Suppose we look at a red soil a few miles north of Matadi near the Congo River in Central Africa. The annual rainfall is about 46 inches. Even during the dry season the sky is often cloudy and the air quite humid. Some 1,100 feet above sea level on a long gentle slope under a dense tropical rain forest we cut our pit. We do not find a thick mat of organic matter such as one might expect under trees. Although the annual fall of leaves is high, they decompose so rapidly that only a very thin litter rests on the surface at any time. The top inch of the soil itself is a highly granular, very dark, reddish-brown clay; below this the main A horizon to a depth of eight inches has the same color but a more irregular nut structure, well permeated with roots and insect holes. At 20 inches, below some transitional layers, we come to the main B horizon: red clay in a firm blocky structure that crushes easily in the hand. Below 40 inches this gradually fades in color and structure through a transitional horizon to the C at six or seven feet. The

latter is a mottled mixture of red and yellowish-red clay with quartz pebbles and stones. The deep rocks underneath are chemically basic.

This general type of soil is known as Red Latosol. It has hundreds of variations. At varying depths there is often a heavy band of quartz pebbles and stones, marking an old stone line once uncovered by soil erosion and later covered by wash from higher land. Occasionally there are rocky hardpans, thought to have been formed in earlier geological ages under the influence of a fluctuating water table.

The clay in this soil is less sticky and more porous than that in the Gray-Brown Podzolic. Because water enters the Latosol more readily, it is less subject to erosion. In the place where we have dug our pit the lower soil is strongly acid (pH 5.3) and very low in plant nutrients, but the surface soil is neutral and fairly rich in mineral nutrients and organic matter released from the rotting plant remains. With clean cultivation, the soil rapidly loses its fertility. After it is abandoned, it first grows a cover of savanna grasses. Unless carefully protected, this cover will burn off each year and the soil will remain poor. But if it is protected the forest gradually returns and the soil becomes productive again.

Under cultivation this soil can be kept productive by the use of chemical fertilizers, shade and organic matter. Indeed, it can be cropped successfully even without fertilizers. The forest is cleared only in part, making space for rubber, cocoa, coffee, bananas and other crops that tolerate some shade. Then, as the crop-bearing trees grow larger, more forest trees can be removed. For food crops, the forest is cut in strips or corridors for plantings of mixtures of corn, bananas, cassava, rice, and so on; after four or five years the exhausted soil is allowed to return to forest. In another 10 to 15 years the soil will be ready for crops again. This corridor system is simply a scientific substitute for the method commonly practiced by natives of the tropics, who clear a patch, farm it until it is worn out, and move on to clear a fresh patch.

On these soils the rejuvenating effect of normal erosion is very important. The best soils in the tropics are the very young ones formed from basaltic lava, those renewed by showers of volcanic ash, those renewed by silty stream overflows and those on steep slopes where normal erosion removes the leached surface soil but leaves enough for good root growth and water storage. In these situations the soil often can be used for crops without fertilizers and even without lying fallow in forest, especially if organic matter is carefully conserved. The trouble is that a great deal of the organic matter is burned. Fire is the great enemy of soil productivity in the tropical areas.

While we are in Africa, we should inspect a peculiar soil that has a tendency to harden and form a substance called laterite, a word that comes from the Latin for brick. Near Elisabethville, for example, is a soil of this type that has developed from schist on an ancient, high, nearly level plateau. During the rainy season the water table is within a foot or two of the surface; in the dry season it drops to 30 feet or lower. At two feet or so below the surface there begins a thick mass of heavy clay which very gradually merges into the unweathered rock at 30 feet or deeper. When thoroughly dried from exposure, say in a ditch or deep pit or after removal of the upper soil by erosion, the clay hardens irreversibly. This hard, slaglike laterite is very resistant to weathering and reconversion to soil. People sometimes use the laterite clay to make bricks, building stones and even statues. It can be carved when fresh and then allowed to turn into hard stone. But in the present state of the agricultural arts, these Ground-Water Laterite soils are hardly usable as soil.

Desert Soils

The same is true, of course, of most desert soils. Most people would not consider the desert to be soil at all, and in extreme deserts, where almost no living matter exists, there can be little true soil. It is only near the margins, where widely spaced shrubs find a foothold and hold the soil material in place, that true soils form. Such soils can be seen in many parts of the U. S. Southwest. Often the surface has a pebbly or cobbly pavement, formed as the high winds swept away the finer materials until a protective covering accumulated. The surface soil coat is usually more or less crusted from beating rains and the hot sun, but the soil underneath is more loose and porous. Since there is little leaching, these soils are limy and often salty. Some have a concentration of clay in the B horizon. Deep down, old desert soils often have a hardpan, cemented with lime or silica.

In low places in the desert, which would be swamps in the Podzol region, the soil accumulates salts. Some of these salty soils, like the salt flats near Great Salt Lake, are practically sterile and useless. Others can be reclaimed.

It is not easy to make "the desert blossom as the rose" through irrigation. Water is usually scarce and not always itself free of salts. More often than not at least some drainage works are necessary. Salts are nearly always a problem; even if there is not altogether too much salt, its presence may upset the optimum balance of the various salts needed for plant nutrition.

Most true Desert soils are on relatively high plateaus and old river deltas—too high to irrigate. But in the same regions there are likely to be much younger and lower soils formed from recent alluvial deposits. These vary from coarse gravel to heavy clay. It is mainly these alluvial soils that are irrigated. For successful irrigation, soils must not be too sandy or gravelly, so that they may hold water at least fairly well; yet they must be well drained so some excess water can leach away. A soil that is well drained in the natural desert under four to five inches of rainfall may be swamped under irrigation with 30 to 40 inches. If drainage is poor, the soil is almost bound to accumulate enough salt to kill any cultivated plants.

Great successes and dramatic failures have attended irrigation projects since the dawn of history. Modern soil science and engineering now make possible much better predictions of the outcome of proposed projects and much better systems of water use and soil management. Yet even now enthusiastic promotion sometimes gets too far ahead of scientific prediction, to the sorrow of new settlers.

Prescriptions

We have looked at specimens of nine or so of a total of some 40 great soil groups found on our planet. Within each of these major groups are many local soil types, varying from one another in ways that affect their use. To make a detailed map of local soils for just one average Midwestern county of 500 square miles, a map sufficiently detailed so that results of research on the various soils can be applied effectively to individual fields, involves the identification of from 100 to 200 significantly distinct soil types and phases of soil types.

It is on the basis of these soil types and phases that detailed recommendations about crop varieties, fertilizers, rotations, erosion-control devices and the like are developed. Of course any single recommendation, such as a specific fertilizer, may apply to a large group of soil types, but for each recommendation the soils have to be grouped differently. A statement covering the essentials for a single county needs to be much longer than this article.

Gradually our knowledge about soils is being expanded and classified into a system. The system is still very imperfect, but ultimately all the local soil types of the world will be known and understood. Maps will give their location. Then every farmer, gardener and forester will have the benefit of what soil science has learned to enable him to use his own soil most effectively.

GRAY-BROWN PODZOLIC SOILS have been intensively studied. On the similar soil of the Rothamsted Experimental Station in England wheat has been grown without fertilizer for 106 years (*left*). Manured plot is at right.

TROPICAL SOILS occur in great variety. On this tropical soil near the Congo River in Africa grow cultivated cassavas and bananas. If left uncultivated after exhaustion, the soil will be invaded by forest and renewed.

DESERT SOILS can sometimes be turned to agriculture by irrigation, but only with a detailed knowledge of their composition. On this soil in the Mohave Desert grow Joshua trees, which properly are not trees but yuccas.

12

WATER

ROGER REVELLE
September 1963

*Did you ever hear of Sweet Betsy
    from Pike
Who crossed the wide prairie with her
    lover Ike?
The alkali desert was burning and bare
And Ike got disgusted with everything
    there.
They reached California with sand
    in their eye,
Saying, "Good-by Pike County, we'll
    stay till we die."*

This bleary and partly unprintable ballad of the 1850's marks the time when most Americans first became aware of the problems of water in national development. In northern Europe, where most of their ancestors had lived, there had always been plenty of water; in the eastern U.S., where they had learned to farm, abundant rain supplied all the water needs of their crops. But when the pioneers crossed the Missouri River, they came to an arid country where water was more precious than land: its presence meant life, its absence death.

Today water problems are part of the national consciousness, and most Americans are aware that the future development of their country is intimately related to the wise use of water resources. The same obviously holds true for the less developed countries. The water problems of the U.S. and the poorer countries are fundamentally similar, but they also differ in significant ways.

Water is both the most abundant and the most important substance with which man deals. The quantities of water required for his different uses vary over a wide range. The amount of drinking water needed each year by human beings and domestic animals is of the order of 10 tons per ton of living tissue. Industrial water requirements for washing, cooling and the circulation of materials range from one to two tons per ton of product in the manufacture of brick to 250 tons per ton of paper and 600 tons per ton of nitrate fertilizer. Even the largest of these quantities is small compared with the amounts of water needed in agriculture. To grow a ton of sugar or corn under irrigation about 1,000 tons of water must be "consumed," that is, changed by soil evaporation and plant transpiration from liquid to vapor. Wheat, rice and cotton fiber respectively require about 1,500, 4,000 and 10,000 tons of water per ton of crop.

When we think of water and its uses, we are concerned with the volume of flow through the hydrologic cycle; hence the most meaningful measurements are in terms of volume per unit time: acre-feet per year, gallons per day, cubic feet per second. An acre-foot is 325,872 gallons, the amount of water required to cover an acre of land to a depth of a foot. Eleven hundred acre-feet a year is approximately equal to a million gallons a day, or 1.5 cubic feet per second. A million gallons a day fills the needs of 5,000 to 10,000 people in a city; 1,100 acre-feet a year is enough to irrigate 250 to 300 acres of farmland.

The total amount of rain and snow falling on the earth each year is about 380 billion acre-feet: 300 billion on the ocean and 80 billion on the land. Over the ocean 9 per cent more water evaporates than falls back as rain. This is balanced by an equal excess of precipitation over evaporation on land; consequently the volume of water carried to the sea by glaciers, rivers and coastal springs is close to 27 billion acre-feet per year. About 13 billion acre-feet is carried by 68 major river systems from a drainage area of 14 billion acres. Somewhat less than half the runoff of liquid water from the land to the ocean is carried by thousands of small rivers flowing across coastal plains or islands; the area drained is about 11 billion acres, but part of this is desert with virtually no runoff.

Eight billion acres on the continents drain into inland seas, lakes or playas. This includes most of the earth's six billion acres of desert and also such relatively well-watered areas as the basins of the Volga, Ural, Amu Darya and Syr Darya rivers, which transport several hundred million acre-feet of water each year into the Caspian and Aral seas. The remainder of the land surface, about four billion acres, is covered by glaciers.

Even agriculture, man's principal consumer of water, takes little of the available supply. A billion acre-feet per year —less than 4 per cent of the total river flow—is used to irrigate 310 million acres of land, or about 1 per cent of the land area of the earth. Roughly 10 billion acre-feet of rainfall and snowfall is evaporated and transpired each year from the remaining three billion acres of the earth's cultivated lands and thus helps to grow mankind's food and fiber. Most river waters flow to the sea almost unused by man, and more than half of the water evaporating from the continents—

MULTIPURPOSE DAM at Watts Bar, Tenn., appears in the aerial photograph on the opposite page. The blue lake at left is formed where the dam halts the Tennessee River. Spilled back to river depth beyond the dam, the water regains a slate-gray hue. The turbulent area at lower right is caused by the flow of water through the square-capped hydroelectric generators. At bottom cente. is a power-distribution station. At top is the lock that enables shipping to pass the dam. The ribs in the dam are spillways, which here are closed. They can adjust the lake level to help control flooding.

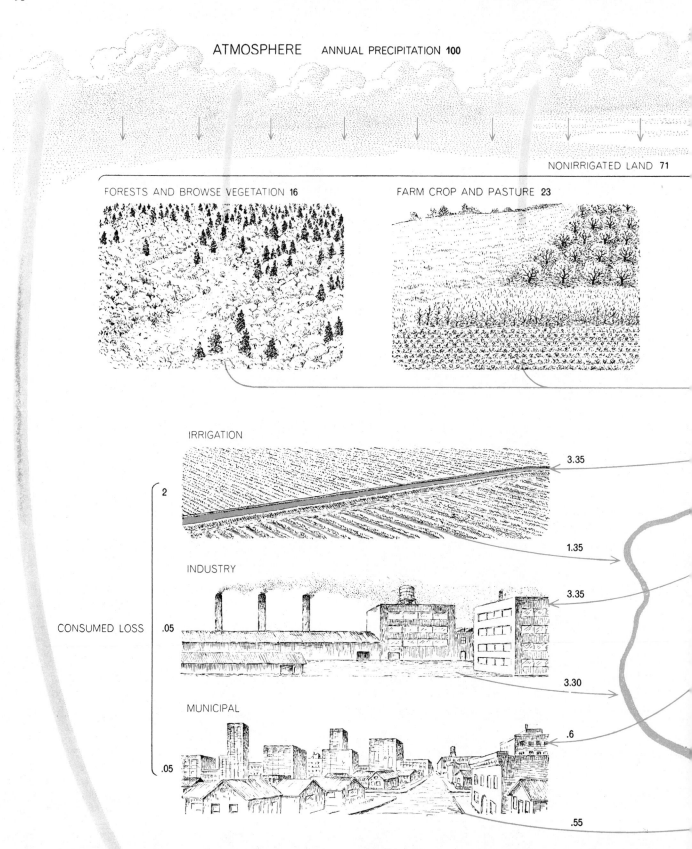

ATMOSPHERE   ANNUAL PRECIPITATION **100**

NONIRRIGATED LAND **71**

FORESTS AND BROWSE VEGETATION **16**

FARM CROP AND PASTURE **23**

IRRIGATION

3.35

2

1.35

INDUSTRY

3.35

CONSUMED LOSS   .05

3.30

MUNICIPAL

.6

.05

.55

OCEAN RESERVOIR

HYDROLOGIC CYCLE for the U.S. shows the fraction of annual precipitation used in a highly developed nation. Twenty-nine per cent of the rainfall arrives at the oceans (*bottom*) via stream flow; 71 per cent falls on various types of nonirrigated land, returning

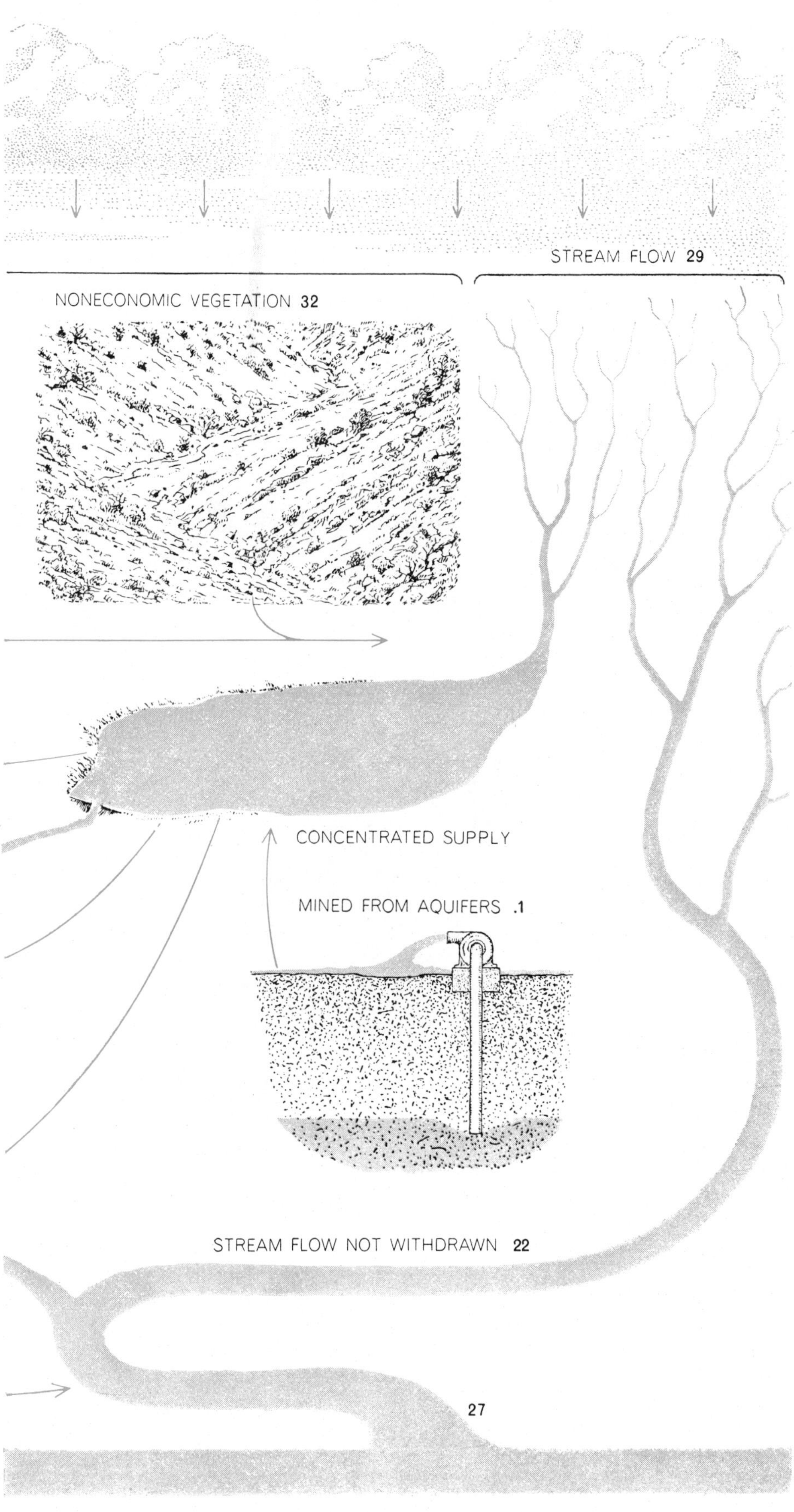

STREAM FLOW 29

NONECONOMIC VEGETATION 32

CONCENTRATED SUPPLY

MINED FROM AQUIFERS .1

STREAM FLOW NOT WITHDRAWN 22

27

directly to the atmosphere (*top*) by transpiration and evaporation. Water withdrawn for irrigation, industry and municipal use is shown at left to constitute only 7.3 per cent of total.

particularly that part of the evaporation taking place in the wet rain forests and semihumid savannas of the Tropics—plays little part in human life.

Although it is not usually reckoned as such in economic statistics, water can be considered a raw material. In the U.S. the production of raw materials has a minor role in the total economy, and water costs are small even when compared with those of other raw materials. The cost of all the water used by U.S. householders, industry and agriculture is around $5 billion a year: only 1 per cent of the gross national product. The less developed countries, where raw materials are a major component of the economy, cannot afford water prices that would be acceptable in the U.S.

In the U.S. water costs $10 to $20 an acre-foot, compared with wholesale prices of $22,000 an acre-foot for petroleum, $100,000 an acre-foot for milk and $1 million an acre-foot (not counting taxes) for bourbon whiskey. The largest tanker ever built can hold less than $1,000 worth of water. Yet Americans use so much water—about 1,700 gallons a day per capita—that capital costs for water development are comparable to other kinds of investment. Although the water diverted from streams and pumped from the ground is equivalent to only about 7 per cent of the rain and snow falling on the U.S., this is still an enormous quantity: 200 times more than the weight of any other material used except air. The annual capital expenditure for water structures in the U.S.—dams, community and industrial water works, sewage-treatment plants, pipelines and drains, irrigation canals, river-control structures and hydroelectric works—is about $10 billion.

One of the most critical water problems of the U.S. is represented by the vast water-short region of the Southwest and the high Western plains. In some parts of the Southwest water stored underground is being mined at an alarmingly high rate, and new sources must soon be found to supply even the present population. The average annual supply of controllable water in the entire region is 76 million acre-feet. If agriculture continued to develop at the present rate, 98 to 131 million acre-feet would be required by the year 2000. Provided that the neighboring water-surplus regions could be persuaded to share their abundance, this deficit could be met by long-distance transportation of 22 to 55 million acre-feet per year. But the annual cost would be $2 billion to $4 billion, or

$60 to $100 per acre-foot of water, including amortization of capital costs of $30 billion to $70 billion. The cost per acre-foot would be too high for most agriculture, although not too high for municipal, industrial and recreational needs.

Nathaniel Wollman of the University of New Mexico and his colleagues have shown that the average value added to the economy of the Southwest through the use of water in irrigation is only $44 to $51 an acre-foot, whereas the value gained from recreational uses could be about $250 an acre-foot and from industrial uses $3,000 to $4,000 an acre-foot. Because the quantities of water consumed by city-dwellers and their industries are much less than those in agriculture, the arid Western states would not require such a vast increase in future supply if they shifted from a predominantly agricultural to a predominantly industrial economic base.

The value of water in the water-short regions of the U.S. that are in a phase of rapid economic development increases more rapidly than the cost. Even high-cost water is a small burden on the gross product of a predominantly industrial and urban economy, and high water costs are only a small economic disadvantage. This is easily overcome if other

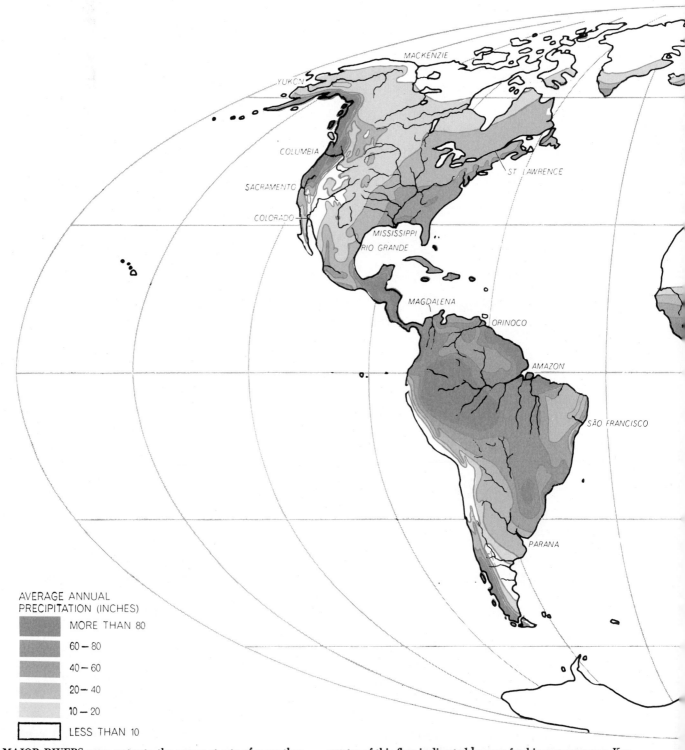

AVERAGE ANNUAL
PRECIPITATION (INCHES)

MORE THAN 80

60 — 80

40 — 60

20 — 40

10 — 20

LESS THAN 10

MAJOR RIVERS carry water to the oceans at rate of more than 11 billion acre-feet every year, but even in the U.S. less than a quarter of this flow is diverted by man for his own purposes. Key at lower left distinguishes areas by amount of precipitation they

conditions, such as climate, happen to be propitious.

Throughout the country favorable benefit-to-cost ratios can usually be attained from relatively high-cost multipurpose water developments for city residents, industry, irrigation agriculture, the oxidation and dispersal of municipal and industrial wastes, the generation of hydroelectric power, pollution control, fish and wildlife conservation, navigation, recreation and flood control.

In the less developed countries water development by itself does not produce much added value for the present economy. Municipal and industrial water requirements are much smaller than they are in the U.S., and the immediate water needs are chiefly for agriculture, which calls for about the same amount of water in any warm region. Most of these countries have a low-yielding subsistence agriculture that brings in very little cash per acre-foot of water, and their farmers can afford to pay only a few dollars per acre-foot. Development of water resources must be accompanied by other measures to raise agricultural yields per acre-foot and per man-hour, and in general to increase the economic value of water.

One means of coping with water problems in both the U.S. and the less de-

receive each year. Several regions where the average annual precipitation is less than 10 inches (*white*) can be seen to lie close to large rivers. Plans for overcoming the aridity of Egypt and central Russia call for giant dams on the Nile and the Ob.

veloped countries is to improve the present rather low efficiency of water use. Here much could be done by effective research. For example, about half the water provided for irrigation is lost in transport, and less than half the water that reaches the fields is utilized by plants.

New mulching methods are already being applied to reduce evaporation from soil surfaces, thereby making more water available for transpiration by the plants. Through research on the physiology of water uptake and transport in plants, and on plant genetics, transpiration could probably be lowered without a proportional reduction in growth. Development of salt-tolerant crops would reduce the amounts of irrigation water needed to maintain low salt concentrations in the solution around the plant

roots. The loss of water by seepage from irrigation canals and percolation from fields would be lowered by the development of better linings for canals and better irrigation practices. Losses from canals would also be reduced if we could learn how to control useless water-loving plants that suck water through the canal banks and transpire it to the air.

In arid regions the runoff from a large area must be concentrated to provide water for a relatively small fraction of the land, and techniques are needed to increase the proportion of total precipitation that can be concentrated. Development of such techniques requires research on means of increasing the runoff from mountain areas (for example, by reducing evaporation from snow fields and modifying the plant cover in order to

reduce transpiration) and on methods for accelerating the rate of recharge of valley aquifers.

Finally, water problems could be dealt with by steps that—in contrast to those seeking to make better use of existing supplies—sought to increase the total volume of fresh water. Here research moves on two fronts: attempts to modify precipitation patterns by exerting control over weather and climate, and development of more economical methods of converting sea water or brackish water to fresh water. The ability to control weather and climate, even to a small degree, would be of the greatest importance to human beings everywhere. Whether or not a measure of control can be obtained will remain uncertain until we understand the natural proc-

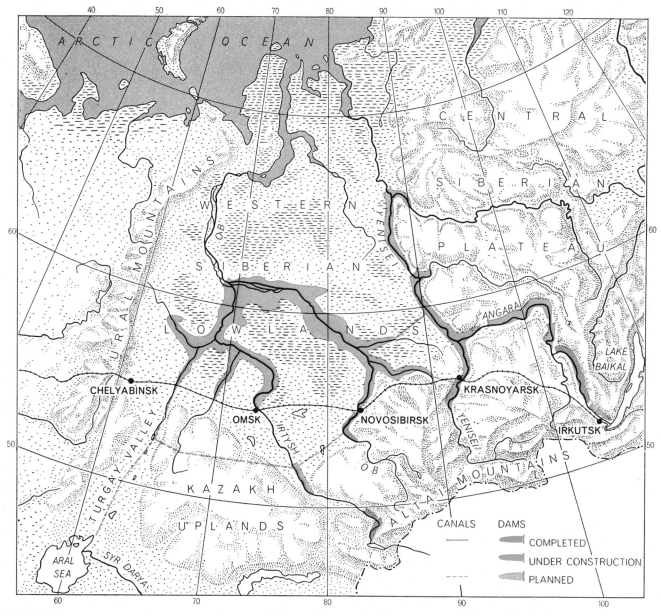

OB-YENISEI PROJECT calls for huge dam on the Ob, creating an inland sea five-sixths the size of Italy. Canal would link Ob and Yenisei rivers so that 12.5 per cent of the water now flowing unused to the Arctic would irrigate the central Soviet steppes.

esses in the atmosphere much better than we do now. As for desalination, this could be accomplished more economically than at present if the amount of energy required to separate water and salt could be reduced or the cost of energy lowered. Research on the properties of water, salt solutions, surfaces and membranes is fundamental to the desalination problem. So is research aimed at lowering energy costs.

We know too little to be able to make more than a rough appraisal of the potentialities of water-resources development for agriculture in the less developed countries. The modern technology of irrigation engineering, drainage, sanitation and agricultural practice is quite different from that which determined patterns of land and water use in the past. At the same time technology is almost completely lacking for expanding productive agriculture in the areas of most abundant water and almost unused land: the humid Tropics. Our concern should be not only to find ways of increasing total production in order to feed and clothe the world's expanding human population but also to raise production per farm worker, that is, to raise living standards. A world-wide strategy for development of land and water will require a careful analysis of existing knowledge, region by region, together with field surveys and experimental research in each region by experienced and imaginative specialists.

In humid areas agriculture is limited only by the extent of good land; in arid lands water is the absolute limiting factor. Unless climates can be modified or sea water can be cheaply converted and economically transported, the area of arable land in the arid zone will always exceed the available water. At present, however, neither surface nor underground waters are fully utilized, either for double-cropping in presently cultivated lands or for bringing new land under cultivation.

In addition to improving the utilization of water and increasing agricultural yields other problems that contributed to the destruction of desert civilizations in the past must still be overcome in arid land development. Among them is the fact that the spreading of water over large areas provides a fertile ground for human diseases, such as malaria and bilharzia, and for plant pests. Egyptian records show an average of one plague every 11 years. Uncontrollable malaria might well have been the cause of the mysterious disappearance of the great civilization of the cities Mohenjo-Daro and Harappa, which flourished 4,500

WATERLOGGED FIELDS near Sargodha in Pakistan reflect leaky canal system and inadequate drainage. Cultivated plots can be seen under water in center of photograph. When it evaporates, the water will deter renewed cultivation by leaving salts in topsoil.

SALINE FIELDS stand out against darker cultivated land in this aerial photograph. The related problems of waterlogging and salt accumulation in the soil have made five million acres of West Pakistan's irrigated farmland either impossible or unprofitable to cultivate.

years ago in the Indus valley of Pakistan.

Soil drainage in a nearly level flood plain is very difficult and is usually neglected, with the result that the water table comes close to the surface and drowns the roots of most crop plants. Water rises through the soil by capillary action and evaporates, leaving an accumulation of salt that poisons the plants. The related disasters of waterlogging and salinity may have caused the ruin of the Babylonian civilization in the valley of the Tigris-Euphrates, and they are a frightening menace today in West Pakistan.

Another threat is the conflict between the sedentary farmers of the plain and nomadic herdsmen. The present-day Powindahs of West Pakistan remind us of this ancient conflict. In our own West

the feuds between cattlemen and farmers are still a vivid memory.

In considering the possibilities of agricultural development in the world's arid lands one thinks first of the famous rivers that have played so large a role in human history: among them the Nile, the Indus and its tributaries and the Tigris-Euphrates.

For thousands of years the Egyptians carried out irrigation by allowing the Nile waters during flood stage to spread in ponded basins broadly over the delta and the valley. When the flood subsided, the basin banks were cut and the ponded water flowed back to the river. The Nile and the sun were said to be the prime farmers of Egypt. It was thought that the river's silt, deposited during the annual flood, fertilized the soil. Sun-

drying and -cracking, during the fallow season before the flood, deeply furrowed the soil and killed off weeds and microorganisms, making plowing unnecessary. The flood arrived in July, reached its height in September and subsided quickly. The fields were sown in early winter with wheat, barley, beans, onions, flax and clover. Summer crops were grown only on the river levees and in areas that contained a shallow water table, where water could be lifted by hand from the river banks or from wells. High floods left the basins pestilential morasses that brought plagues and epidemics. Low floods brought famine.

During the past 140 years this ancient system has been transformed. In 1820 Egypt had reached a nadir, with a population of only 2.5 million and with three

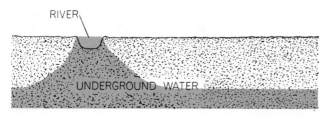

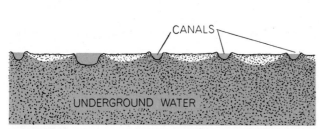

LACK OF DRAINAGE has caused the underground water table to rise disastrously in parts of Pakistan. Before construction of leaky canals, water approached ground level only near rivers. Now water table in many areas is high enough to drown crop roots.

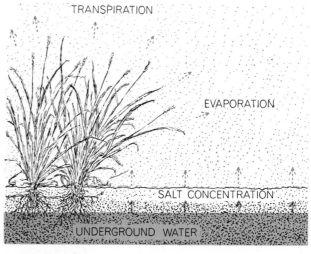

SALT ACCUMULATES on topsoil in two ways. Underground water rises by capillary action, lifting dissolved salts that will be left behind after evaporation. If farmer uses thin layer of water to irrigate topsoil, it will evaporate before percolating down.

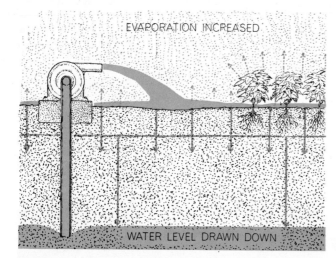

VERTICAL DRAINAGE might solve related problems of salinity and waterlogging. Pumped through cased well from underground table, as illustrated here, enough water could reach surface for salts to percolate down beneath topsoil before full evaporation occurs.

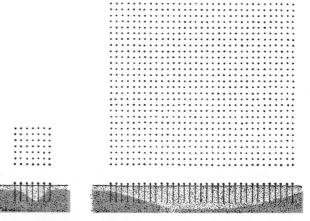

LARGE-SCALE PROJECT will be required to lower the water table. Drainage of a field of one million acres (*right*) could negate seepage from adjacent land. But pumping out a small field (*left*) would have no appreciable effect on underground water level.

million cultivated acres. This date marked the beginning of perennial canal irrigation and widespread planting of summer crops, including cotton, corn, rice and sugar cane as well as the traditional winter crops. Low dams called barrages were built across the river; the water backed up behind these structures was diverted through large new canals that flowed the year round. By 1955–1956 the cultivated area had increased to 5.7 million acres and the intensity of cultivation to 177 per cent; that is, more than 10 million acres of crops were harvested. Salinity and waterlogging became serious menaces in the early part of this century, but they have been fairly well controlled by an extensive drainage system. Chemical fertilizers are used in large amounts and crop yields per acre are high, even though the Nile silt no longer settles on the fields but is deposited back of the barrages. Sufficient food is grown to feed the present population of 27 million. From the standpoint of crop yields per acre, although not per man, Egypt is a developed country.

The average annual flow of the Nile is 72 million acre-feet, but it is occasionally as high as 105 million acre-feet or as low as 36 million. If all the average flow could be utilized, it would be enough to irrigate 12 to 15 million acres on a year-round basis. At present the area of irrigated land in Egypt is less than half that. During the flood season much of the water flows to the sea unused, and during the rest of the year a shortage of surface and underground water limits the size of the cultivated area.

Now the construction of the Aswan High Dam promises to bring the river under complete control. The dam will have a storage capacity of 105 million acre-feet, equal to the highest annual flow during the past century. There will no longer be a Nile flood; the tamed river will become simply a huge feeder canal for irrigation. With the average of 55 million acre-feet per year available to Egypt (17 million acre-feet from the reservoir is allocated to the Sudan), it will be possible to increase the cultivated area in the delta and the valley floor by 2.2 million acres, or nearly 40 per cent, and to convert .7 million acres from flood to perennial irrigation. Hydroelectric power generation of more than a million kilowatts will make power available for pump drainage, which may increase crop production by 20 per cent, and for the manufacture of chemical fertilizers. The electric power will also be used to lift water to the desert margins of the valley, where it is

hoped that an additional one to two million acres can be brought under the plow. If all these benefits can be realized, total agricultural production in Egypt can be increased by 90 per cent, enough to feed almost twice the present population and at the same time provide crops for export.

When Alexander the Great pushed his tired armies eastward some 2,300 years ago, they came at last to an old desert civilization on the banks of the mightiest river they had ever seen. The Aryans, who had preceded Alexander by 1,000 years, did not give the river a name; they called it simply the Indus, which was their word for river, and they named the subcontinent they had invaded "India": the land of the river.

The Indus and its five tributaries of the Punjab, together with the flat plain through which they flow, are one of the major natural resources of the earth. In the Punjab and Sind regions of West Pakistan 30 million persons dwell on the plain; 23 million make their living from farming it. They produce most of the food and fiber that feed and clothe nearly 50 million people.

The rivers carry more than twice the flow of the Nile. Half this water is diverted into a highly developed system of irrigation canals and is used to irrigate some 23 million acres—by far the largest single irrigated region on earth. Underneath the northern part of the plain lies a huge reservoir of fresh ground water, equal in volume to 10 times the annual flow of the rivers.

In spite of the great potentialities of the plain, the fact is that poverty and hunger, not well-fed prosperity, are today the common lot of the people of West Pakistan. These afflictions are nowhere more desperately evident than in the farming villages of the countryside. In a country of farmers food must be imported to provide the most meager of diets; the gap between food production and the number of mouths to be fed is widening.

The problem of agriculture in West Pakistan is both a physical and a human one. It is a problem of land, water and people and of the interactions among them. One of its aspects is the waterlogging and salt accumulation in the soil, caused by poor drainage in the vast, nearly flat plain, that are slowly destroying the fertility of much of the irrigated land. The area of canal-irrigated and cultivated land already seriously damaged by waterlogging and salinity is close to five million acres, or about 18 per cent of the gross sown area. Three

other difficulties also beset agriculture: shortage of irrigation water, problems of land tenure and poor farming practices.

Although crops can be grown throughout the year, and both a winter and a summer growing season are traditional, the irrigation canals lose so much water by seepage that the amount carried to the fields is sufficient to irrigate only about half the land during each season. Even so, the crops are inadequately irrigated, particularly in summer. Much of the cropped area receives insufficient water to prevent salt accumulation.

Many of the farmers are sharecropping tenants who have little incentive to increase production. Nearly all of them struggle with small and widely separated plots that multiply the difficulties of efficient use of irrigation water and farm animals and gravely inhibit change in traditional practices.

In West Pakistan we have the wasteful paradox of a great and modern irrigation system pouring its water onto lands cultivated as they were in the Middle Ages. Plowing is done by a wooden plow of ancient design, pulled by undernourished bullocks. Unselected seeds are sown broadcast. Pakistan uses only a hundredth as much fertilizer per acre as Egypt.

Careful investigation shows that in most of the Punjab the problems of waterlogging and salinity could be cured, and at the same time adequate water could be supplied to the crops, by sinking fields of large wells to pump the underground water and spread it on the cultivated lands. Part of the pumped water would be carried off by evaporation and transpiration and part would percolate back into the ground, in the process washing the salt out of the soil.

If the well fields are too small in area, lateral infiltration of underground water from the surrounding land will be large compared with the rate at which the pumped water can evaporate, and the process of dewatering will be retarded or completely inhibited. For this and other reasons each Punjab project area should be about a million acres in size.

Removal of salt and provision of additional water are necessary, but by no means sufficient, measures to raise agriculture in West Pakistan from its desperate poverty. Equally essential are chemical fertilizers, higher-yielding seeds, pest control, credit and marketing facilities, and above all incentives and knowledge to adopt better farming practices. The job cannot be done all at once; it is necessary to concentrate on project areas of manageable size. Initial capital costs for a million-acre project in the Punjab would be of the order of $55 million, including costs of wells and electrification, nitrogen-fertilizer plants, pest-control facilities and filling of administrative, educational and research pipelines.

In the Sind region initial capital costs would be considerably higher, probably between $130 million and $165 million per million acres. That is largely because the underground water in most of the Sind is too salty to be used for irrigation, and drainage is therefore a more difficult matter than in the Punjab.

After a few years the minimum net increase in crop value in each million-acre project in the Punjab could be $55 million to $60 million a year, equal to the capital costs and to twice the present gross production, excluding livestock. In the Sind the net increase, including livestock, could probably be at least equal to the present output.

The same interrelated problems of water, land and people that afflict the Indus plain also exist in the valley of the Tigris-Euphrates, but on a much smaller scale. Salty soil is found over large areas; because of waterlogging it is possible to cultivate only about a third of the seven million acres of irrigated land each year. The remainder is left fallow and unirrigated to dry out the subsoil and to build up a little soil nitrogen. Great damage was done long ago when the ancient canal systems were destroyed and the land was depopulated by waves of nomadic invaders. But the nomads merely hastened the salt accumulation and waterlogging that were the seeds of destruction. These had begun centuries earlier as a result of inadequate drainage and inability to control floods.

If the flow of the Tigris-Euphrates could be fully utilized, through combined development of surface and ground water, and if the soils were adequately leached and drained, the irrigated area cultivated each year could be increased to 10 to 12 million acres. If greater water usage were combined with perennial cropping, better farming practices and the application of chemical fertilizers, total agricultural production could be raised at least fivefold.

The largest opportunities for expansion of the area of irrigated arid and semiarid lands exist in the U.S.S.R. Between 1950 and 1960, 15 million acres in the neighborhood of the Black and Caspian seas were provided with irrigation water from the Volga, Dnieper, Amu Darya and Syr Darya rivers. The total flow of these rivers is more than 300 million acre-feet, sufficient, under the cold-winter and warm-summer cli-

mate of the steppes, to supply all the water needed to irrigate 70 to 100 million acres.

Because of the relatively advanced economic level of the country, large multipurpose water developments in the U.S.S.R. are economically feasible, and a high percentage of the capital invested goes for power, transportation, industrial water supplies and flood control.

Soviet engineers have outlined a plan to build an immense dam on the Ob River, creating an inland sea five-sixths the size of Italy, and to dig a canal connecting the Yenisei with the Ob above the dam [see illustration on page 102]. The impounded waters would be transported through a giant system of canals, rivers and lakes to the Aral Sea and thence by canal to the Caspian Sea. Several hundred million acre-feet of water that now goes to waste each year in the Arctic Ocean would be conserved. This water would be used to irrigate 50 million acres of crop lands and a somewhat larger area of pasture in arid western Siberia and Kazakhstan. Accompanying hydroelectric power installations would have a capacity of more than 70 million kilowatts. Major storage, irrigation and hydroelectric works are also under construction or planned in the northern Caucasus and in the Azerbaijan, Georgian and Armenian Soviet Socialist republics. These will bring additional tens of millions of acres under irrigation.

In some parts of the arid zone both surface and ground water are so scarce that it is difficult to see how irrigation agriculture can be developed to support the rapidly expanding population. In the Maghreb countries of North Africa—Tunisia, Algeria and Morocco—there is probably not enough water in the region north of the Sahara to irrigate more than 3.5 million acres of land, yet the combined population of these three countries is already 26 million (equal to Egypt's) and will double in 20 to 25 years. Elaborate systems of dry farming have been developed in the Maghreb; for example, the planting of olive trees far apart in light, sandy soils that catch and hold the nighttime dew. With this technique it has been possible to grow olive and other fruit trees on more than a million acres in Tunisia. In the long run it may be necessary to employ most of the available water in the Maghreb countries for industrial purposes, because these can provide a tenfold to hundredfold higher marginal value for water than agriculture can.

A new possibility for water development has recently been opened, how-

ever. During the past few years evidence has been obtained that large areas in the Sahara may be underlain by an enormous lake of fresh water. In some places the water-bearing sands are 3,000 feet thick, and they appear to extend for at least 500 miles south of the Atlas Mountains and perhaps eastward into Tunisia and Libya. If this evidence is correct, the amount of useful water may be very large indeed—of the order of 100 billion acre-feet, sufficient to irrigate many millions of acres for centuries.

In general the possibilities of expanding the area of irrigated land in the arid zone outside the U.S.S.R. are not large when measured in numbers of acres. But crop yields under irrigation in the arid lands are high and assured if all the factors of agricultural production are properly applied. In fact, irrigation agriculture in arid regions can be successful only if it is intensive and high-yielding; it is costly to construct and maintain drainage systems that will keep the water table from rising too close to the surface, and to provide enough water on each acre to leach the salts out of the soil. In hot, arid lands some kinds of irrigation agriculture can be so productive that very expensive irrigation water, such as could be produced by sea-water desalination, may soon become economical.

Much greater possibilities (and also greater difficulties) exist for agricultural expansion in the regions of savanna climate, which are characterized by an annual cycle of heavy rainfall during one season, followed by drought the remainder of the year, and by warm weather at all seasons. In Africa, for example, many millions of what are now barren acres could be brought under irrigated cultivation, provided that interested farmers could be found, in the neighborhood of the great bend of the Niger River in former French West Africa, in the basin of the Rufiji River of Tanganyika and near Lake Kyoga in Uganda. Similarly, in the area extending from India east through Burma, Thailand and Vietnam to the northern Philippines, air temperature and solar radiation are suitable for year-round crop growth, and water and land are the limiting factors [see "The Mekong River Plan," by Gilbert F. White; SCIENTIFIC AMERICAN, April, 1963].

In the lower basin of the Ganges and Brahmaputra rivers, comprising East Pakistan and the Indian states of Bengal, Bihar and Assam, some 140 million people live on 70 million cultivated acres. The basic resources of soil and water are grossly underutilized in this land of ancient civilization, extreme present poverty and strong population pressure. Each year the rivers carry about a billion acre-feet to the Bay of Bengal, and in the process they flood most of the countryside. Yet only one crop is grown a year. The land is left idle half the year because of the shortage of water and there is a lack of useful occupation for the people six to eight months of the year. Agricultural practices are adjusted to the rhythm of the monsoon.

The opportunities for increasing production are enormous in this region of land shortage and overabundant water. Through surface and underground storage of a portion of the flood waters, water could be provided for three crops each year over more than half of the cultivable area in the alluvial plain, and a considerable additional area could receive sufficient water for two crops. An assured year-round water supply would also provide favorable conditions for intensive use of fertilizers, higher-yielding plant varieties and better farming practices, which could result in a tripling of yields per crop and per acre for cereals, pulses (the edible seeds of leguminous plants such as peas and beans) and oilseeds.

A well-fed livestock industry could be developed in addition to improvements in field crops, and a balanced diet, instead of the present completely inadequate one, could be provided for twice the present population. Expansion of agricultural production here, based on irrigation, would raise few basic problems of land and settlement, but it would require a reorientation of thinking regarding patterns of land and water use. Because of the enormous volumes of water involved and the flatness of the alluvial basin, the cost of water storage and distribution and of flood control and drainage would be high, but the returns through increased farm and livestock production could be several times higher than the cost. The yields per worker must also be increased, however, and a large degree of industrialization accomplished if the project is to finance itself.

Development of water resources is not an end in itself. The investment can be justified only if it leads to higher agricultural or industrial production, or in other ways to an increase of human well-being. To gain these objectives water development must be accompanied by other actions needed to use the water effectively. This is well illustrated in agriculture. One of the basic principles of agricultural science is the principle of interaction: the concurrent use of all the factors of production on the same parcel of land, which will give a much larger harvest than if these factors are used separately on different parcels. Adequate water and water at the right time are essential if seeds of a particular crop variety planted in a given soil are to yield a good crop. But a much larger crop is possible if seeds of a higher-yield variety are planted. This potential increase in the harvest will be realized, however, only if the soil contains sufficient plant nutrients. Usually nitrogen fertilizers and phosphate fertilizers must be added in large amounts to provide the maximum yield. Increased soil fertility will be drained off by weeds unless these are rigorously controlled, and an eager host of insect pests and plant diseases will fight to share the crop with the farmer unless he can combat them with pest-control measures. Improved seed varieties planted without adequate water, abundant fertilizer and rigorous pest control may not do even as well as the traditionally planted varieties. The potentialities for double- or triple-cropping in a perennial irrigation system cannot be achieved if the farmers do not have tractors and efficient tools to enable them to prepare their fields in the short interval between harvest and planting.

To meet the cost of new irrigation systems the farmer must produce much more per acre-foot of water than he has in the past, and this can be done only if all the factors of production are made available to him and if he is taught how to use them effectively. The human, educational, social and institutional problems of bringing the necessary knowledge to millions of farmers are immense. The task of remaking methods of production that are intimately tied to ways of living and of overcoming institutional and political resistance to change is more difficult than any of the engineering problems. Illiteracy, malnutrition and disease; poverty so harsh that the farmer does not dare risk innovation because failure will mean starvation; small and fragmented farm holdings; land-rental and taxation systems that destroy incentive; extreme difficulties in obtaining a farm loan promptly at a reasonable interest rate; poor marketing and storage systems; administrative inefficiency and corruption; the shortage of trained teachers and farm advisers; inadequate government services for agricultural research, education and extension and for control of water-borne diseases—all must be overcome if investments in water resources in the developing countries are to produce really beneficial results.

13

CLIMATE AND AGRICULTURE

FRITS W. WENT
June 1957

In our elaborately industrialized country we tend to lose sight of the fact that modern man's life still depends fundamentally on agriculture. And it is difficult to appreciate how insecure this foundation is, from the standpoint of feeding a growing population. Only by prodigies of toil and invention has civilized man been able to wrest enough food from the soil to keep pace with his increasing needs. The invention of systematic agriculture was itself a remarkable technical achievement: to make it possible the original farmers had to develop crop plants (wheat, rice and so forth) and techniques of plowing, sowing, irrigation, weeding and fighting pests and diseases. Nowadays farming is a highly sophisticated technology. By mass production methods each farm worker now can produce enough food for 17 persons; research has brought most of the important crop diseases and pests under effective control; scientific soil management makes it possible to get high yields from the same soil year after year; transportation facilities mitigate local crop failures by distributing food quickly over whole continents; modern preserving methods make fresh food available all year round.

The one great factor that man has not yet learned to control is climate. Droughts, floods, freezes, tornadoes, hailstorms still make farming an uncertain enterprise, even in the U. S. And more subtle climatic aberrations may work even more havoc upon our crops.

The latter point is brought into sharp focus by a look at the annual figures for production of tomatoes in the U. S. The yield of tomatoes per acre in many states fluctuates enormously from one year to another. It is less variable in California than in states with more changeable climates. There cannot be much doubt that differences in weather from year to year are mainly responsible for these variations in yield. A number of other crops, such as beans, peas and various fruits, also suffer from climatic vicissitudes. What are these damaging climatic influences, so unobtrusive that it is difficult to detect them, let alone measure their effects? Plainly they must have to do with factors such as the amount of sunlight, temperature and humidity.

For nearly a decade we have been carrying on an investigation of the effects of climatic factors on plant produc-

GREENHOUSE in the Earhart Laboratory is one of 54 separate rooms in it where plants grow. Outside the laboratory it is night; here daylight is artificially lengthened for the purposes of an experiment. The roof of the greenhouse is cooled by a steady flow of water.

TOMATO PLANTS were grown in the phytotron (the Earhart Plant Research Laboratory of the California Institute of Technology). During the day both plants grew at 86 degrees Fahrenheit, a temperature at which tomato plants grow well. During the night the plant at left was subjected to a temperature too high for fruit set. The plant at right was grown at the proper night temperature.

CONTROL ROOM of the Earhart Laboratory centralizes the regulation of light, temperature and humidity in all of its rooms.

TRAYS OF CHARCOAL remove contaminants from air taken into the laboratory. Air in greenhouses is replaced twice a minute.

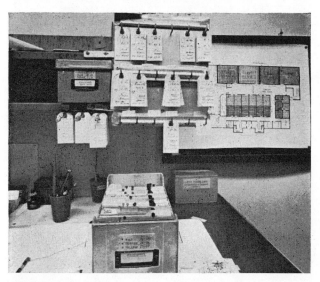

CHART at right has a space for each wheeled table in the entire laboratory. Tags to the left of the chart are reservations for tables.

CIGARETTES smoked by workers in the laboratory are sterilized in autoclave to inactivate plant disease organisms in the tobacco.

NIGHT ROOM is one of several used to provide a controlled dark period for greenhouse plants. Door was opened to make photograph.

WIND TUNNEL is used to grow plants under various wind conditions. Artificial daylight is provided by the fluorescent lamps at top.

tion in our "phytotron" at the California Institute of Technology. It all began with a study of the growth of tomato plants under controlled conditions. The generosity of Miss L. Clark and the technical and scientific knowledge of H. O. Eversole had provided two fully air-conditioned greenhouses, and I decided to study first the response of plants to the relative humidity of the air. Since the temperature had to be high to produce a low enough humidity, I chose the tomato, a warmth-loving plant, as the subject of the experiments.

We kept both greenhouses at a constant temperature of 79 degrees Fahrenheit but set the relative humidity in one at 70 per cent and in the other at 40 per cent. The humidity apparently made no difference: the tomato plants grew at about the same rapid rate in both greenhouses. But we found that all the plants did rather poorly under the conditions we had established. They failed to develop a rich green color, were a little spindly and, worst of all, produced no fruits. From the several hundred plants we raised during the first year in these greenhouses we got only four ripe tomatoes!

Tomato-growing experts who saw the plants offered all sorts of explanations for the lack of fruit set, but none of their suggestions for changes in the culture of the plant helped. Finally we tried lowering the temperature of one of the greenhouses to 64 degrees F. The tomato plants in that greenhouse immediately started to set fruit, and it ripened normally. This seemed very strange, considering that tomatoes in the field grow and bear well at average temperatures much higher than the 79 degrees we had maintained in our greenhouses. We soon determined by experiments that the plants needed a daily cycle of temperature change, and that they did best if the night temperature (*i.e.*, during the dark period of the plant's growth) was near 64 degrees.

The Phytotron

It was plain that we needed much more detailed information to understand the effects of individual climatic factors on plants. The Earhart Foundation made a generous grant of $407,000 (to which it later added $74,000) for the construction and operation of a set of greenhouses in which all sorts of climatic factors can be controlled. This building, known as the Earhart Plant Research Laboratory, is popularly called the "phytotron" from *phyton* (the Greek word

COOLING WATER is sprayed on the roof of a greenhouse in the laboratory. Most of the solar heat which accumulates in the greenhouses is removed by the rapid circulation of air.

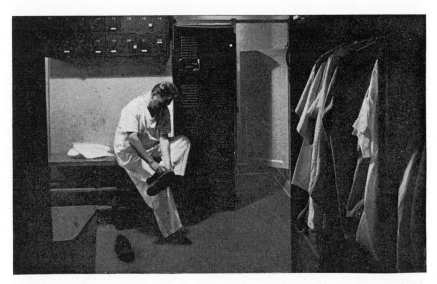

CHANGE ROOM is used by men who enter the laboratory to remove their street clothes (which may harbor insects or other disease organisms) and to don laboratory garments.

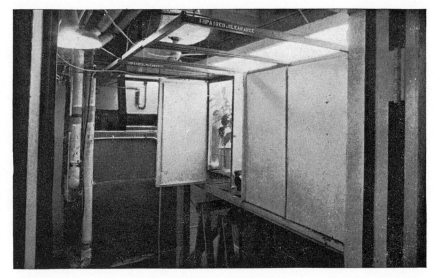

ARTIFICIALLY LIGHTED CABINETS are used to grow plants at the same temperature but under different light conditions. They may be darkened by electrically operated doors.

for plant) and *tron* (suggesting the contribution of physics to its remarkable equipment).

To give an idea of what the Earhart Laboratory is, let us make a quick tour through it. From the outside it looks like a big, pleasant house with large windows. But just inside the front door you are at once confronted with two doors, marked "Ladies" and "Gentlemen." Before you can enter the laboratory, you are required, like every visitor and worker there, to go to a dressing room to wash your hands, comb your hair and change your street clothes for freshly laundered laboratory garments, to make sure that you will bring no insects or diseases into the laboratory.

Inside the building you find yourself in a remarkable atmosphere, free of any trace of insects, dust or smog. When you pass through the door into a greenhouse, you are suddenly in a new world, filled with plants on which a shimmering light plays. The shimmer is the result of water spraying on the glass roof, the purpose of the water being to absorb most of the

sun's infrared radiation, which is not used in plant growth. The air feels perfectly fresh and pleasant, not muggy as in conventional greenhouses. Conditioned air circulates constantly throughout the greenhouse, entering through slots in the floor and leaving by ventilators in the walls. The air is completely replaced twice each minute. It removes most of the solar heat; at noon the greenhouse absorbs so much heat from the sun that during the half minute the circulating air spends in the room its temperature is raised by seven degrees. The ventilating system keeps the air moving so evenly that there are no stagnant spots, and all the plants within a greenhouse are subjected to exactly the same temperature and humidity.

The warmest greenhouse in the group is kept at 86 degrees during the day, summer and winter, but it does not feel too warm and is pleasant to walk in. You are, in fact, surprised to learn the actual air temperature. As every skier knows, the temperature of the air is less important, so far as the feeling of cold or

warmth on our skin is concerned, than radiation. Moreover, the relative humidity of the air has a large influence on our perception of heat: dry air feels much cooler than humid air at the same temperature. And thirdly, air in motion feels cooler than quiet air at the same temperature.

Factors of Climate

A complete tour of the Earhart Laboratory would take us to 54 separate rooms where plants grow: individual greenhouses, darkrooms, artificially lighted compartments. All the plants stand on wheeled tables, so that they can easily be moved to any one of 54 different environments. We can combine a high daytime temperature with a high or low night temperature, vary the length of illumination, subject plants to artificial rain, to wind or to special gases. The particular variables that we manipulate in the laboratory are day temperature, night temperature, light intensity, duration of daily illumination, light quality, relative humidity of the air, wind, rain and gas content of the air. All other variables are excluded as much as possible, so that we can concentrate on the ones we have chosen. We keep the nutrition and soil conditions uniform by growing all the plants in vermiculite or gravel or a mixture of the two and watering them with a standard nutrient solution, which is piped into all greenhouses and growing-rooms. We can, of course, change the feeding of plants for special studies.

We grow many different plants together in the same greenhouse. Commercial growers generally use separate greenhouses for their different crop plants, each adjusted to the best growing conditions for its inhabitants. But our purpose is to test given plants under a wide variety of conditions. We may have tomatoes, peas, potatoes, African violets and a number of varieties of orchids all in the same greenhouse, measuring their various responses to the same temperature. In another greenhouse you may see desert plants growing next to coffee, barley, spinach, carnations and dozens of other plants—each in an experiment with different objectives.

Having walked through the greenhouses, let us look into the artificially lighted rooms. There are 13 groups of these, each serviced by a separate air-conditioner. A group is divided into separate compartments by sliding doors, making it possible to keep plants in the different compartments at the same temperatures but under different light treat-

VARIATION OF NIGHT TEMPERATURE for 24-hour periods over the year increases with latitude. In Fairbanks it varied over several years from —25 degrees F to 54 degrees; in San Juan, only from 72 degrees to 79 degrees.

ments. The plants are on tables of adjustable height, so that as they grow taller they can be kept at precisely the same distance from the light by lowering the table. Time clocks turn the lights on and off automatically. The plants can be wheeled into adjoining dark compartments on their trucks.

The temperature range in the artificially lighted rooms is from 38 to 86 degrees F.—a range which includes the temperature extremes at which tropical

plants and alpine plants grow best. In the greenhouses we maintain day temperatures between 63 and 86 degrees and night temperatures between 53 and 73. The temperature is always lowered during the night, because most plants need this daily change to achieve their best growth.

When we come to the control and machinery areas, you will get a vivid idea of the complexity of the laboratory and the multiplicity of operations need-

ed to keep it going. Located strategically in the center of the main floor is a control room with long panels covered with dials and recorders, showing exactly what the conditions are in every room. Here the superintendent of the building reigns over the controls, and here also the complicated administration to insure the best use of the growing space is carried out. In the basement we find labyrinths of air ducts and pipes for conveying hot and cold water, compressed air,

RESULTS OF EXPERIMENT on coleus plants in the Earhart Laboratory are recorded by photograph. The plants in the top row were subjected to various night temperatures ("nyctotemp"). The temperatures are given in degrees centigrade below each plant. The day temperature ("phototemp") and the length of the day ("photoperiod") were kept constant. The plants in the middle row were subjected to various day temperatures; the plants in the bottom row, to various photoperiods. Date of experiment is at upper right.

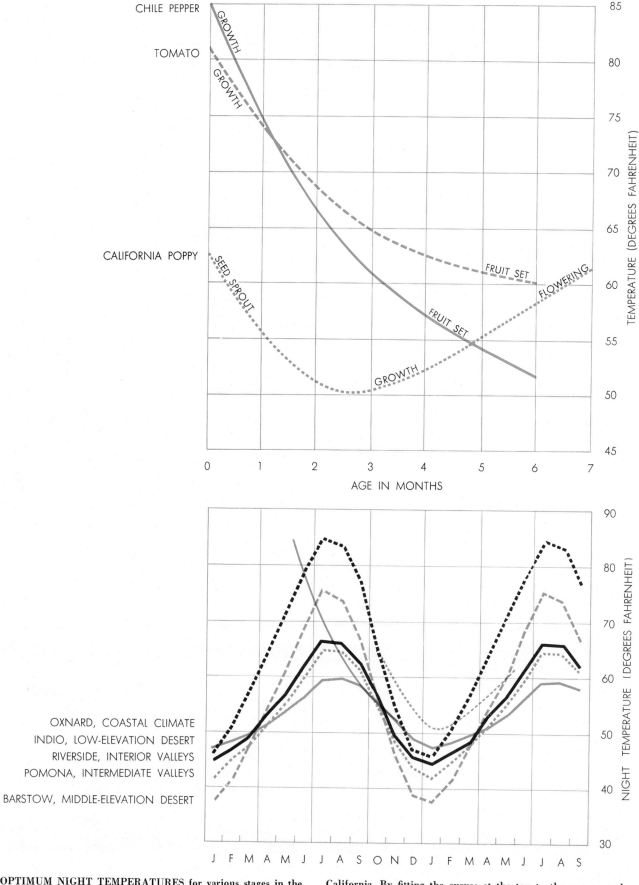

CHILE PEPPER

TOMATO

CALIFORNIA POPPY

OXNARD, COASTAL CLIMATE
INDIO, LOW-ELEVATION DESERT
RIVERSIDE, INTERIOR VALLEYS
POMONA, INTERMEDIATE VALLEYS

BARSTOW, MIDDLE-ELEVATION DESERT

OPTIMUM NIGHT TEMPERATURES for various stages in the life of three annual plants are given by the curves in the diagram at the top. The curves in the diagram at the bottom give the average night temperatures during 1950 and 1951 for four locations in California. By fitting the curves at the top to the curves at the bottom it is possible to determine when it is best to plant these species. The curves are fitted for the chili pepper (*in color at left in the bottom diagram*) and the California poppy (*center*).

nutrient solutions, deionized water, ordinary tap water and so on. There is a separate air conditioner for each of the rooms upstairs. Also in the basement are a general laboratory, shops, photographic rooms, etc.

One of the major problems that had to be solved to make the laboratory possible was filtering of the incoming air. Since much of the air conditioning is carried out through evaporative cooling, large volumes of air have to be taken into the building. All this air has to be filtered clean of dust, insects, disease spores and smog—no small matter in a city such as Pasadena where the air is contaminated with oxidation products of gasoline vapors which are toxic to plants. The capacity of our air-filtering system is approximately one ton of air per minute. Gradually all the difficulties involved in the filtering process were ironed out, and we now have completely clean air in the laboratory.

Plants and Temperature

Now that we have looked over the laboratory, let us review some of the things that have been learned about plants and climate by experiments during its seven and one half years of operation. We shall start with the daily alternation of warm and cool temperatures that plants need, as we found in our original experiments with tomatoes. From extensions of those experiments we now know that most varieties of tomato plants require almost the same night temperature to set fruit. This explains why tomato production is usually very low in the tropics: the night temperatures there are above the optimal fruit-setting range. On the other hand, in cool climates the night temperatures are likely to be too low. The tomato yield is most reliable in California, where a relatively stable air mass during the summer insures fairly consistent night temperatures in the appropriate range. In areas such as the Southwest and the East, the great fluctuations in summer night temperatures make tomato production very variable. In some years there are only a few weeks of optimal night temperatures, and the crop is small. We are testing in our laboratory new varieties of tomatoes which breeders hope will tolerate a wider range of night temperatures, and indications are that the effort to breed a less sensitive tomato will be successful.

The potato plant, a close relative of the tomato, has in general the same temperature response; that is to say, it will form tubers only if the night temperatures fall within a rather narrow range. The optimal range is between 50 and 57 degrees F., about 10 degrees below the best range for tomatoes. This explains why the most successful potato-growing areas are mainly in the northern regions, such as Idaho, Maine, Ireland and northern Europe. In the central valley of California potatoes can be grown in the spring and late fall, but not during the middle of summer. In the tropics potato production is possible in the mountains, where the night temperatures are within the proper range. We have found that the important thing is the temperature to which the top of the potato plant is exposed: artificial heating or cooling of the soil has very little effect on the formation of tubers.

The response of sugar beets to temperature is somewhat more complicated. The plant grows best, at least in the case of warm-climate varieties, when the night temperature is about 68 degrees, but this is poor for sugar production. Sugar beets grown on a regime of 68-degree nights produce monstrous tubers with a very low sugar content. They develop the highest sugar content when they are exposed to considerably lower night temperatures and also get comparatively little nitrogen nutrition. This suggests that the best sequence of conditions for beet sugar is a warm summer and early nitrogen feeding, while the plant is growing, followed by sunny autumn weather (for photosynthesis of sugar) with cold nights near freezing.

Peas and sweet peas grow mainly during the day and are little affected by the night temperature. They do best at daytime temperatures below 70 degrees; as soon as the regular daily temperature goes above 80 or 85 degrees, they start to die. In warm climates they can be grown only as winter crops.

The responses of a large number of other plants to day and night temperatures have been investigated, and for each of them an optimal range was established. Representative ranges for a few garden plants are shown in the accompanying chart [see page 116]. As temperatures depart from the optimum, the growth of a plant becomes poorer and poorer, until it fails entirely. For example, African violets will die within weeks or months when they are kept continuously at the growing conditions which are best for the English daisy, whereas the English daisy dies when it is kept under the conditions best for the African violet.

If we lay such a chart over the climatic chart compiled by meteorologists for a particular locality, we can readily see the most favorable growing season for any given plant [see opposite page]. Within California we have a considerable range of climates: near the coast, because of the effect of the ocean, the daily and seasonal temperature variations are comparatively small; farther inland we have a drier climate and greater temperature fluctuations.

Perennial Plants

Most perennials of temperate climates depend upon cycles of temperature change. For instance, tulips will not flower in a completely even climate. There is a different optimal temperature for each successive stage of the plant's growth. First there must be a period of temperatures under 50 degrees to prepare for the growth of the flower stalk; the stalk then requires 63 degrees for its best growth; finally, if it is to produce new leaves and flowers in the following season, the tulip needs about 80 degrees. For the hyacinth these requirements are all approximately 10 degrees higher. The onion starts its flowers at a low temperature, but it needs higher temperatures for other growth processes.

The biennial plants, such as beets, carrots and foxgloves, spend their first year making a rosette of leaves and a tap root which becomes filled with storage food. They are more or less dormant over the winter. Then in the following spring a long stem develops from the center of the rosette, and this produces flowers and fruits. They must have the cold winter period: if a beet, for instance, is kept instead at continuously high temperatures, it will live for years and grow to enormous size, but it will never form flowers.

Most of our deciduous trees also require a sequence of warm summers and cold winters. A peach or pear will not open its leaf or flower buds in the spring unless the tree has passed through a sufficiently cold winter. This chilling requirement has not been worked out in great detail, but in general it seems that the temperature must be below 40 degrees for several months. The requirement varies with the tree and the variety: the peach varieties that normally grow in cold climates require a longer period of low temperatures than peaches normally growing in warmer climates. Thus a peach that does well in the St. Louis area will not leaf out in southern California, whereas a southern California peach transplanted to St. Louis is apt to sprout during a warm spell in the middle of winter because its moderate cold requirement has been fulfilled too soon.

The deciduous trees are controlled by climate in two ways: by changes in the temperature and by changes in the length of the day [see "The Control of Flowering," by Aubrey W. Naylor; SCIENTIFIC AMERICAN Offprint 113]. A peach or pear tree senses the approach of autumn through the decrease in day length; its buds then begin to go into the dormant state. If a peach tree is kept continuously on long days in the laboratory, it will go on growing vegetatively without forming resting buds. In nature the sequence of long days, short days, cold temperatures and warm temperatures synchronizes the peach tree with the progression of seasons. The tree's response is complex. Two successive cold winters, separated by a warm summer, are needed—the first for the initiation of new flower buds during the following summer and the second to prepare for the flowering of these buds.

Warm-climate evergreen shrubs such as camellias also have a dual control by seasonal fluctuations in temperature and day length. In their case the flower buds are formed during high summer temperatures and the flowers open during the next winter. For vegetative growth they must have long days. Therefore camellias do well only when subjected to sufficient seasonal variations.

Many tropical plants, e.g., palms and hibiscus, develop leaves and flowers all the year around. The reason they cannot be grown in a temperate climate is that their vegetative growths would be killed

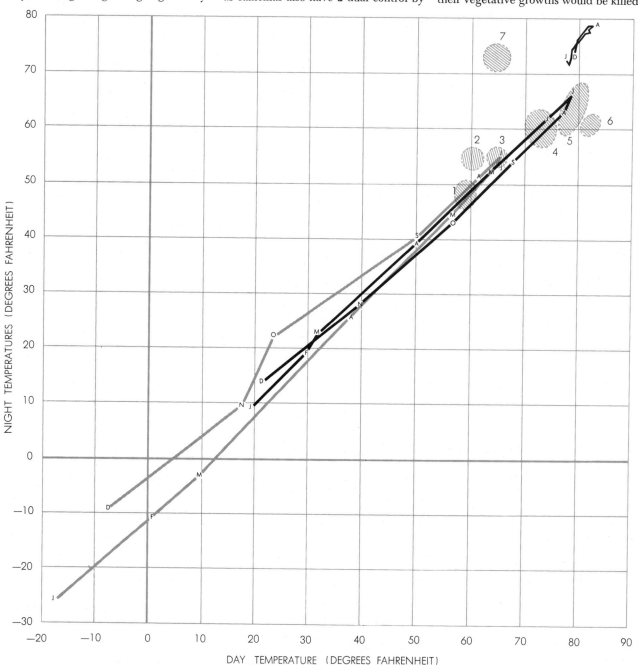

OPTIMUM DAY AND NIGHT TEMPERATURES for seven plants are indicated by the areas in color on this diagram. The plants are the English daisy (1), stock (2), ageratum (3), China aster (4), zinnia (5), petunia (6), African violet (7). The average day and night temperatures for three locations are also plotted. The long gray curve plots the temperatures for Fairbanks, Alaska; the long black curve, for Sioux Falls, S.D.; the short black curve at upper right, for San Juan, Puerto Rico. The letters on the curves indicate months of the year. Where the curve for a locality intersects the colored area for a plant, the plant will grow well there.

by frost. Yet even some tropical plants, such as the royal poinciana, respond to a yearly cycle—namely, the cycle of rainy and dry seasons.

Strawberry Flavor

Obviously climate exerts its influence upon plants by affecting their biochemistry. Concerning this we do not yet have much specific information. But we do find definite effects of climatic factors upon the taste and other qualities of fruits. Using the tongue as an analytical instrument, we have investigated the influences of various factors upon the flavor of strawberries.

When strawberry plants are grown in warm or moderate temperatures, the fruits are red, sweet and slightly acid, but they have no strawberry flavor! To develop flavor they must ripen at daytime temperatures of about 50 degrees. By various experiments we learned that the plants have to be exposed to the right light and temperature conditions for at least a week to acquire the full strawberry aroma.

These results explain why generally the first strawberries of the season taste best. They ripen while the early morning temperature is about 50 degrees. Later in the spring and during the summer the strawberry crop is practically without aroma, because the ripening fruit does not receive the proper temperature at any time during the day. At high altitudes, or far north, low morning temperatures occur even during summer, and strawberries from Alaska, northern Sweden or the high Rockies taste marvelous at any time.

With this information it should be possible to grow strawberries deliberately under conditions in which full flavor develops, producing a product of reproducibly high quality. For it is after all the flavor for which we pay 40 cents a basket; the amount of sugar and acid contained in a basket of strawberries (if bought separately) would be worth less than one cent.

It is possible that in many other fruits the same factors are important for flavor development. Probably the excellent taste of northern apples is due to the low temperatures during the last days of ripening, especially in the morning. Breeders of fruit give the main emphasis to its keeping quality and appearance, but growers ought to pay as much attention to flavor, which is the basis on which we buy fruits in the first place. With more experiments we may obtain knowledge of the conditions under which flavor can be developed most effectively.

Control of Climate

Since agriculture is the backbone of our existence, and since climate is a decisive factor in crop production, what can we do about it? This is a problem of such magnitude that it certainly should occupy many of our best brains, and research and development groups in agroclimatology should be active all over the world. Every agricultural experiment station and agricultural college has groups of scientists studying plant diseases and pests, breeding new varieties, determining proper soil treatments, developing agricultural implements and machinery. But only a few places have research teams in agroclimatology. Among these are the Drexel Institute of Technology's Research Laboratory of Climatology, the Department of Agronomy at the University of Wisconsin and the College of Agriculture at the University of California.

Such a team should comprise meteorologists, climatologists, agriculturists and plant physiologists. A phytotron of the kind we have at Cal Tech is a necessary instrument for the investigation. Now that the need for such a laboratory has been recognized, we can expect that in the near future a number will be built, not only for basic research but also for the solution of many problems in agriculture, horticulture and forestry.

Not much can be done about improving the selection of climate for the great staple crops, such as wheat, rice and cotton, for these crops are already generally grown in the most favorable areas, as a result of experience, economic forces (the profit margin is narrow) and the development of varieties adapted to local conditions. The specialty crops, on the other hand, are tried everywhere, with more or less (usually less) success. Agroclimatology can advise as to whether such crops are likely to succeed in a given locality and which varieties are the best to plant.

It can also be helpful with weather warnings. One of the best examples is the frost warning system for citrus growers in southern California. During the winter farmers receive a daily report by radio on the expected minimum temperature in each area. They are told at what hour temperatures will drop to the point where their orchard heaters need to be fired. This service is so reliable that farmers know when they can sleep soundly through the night and when they must organize crews to light the heaters. In the Netherlands the Meteorological Institute has an efficient warning service against outbreaks of potato blight. It tells potato growers when the climatic factors threaten spread of the disease, so that they can spray their fields with Bordeaux mixture in time.

Similar warning services are being developed for many other farming operations. For instance, calculations of the rate of water loss by the soil make it possible to tell farmers when and how much they must irrigate their fields. If reliable long-range weather forecasting can be developed, whole new possibilities will open up for agroclimatology. Experts will then be able to recommend what varieties of a plant (*e.g.*, tomato) should be planted for the best yield.

There are already a number of ways in which farmers can control climatic factors in the field to some extent. They can accelerate the flowering of chrysanthemums and other plants by means of electric lights or curtains which artificially regulate the day length. We have found that it is possible to induce tomato plants to set fruit as early as May or June in Pasadena by covering them with dark cloth during the late afternoon, when the temperature is in the proper range for fruit set. These plants start to form tomatoes at least a month ahead of uncovered plants. A more inexpensive way to obtain early fruit set is to plant the tomatoes along the east side of a wall or a shade tree, so that they are shaded and start their night activities early in the evening, while the temperature is still high enough.

As we learn more about plant responses to climate, we undoubtedly will find more ways to control their growth, and it is a comforting thought that technology is so advanced in our present world that almost any technical problem can be solved if it is urgent enough.

Climate-Tailored Plants

We have considered the possibilities of adjusting climate to plants; to what extent can we adjust plants to a climate? There is very little scientific evidence to support the idea that over the course of time a given variety of plant can become acclimated to conditions for which it is not originally suited. The soundest scheme is to attempt to breed new varieties which are fitted to the climate. This of course has been tried by many plant breeders and in many experiment stations. Every state of the Union has an experiment station with its own breeding program seeking to adapt cereals and other plants to the particular climatic conditions in that state. One of the great difficulties in these breeding programs is that there are minor differences in climate from year to year. The plants selected in one year may not be appropriate for typical conditions.

With the phytotron, in which the critical climatic factors can be controlled, it should be possible to select varieties which are specifically adjusted to particular climates. We are conducting such a program at the present time to breed tomato plants which will set fruit at comparatively high night temperatures in places such as Texas; the research program is financed by the Campbell Soup Company, the largest single grower of tomatoes in the U. S. Starting with varieties which are able to set at high temperatures but produce inferior tomatoes, we have found that by cross-breeding we can transfer their high-temperature tolerance to good tomato varieties. By such breeding it should be possible to develop a series of forms suitable for different climates.

The examples discussed in this article should make it clear that a great future lies ahead for the field of agroclimatology. The Earhart Laboratory has already shown in a number of cases that experiments under carefully controlled conditions can greatly cut down the expense of field testing. For example, it has assisted the U. S. Forest Service in selecting cover plants which are likely to grow well on mountainsides in southern California. In another project we found that the herb called the American hellebore (*Veratrum viride*), which had failed in all field experiments, needed six months of freezing temperatures followed by six months of very cool weather. The plant was then tried at high elevations in the mountains of northern Washington and grew successfully there. In these cases climate turned out to be the most important selective factor, and this is probably true of many plants.

More from the Sun

Basically the most important climatic factor that man seeks to put to use in agriculture is the energy of sunlight. Our present methods capture only a small part of the available energy. Laboratory experiments show that a plant can convert 10 per cent or more of the incident light energy into chemical energy in the form of its organic products. But in practice we harvest as plant material no more than 2 per cent of the solar energy that falls on a field of, say, corn or sugar beets. One reason for this relative inefficiency is that an annual, starting anew from seed, covers only a small percentage of the surface of the field in the early part of the season. Perennial crop plants provide more coverage and absorb more solar energy over the season, but most of their production goes into unusable leaves and branches.

Obviously these facts offer a vast field of research for students of climate and plant production. The objectives must be to get better coverage of the earth's surface with plants and to improve the efficiency of conversion of sunlight into chemical energy. Already it has been found that algae in a nutrient solution can transform light into organic energy two or three times more efficiently than the higher plants. However, a rather elaborate setup is needed to cultivate algae effectively; the expenditure of an equal amount of ingenuity and technological effort could make the higher plants just as efficient in absorbing the sun's energy.

Climate is one of our major natural resources. By learning exactly what the relationships are between climatic factors and plants we should be able to make far greater use of this resource.

IV

PRODUCTION TECHNOLOGY

IV

PRODUCTION TECHNOLOGY

INTRODUCTION

Present-day crop production is based on a great number of technological advances—some ancient, others quite new. The progress that these advances have helped to bring about is manifested in two distinct ways: by increases in production per acre and by increases in efficiency, which permit a reduction in the labor force. The explosive period of growth that agricultural technology underwent in the mid-twentieth century seems to be continuing still.

The use of technology to produce more where little was previously obtained is detailed by Walter C. Lowdermilk in "The Reclamation of a Man-Made Desert." The restoration of worn-out land in arid Israel provides a striking example of what technology can accomplish. To survive, the new state first had to become self-sufficient agriculturally. This was achieved through the adoption of progressive agricultural practices, foremost among them the efficient management of water resources, and the use of intensive soil-conservation measures to prevent the land erosion that for thousands of years had depleted Israel's soils. As a result, a ruined land has been salvaged.

Restored land means little without adapted crops. The implications for world food supply of rice varieties developed in the Philippines, maize developed in Mexico, and wheat based upon Japanese germplasm are tremendous. Perhaps the greatest success story of all is detailed by Paul C. Mangelsdorf in "Hybrid Corn." Heterosis, or "hybrid vigor"—the increased vigor associated with wide crosses—has been known for thousands of years (the mule and its virtues, for instance). American Indians planted various kinds of corn in the same field, and benefited from their crossing. Charles Darwin clarified cross- and self-pollination, and W. J. Beal put theory to work by producing crosses through detasseling. Nevertheless, hybrid corn became possible only when George H. Shull developed pure inbred lines; it became practical when D. F. Jones produced his high-seed-yielding double cross in the second decade of this century. In the 1940's open-pollinated corn had been almost entirely supplanted by hybrids in the United States' corn belt.

Probably no one recent technological change has had more far-reaching consequences for agriculture. Hybrid corn is in many ways a new crop, compared with the open-pollinated varieties. Its great success has led to more efficient practices and to changes in concepts of plant improvement that have had a great impact on the production of other crops. The incorporation of cytoplasmic male sterility has eliminated the laborious detasseling procedure used in the production of hybrid corn. This same technique has made hybrid breeding possible in crops where hand emasculation is not feasible, such as sorghum, and has transformed wheat from a normally self-pollinating crop into a cross-pollinating one, thus making possible the development of hybrids. The details of this bit of genetic engineering are chronicled in "Hybrid Wheat," by Byrd C. Curtis and David R. Johnston. They suggest that the eventual widespread introduction of wheat hybrids will have a far greater economic and nutritional impact than the introduction of any previous hybrid crop. Today hybrid breeding is becoming increasingly used in animals as well as in plants, for in addition to increased production, hybridization confers a bonus of uniformity of individuals, an indispensible advantage in this age of mechanization.

Fertilizer technology offers great promise for immediate increase in agricultural production. Christopher J. Pratt, in "Chemical Fertilizers," points out that fertilization can increase crop yield as much as 200 percent. The metabolic train of events that results from chemical fertilization is not entirely clear; yet the great bulk of empirical research has proved the usefulness of both macro- and micronutrients, related specifically to soil types and soil conditions. Much is known, too, of such things as leaching loss, salt injury, nutrient balance, and deficiency diseases, and of manufacturing techniques that make once-scarce nutrients abundant today (for example, synthetic ammonia, processed phosphates and potassium, trace elements). Of promise for the future is the economical production of fertilizers that release nitrogen gradually (either from particular molecular combinations or from coated soluble particles). Also promising is minimum tillage, made practicable by the use of herbicides, special seeding techniques, and residues of previous vegetation. Perhaps most outstanding is the capacity of the fertilizer industry to make readily available inexpensive substitutes for nature's age-old organic residues—substitutes tailored to particular soils, farms, and crops.

To bring home any crop successfully requires control not only of the environment, but control of the biological hazards that interfere with production. In "Third-Generation Pesticides," Carroll M. Williams describes efforts to control insect pests chemically as a series of technological changes. The chemical club of the broad-spectrum poisons, such as lead arsenate, which killed insects when consumed or contacted, constituted the first generation. About the time of World War II, DDT ushered in a new group of especially potent organic compounds that could be used in small concentrations and killed almost all insects on contact. Although this "second generation" of insecticides greatly increased effectiveness, they failed to eliminate a basic problem of chemical control; the cure became as hazardous as the malady. The complications that followed are well known: the development of resistance by the insects—which rendered many pesticides impotent—and the widespread pollution of the environment. Research indicates that more specific biological regulators seem to be available in the hormones that control an insect's life cycle. Substances that derail the orderly progression of insect maturation have already been synthesized. As a result of their extremely high specificity these "third generation" pesticides are not only generally nontoxic to other organisms, but the possibility that insects will develop resistance to them seems extremely unlikely.

The sophistication of crop-production technology varies widely around the world. Its *raison d'etre* is primarily the conservation of labor. In lands with inexpensive labor an investment in labor-saving equipment may seem questionable, saving little and idling many. But as Clarence F. Kelly has demonstrated in "Mechanical Harvesting" the development of agricultural mechanization has implications that go beyond crop production, for widespread mechanization triggers the release of labor in economies based upon hand agriculture, and thus becomes one of the keys to industrialization.

A study of how technology has enabled the United States to achieve its high level of agricultural mechanization should reveal principles applicable to crop production elsewhere. Many disciplines in addition to those traditional to agriculture are brought to bear on the problem, and the engineer and the crop breeder work together to achieve desired ends. California agriculturists set as a goal the development of mechanically harvestable tomatoes, which not only required the invention of a picking machine but the genetic restructuring of the tomato plant for fruits that would all ripen at about the same time and thus be suitable for machine handling. Confidence that such a tomato could be bred permitted a decade's engineering work before commercial crops were ready.

Kelly outlines the principles behind grain combines, hay harvesters and pelleters, sugar-beet diggers, cotton-gatherers, and a host of ingenious machines for harvesting perishable fruits and vegetables. These machines are of course most economical on large acreages, where their cost is more widely spread. The economics of such mechanization are amazing: one man-hour suffices to harvest an acre of rice in the United States, but 258 hours are required in Japan.

How a particular agricultural industry builds upon ancient practices to evolve a modern technology is documented by Maynard A. Amerine in "Wine." That wine-making is an ancient art is attested to by murals on Egyptian tombs that date as far back as 2500 B.C. When acted upon by the yeasts and molds that coat all grape skins, the sugary juice of the grape ferments naturally. But wines produced from unselected grapes by unregulated fermentation would hardly find a market today. Instead, select *vitis vinifera* grape cultivars are planted to suitable environments in the United States (in California and east of Great Lakes), southern Europe, northern Africa, and the southern parts of South America, Africa, and Australia, and their fermentation is carefully controlled. The variability of the *v. vinifera* grape is such that there would seem to be unlimited possibilities for the development of new cultivars. Even though science has contributed much to our knowledge of the processes of fermentation, aging, and maturation, there are still many opportunities for things "to go wrong." Wine-making is still more of an art than a science. To a degree this is true of all food production, a point worth remembering as technically advanced nations send advisors to less advanced lands.

14

THE RECLAMATION
OF A MAN-MADE DESERT

WALTER C. LOWDERMILK
March 1960

The State of Israel has undertaken to create a new agriculture in an old and damaged land. The 20th-century Israelites did not find their promised land "flowing with milk and honey," as their forebears did 3,300 years ago. They came to a land of encroaching sand dunes along a once-verdant coast, of malarial swamps and naked limestone hills from which an estimated three feet of topsoil had been scoured, sorted and spread as sterile overwash upon the plains or swept out to sea in flood waters that time after time turned the beautiful blue of the Mediterranean to a dirty brown as far as the horizon. The land of Israel had shared the fate of land throughout the Middle East. A decline in productivity, in population and in culture had set in with the fading of the Byzantine Empire some 1,300 years ago. The markers of former forest boundaries on treeless slopes and the ruins of dams, aqueducts and terraced irrigation works, of cities, bridges and paved highways—all bore witness that the land had once supported a great civilization with a much larger population in a higher state of well-being.

Last year, as a finale to the celebration of the 10th anniversary of the founding of the State of Israel, an international convention brought 485 farmers from 37 countries to see what had been accomplished. They found a nation of two million people, whose numbers had doubled in the decade, principally by immigration. Yet Israel was already an exporter of agricultural produce and had nearly achieved the goal of agricultural self-sufficiency, with an export-import balance in foodstuffs. It had more than doubled its cultivated land, to a million acres. It had drained 44,000 acres of marshland and extended irrigation to 325,000 acres; it had increased many-fold the supply of underground water from wells and was far along on the work of diverting and utilizing the scant surface waters. On vast stretches of uncultivable land it had established new range-cover to support a growing livestock industry and planted 37 million trees in new forests and shelter belts. All this had been accomplished under a national plan that enlisted the devotion of the citizens and the best understanding and technique provided by modern agricultural science. Israel is not simply restoring the past but seeking full utilization of the land, including realization of potentialities that were unknown to the ancients.

For the visiting farmers, many of whom came from the newer and less developed nations of the world, the example of Israel was a proof and a promise. Civilization is in a race with famine. The doubt as to the outcome is due not so much to the limitations of the earth's resources, plundered as they are, but to a lag in the uptake of progressive agricultural practices and failure in the distribution of the present output of food. More than two thirds of the people of the world are undernourished. Most of them live in the lands where mankind has lived longest in organized societies. There, with few exceptions, the soil is in the worst condition. The example of Israel shows that the land can be reclaimed and that increase in the food supply can overtake the population increase that will double the 2,800-million world population before the end of this century. Israel is a pilot area for the arid lands of the world, especially those of her Arab neighbors, who persist in their destitution in the same landscape that Israel has brought into blossom.

The achievement of Israel is the more remarkable for the fact that politics showed little regard for the logic of terrain and watershed in setting the boundaries of the state. The 7,815 square miles allocated to Israel in the 1948 partition of Palestine make a narrow strip of land along the eastern shore of the Mediterranean, roughly 265 miles long and 12 to 70 miles wide. It comprises only part of the Jordan River Valley, the principal watercourse of the region, with its three lakes: Lake Huleh, 230 feet above sea level at the northern end; the Sea of Galilee, nine miles to the south and 680 feet below sea level; and the Dead Sea, 65 miles farther south and 1,290 feet below sea level. More than half of Israel's territory is occupied by true desert or near desert, and the principal agricultural acreage lies on the narrow coastal plain, on the northern uplands and on the western slope of the Jordan Valley from Lake Huleh downstream to 25 miles below the Sea of Galilee, where Israel's boundary comes down to the river. This division of territory and the persistent hostility of Israel's Arab neighbors continue to frustrate programs to realize the full benefits of the water supply to all concerned in a region where water is scarce.

Climatically Israel much resembles California. Rains come in winter, and the summers are long and dry. Moreover, the erratic rainfall varies considerably from one end of the country to the other, from an average of 42 inches in the north, to 26 inches at Jerusalem, to less than two inches at Eilat on the Gulf of Aqaba at the foot of the desert of the Negev. Temperatures range to similar extremes over short distances, being cool at high elevations and hot and tropical in the Jordan Valley. In the spring a hot, dry wind, called the khamsin, may blow for days at a time out of the desert to the east, with calamitous effect upon unprotected crops. Harsh as these conditions are, there has been no significant deterioration in climate since Roman times. The same plants still thrive in protected places, and springs recorded in the Bible still bubble from the ground. The "desert" that took over the once-flourishing land was the work of man, not of nature.

Fortunately one geologic feature operates in favor of the conservation of rainfall; the porous limestone of the landscape absorbs a high percentage of the rain and distributes the water widely from the regions of heaviest fall through labyrinthine aquifers underground. The total discharge from springs exceeds the flow of the Jordan: a single great spring near the foothills of Judea gives rise to the Yarkon River. Another important source of water, the heavy summer dew, helps crops to grow in the uplands.

The agricultural restoration of Israel began in the 1880's, with the arrival of the first immigrants brought by the emergent Zionist movement as refugees from the pogroms of Eastern Europe. They were able to buy "useless" marshland on the coastal plain. These marshes had been created by the shoaling of erosion-laden streams and by the damming effect of the inland march of sand dunes. With heroic labor the early settlers succeeded in draining the marshes and farming them successfully. But until

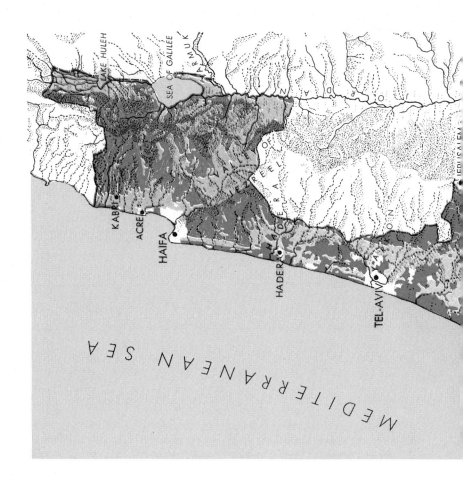

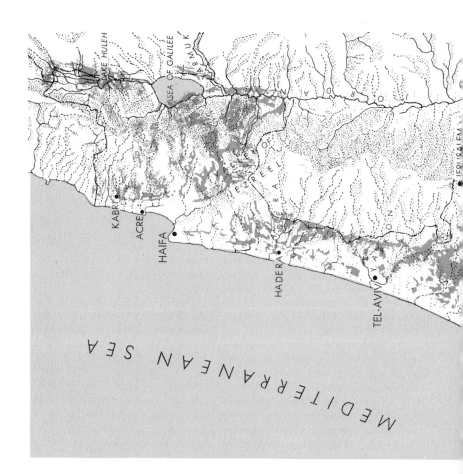

SOIL-RESOURCES SURVEY, covering those areas where water is or can be made available for agriculture, shows the effects of 1,300 years of misuse of the land. Most "very severe" erosion areas are beyond reclamation for cultivation and have been downgraded to forest and pasture or to wasteland. "Severe" erosion areas are reclaimable when watered; "moderate" areas can be restored by using elementary measures of soil management.

MODERATE   SEVERE   VERY SEVERE

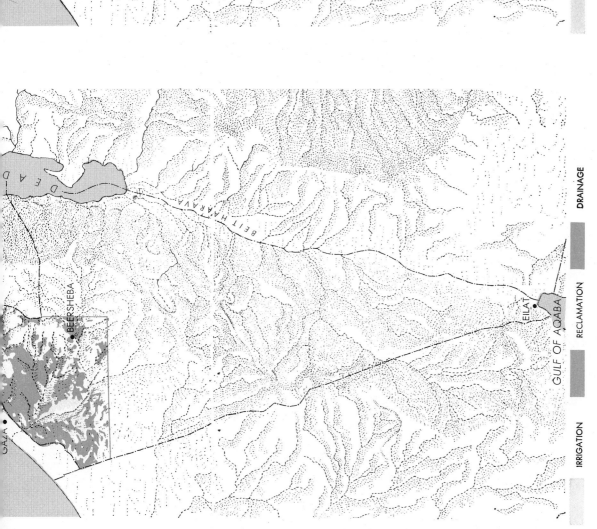

LAND-USE SURVEY shows measures to be taken to achieve maximum productivity. Extensive "drainage" areas reflect problem of salinity as well as of standing water in lowlands after winter rains. Marshes along coast were created by inland drift of sand dunes and erosion outwash from uplands which shoaled stream beds. "Irrigation" areas can be cultivated when watered; "reclamation" areas require repair as well as irrigation.

IRRIGATION   RECLAMATION   DRAINAGE

CONTOUR FARMING in Jezreel Valley is being extended to all slopes that exceed the safe gradient for "safe-line" farming. Steeper slopes are terraced and crops are rotated on contour strips. These practices prevent erosion and conserve the rainfall on fields.

the State of Israel was established, the effort was on a "first aid" basis.

When the new government set out to frame a comprehensive program for the development of the country's soil and water resources, it could call upon a number of outstanding authorities among its own citizens: specialists in forestry, horticulture, soil science, plant breeding and civil engineering who had come as refugees from Germany and Central Europe. But with a major por-

tion of its expanding population coming from the Arab countries of North Africa and the Near East, Israel did not have enough experts in the many disciplines needed to establish a modern agriculture in short order. The government therefore was among the first to draw upon the technical assistance offered by the specialized agencies of the United Nations and by the "Point Four" program of the U. S. I had the rewarding experience of sharing in this work as a

member of missions that served in Israel under the Food and Agriculture Organization of the U.N. from 1951 to 1953, consulting in the establishment of a national program of land development and in building up a staff of men to carry it out; and again from 1955 to 1957 helping to build a department of agricultural engineering at Technion, the Israeli institute of technology.

The first order of business, begun in 1951 and completed in 1953, was the

OLD SETTLEMENT of small-holders at Nahalal is laid out on the eastern European pattern with the village at center, surrounded by individual truck-garden farms, each seven acres in extent.

CITRUS GROVES, managed on a cooperative basis, support a com-

SOIL RECLAMATION has halted gully erosion and brought a limited area of heavily eroded soil under cultivation in the north-ern Negev. The soils in this region developed a "bad land" topography during centuries in which they were stripped of grass cover.

taking of a comprehensive inventory of the land. This comprises the 2.38 million acres north of the 60th parallel, about half the territory of Israel, where major agricultural development is possible. One of the most thorough inventories of its kind in the world, it furnished a secure foundation for land-use policy and for the immense task of reclamation and water development that has followed. Classification of the inventoried land by end-use shows that,

given adequate water supply, about 40 per cent, or a million acres, can be made suitable for general cultivation; about 15 per cent for orchard, vineyard, pasture and other use that will keep a permanent plant cover on the soil; 20 per cent for natural pasture without irrigation; and 25 per cent for forests, parks and wasteland. Outside the area of detailed survey, in the Negev, an extensive reconnaissance has projected a program for range development and

for the cultivation of forage crops in those areas where the scant winter run-off can be diverted or impounded to support irrigation.

A major feature of the land inventory was the classification of the lands according to their relative exposure to erosion by wind and water. In the hands of the Israeli Soil Conservation Service this has served as a blueprint for measures to preserve the best soils and ultimately to reclaim land now

munity on the Sharon Plain. Taller eucalyptus trees shelter groves.

NEW SETTLEMENT has begun to farm land opened up to cultivation by irrigation and soil-conservation measures. Open center of the village is eventually to be occupied by service buildings.

IRRIGATION BY SPRINKLING is the method that has proved most economical in bringing water to the field in Israel. Aluminum pipes are moved from field to field as needed.

unusable. The hazard of erosion increases in geometrical ratio with increase in the gradient of the soil. The first line of defense is directed against the dynamics of the falling raindrop and includes measures of soil management that are also required for sustained crop yields such as the build-up of organic matter to increase the water-holding capacity of the soil and the use of crop litter to absorb the energy and reduce the splash-erosion of the raindrop. Contour plowing and the planting of crops in strips along the contour provide the second line of defense and usually suffice against the hazards of moderate storms. These defenses can be set up by the individual farmer or farm cooperative and are everywhere encouraged through education and demonstration by the Soil Conservation Service. But rains in Israel characteristically come in downpours, in a few heavy storms during the rainy season and in extreme storms every few years. Where such rains overtax the first two lines of defense, more elaborate and costly measures must be designed and laid out by soil-conservation engineers of the Soil Conservation Service. Slopes must be broken by broadbase terraces to pick up and slow storm runoff and the terraces must be interconnected by waterways to keep the accumulated water from cutting gullies through the fields. Storm waters are then available for storage in surface ponds and reservoirs or to recharge ground waters. This line of defense must be accu-

rately and adequately engineered, for running waters do not forgive a mistake or oversight in design.

One of the effects of man-induced erosion in the past was the creation of marshes on the narrow coastal plain, notably at Hadera, Kabri and in the Jezreel Valley. Carrying through the work started by the early settlers, Israel has now fully reclaimed these lands, draining and planting them to eucalyptus trees in the lowest spots and to citrus groves and crops on the higher ground.

A more substantial engineering challenge was presented by the marshlands of the Huleh basin at the head of the Jordan Valley. In Roman times and before, this region was fertile and thickly populated, but it had become a dismal swamp and a focus of malarial infection to the country at large. Sediments from the uplands to the north had progressively filled in the northern end of Lake Huleh, thus creating a marsh that was overgrown with papyrus. The marshes have now been drained by widening and deepening the mouth of the lake to bring down its water level and by a system of drainage canals. With the papyrus cleared away, the deep deposit of peat beneath yields richly to cultivation, much as do the delta peat-lands at the head of San Francisco Bay. The Huleh Reclamation Authority estimates that this little Garden of Eden will support a population of 100,000 in an intensive agricultural economy, cultivating

vegetables, grapes, fruits, peanuts, grains, sugar cane, rice—even fish (in ponds impounded on the old lake bed). The yield of fruits and vegetables will soon require the installation of processing and canning plants on the spot. Another gain achieved by the reclamation of this land is the conservation of water; the reduction of the evaporation surface of the lake and surrounding marshes will save enough water to irrigate 17,000 to 25,000 acres of land, depending on the rainfall of the district to which these waters will be delivered. The Huleh Drainage and Irrigation Project is not great in size, but it symbolizes the determination of Israel to make the most of its resources.

The development of water supplies and irrigation constitutes the most significant achievement of the new nation and differentiates its agriculture most sharply from that which prevails in all but a few areas in the surrounding Arab countries. Since the time of Abraham, when "there was famine in the land," agriculture in this region has been at the mercy of the variable winter rainfall. In ancient Palestine irrigation was limited to small areas that could be fed by gravity from perennial springs. These works had long since fallen into disuse, and at the beginning of this century very little of the Holy Land was irrigated. In 10 years the State of Israel has quadrupled the acreage under irrigation, from 72,500 to 325,000 acres. It was this achievement that made possible the absorption of the great influx of immigrants. Irrigation has increased yields per acre from three to six times and more over those achieved by dry farming in the region and has secured dependable yields from year to year.

With most of the water coming from wells, irrigation in Israel is accomplished by sprinkling, rather than by furrow or border ditch. The grid of pumps and pipes delivers the water under pressure but at low rates of flow. Irrigation engineers soon found that sprinkling was best adapted to this mode of delivery and for application of the water to sandy soils and to rougher, stony land unsuited for leveling. The high investment in pumps and piping has been more than offset by the intensive year-round cultivation made possible by irrigation and by the urgent need to settle immigrants in self-supporting activity on the lands. Each year from 25,000 to 30,000 additional acres are being brought under irrigation, and the prospect is that this will continue until the limit of water supply is reached.

Meanwhile extensive field research

is devoted to achieving the most efficient use of water. In the northern Negev, for example, it has been found that about six inches of irrigation water, applied just before the winter rains to soak the soil to its water-holding capacity down to a depth of about four feet, will make up the equivalent of 20 inches of rainfall, sufficient for winter grain. In many soils irrigation raises a serious drainage problem. Evaporation from the soil and transpiration by crops in the "consumptive" use of the water leave behind the salts it carries in solution. After a few years the accumulation of salts may reach toxic proportions. Certain crops, such as sugar beets, take up some salts and may be planted in rotation to reduce this accumulation. But whatever the crop, drainage must be provided in time to leach away the salts, and the chemical composition of the soil must be kept under surveillance.

To bring much of the land under irrigation and cultivation has required strenuous repair of the damage done by centuries of erosion. The slopes in stony soils are typically covered by an "erosion pavement" made up of stones too heavy to be moved by rain splash and by the sheet flow of the storm runoff that carried away the topsoil. In some parts of the country, farmers have raked these stones from the fields and piled them into great heaps. Where erosion has exposed the rock or gullied the deep soils beyond plowing, the land has been put to some lower use, such as rough pasture or woodlot. In many parts of the highlands modern farmers have been able to take advantage of the soil-conservation works of the ancient Phoenicians. My own investigations indicate that the Phoenicians, 3,000 to 4,000 years ago, were the first people in the Middle East to clear and cultivate mountain slopes under rainfall agriculture and so were the first to encounter soil erosion. They were also the first to control soil erosion by using the principle of the contour and by building stone walls to convert a slope into a series of level benches. Most of these ancient terraces had been allowed to fall in ruins. Today they are being reconstructed and redesigned. Since the terraces are narrow and so suited only to hand labor, the practice is to collect the stones from old terrace walls and to pile them into ridges spaced more widely apart on the contour, creating terraces with gentle gradients for cultivation by tractor-drawn farm implements. Under sprinkler irrigation the new terraces are proving to be favorable sites for vineyards and orchards.

Over the large stretches of the coun-

SAND DUNES carried inland from coast have been encroaching for centuries upon arable land. Dunes are now being held in place by planting and even reclaimed for cultivation.

FLASH FLOOD from winter downpour rushes over ruins of ancient desert irrigation works in northern Negev. Waterfall in foreground demonstrates gully-cutting action of floods.

PLOWING BY CAMEL in Gaza Strip on Israeli border reflects survival of primitive agricultural practices that over the past millennium have wasted the soil resources of the region.

IRRIGATED LAND of small-holders' settlement near Sea of Galilee (*visible through gap in background*) is watered by aluminum sprinkler pipes supplied from wells. Fields are planted to garden crops and to forage for dairy cattle. On slope in foreground, still green from winter rains, is an old olive grove. Land is rented from the state on 49-year leases with rental at 2 per cent of land value.

DESERT LAND in the Negev is made to support forage crops by diversion of flash-flood water, after methods employed by ancient Nabataeans. Water is trapped in basin fields, held long enough to soak the ground and then passed on to the field below over concrete spillways such as that under construction here. Agricultural engineering students inspecting site are armed against border raids.

try that are beyond such reclamation and are too dry for forests, the effort is to develop the land for pasture. Throughout the Near and Middle East and North Africa the land has been overgrazed for more than 1,000 years. What sheep will not eat, goats will, and what the goats leave, camels will graze. By the time these hardy animals have ranged over the land through the long, hot, rainless summer, there is little plant cover left to protect the soil from the winter rains. But if one may judge by the relict species of forage grasses and plants that survive in rocky places and thorn thickets beyond the reach of goats and camels, it may be surmised that this land was once a pastoral paradise. The prompt return of a good cover of grasses and herbs after the goats were removed from the land by the Israeli Government in 1948 confirms this appraisal. The Soil Conservation Service has since been reseeding the range with native plants and with species imported from the U. S. and South Africa. In addition, certain woody bushes and low trees are being planted to hold the soil and furnish browse for livestock; the rich beans of the hardy carob tree, for example, yield as much

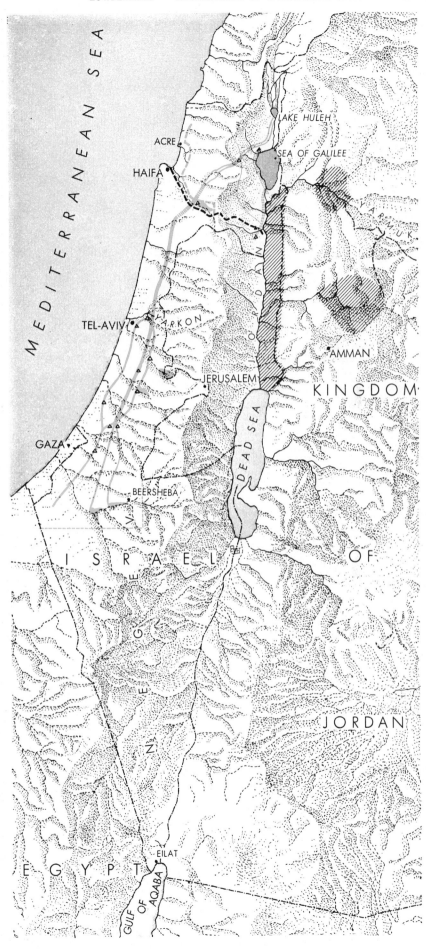

JORDAN-NEGEV CANAL
JORDAN-NEGEV TUNNEL
JORDAN-NEGEV PIPELINE
YARKON-NEGEV PIPELINE
WESTERN GALILEE PROJECT
HULEH DRAINAGE CANALS
OTHER PROJECTS
MEDITERRANEAN-JORDAN POWER PROJECT
DAMS
YARMUK YABOQ SCHEME
▲   POWER PLANT
△   PUMPING PLANT
◆   POWER AND PUMPING PLANT

MASTER WATER PLAN of Israel (*color*) is based upon plan originally designed to maximize water resources for entire Jordan River Valley, including both Israel and what is now the Kingdom of Jordan. Basic scheme calls for diversion of water from the head of the Valley, where rainfall is heaviest, to arid lands in the south. Via the Jordan-Negev canal, a tunnel and a 108-inch pipeline now under construction, water from Lake Huleh region is to be carried to the Negev, with additional water from the Yarkon River being carried by the Yarkon-Negev pipelines. The Mediterranean-Jordan Power Project and Yarmuk-Yaboq scheme await cooperation and action by Jordan.

feed as an equal planting of barley. Measures to divert and spread the storm waters over the pastures are further increasing the yield. Herds of beef and dairy cattle are now beginning to multiply on the restored range.

Early in the Jewish immigration to Israel the planting of trees came to be a symbol of faith in the future. Afforestation now plays a central role in the control of erosion, in reclamation of stony hills and in sheltering orchards and garden plots from the winds, whether from the sea or the desert. Some 250 million trees, both native and imported species selected by the Israeli Forest Experiment Station, are to be planted in the next 10 years. The growing of stock in the nurseries and the planting of trees on uncultivated hillsides, on roadsides, in shelter belts and on sand dunes provides interim employment for new immigrants until they become established. Already the new stands are yielding timber, poles and fuel products—valuable commodities in a deforested land.

The land inventory has served to protect the best agricultural lands from being engulfed by the growing cities and towns of Israel. Along the coast, for example, the communities have been encouraged to expand their boundaries into the sand dunes rather than into surrounding cultivable land. The dunes comprise 10 per cent of the coastal land and, under the drag of the prevailing westerly winds, are overwhelming good land, orchards and even houses. Experiments are under way to hold the shifting dunes by stabilizing the sand surface and by aggregating sand grains into crumb structures. This is accomplished by plantings of hardy shrubs and sand grasses, and of such fibrous-rooted plants as alfalfa, with water supplied to some tracts by sewage effluent and partially rectified sewage water. The rapid growth of the plants where this has been tried converts the sand in a few years into a stable soil-like material suited to the planting of trees and even some crops. But the full reclamation of the dunes to agricultural use is still in the research stage.

Ultimately the expansion of agriculture is limited by the availability of water. The Israeli Water Planning Agency is seeking to double the 1956 water supply by 1966, giving the country a total of 14.5 million acre feet (an acre foot is 12 inches of water per acre) per year. A central feature of the plan derives from a survey that I conducted in 1938

CONCRETE PIPE 70 inches in diameter for Yarkon-Negev pipeline is delivered to location where it is to be set in place in a great trench and buried. This line will carry water from the Yarkon River in the north to the arid lands of the Negev (*see map on page 131*).

and 1939 for the U. S. Department of Agriculture and from the proposal, growing out of that survey, of a Jordan Valley Authority to achieve the fullest development of the surface and underground waters of the valley for the entire original Mandate under the League of Nations, including what is now the Kingdom of Jordan as well as Israel. That proposal called for the development of ground waters and diversion of the upper Jordan waters within Israel to the dry lands in the south and for the diversion of the waters of the Yarmuk River to the eastern side of the Jordan Valley for irrigation of a promising subtropical region in Transjordan. In order to replace the flow of these rivers into the Dead Sea, salt water was to be brought in from the Mediterranean Sea through canals and tunnels to drop through two sets of hydropower stations nearly 1,300 feet below sea level to the Dead Sea. This salt water would not only produce electric power but also would maintain the level of the Dead Sea for the extraction of the minerals and chemicals that are there in fabulous amounts. The plan was declared feasible by an international consulting board of engineers. All parts of the plan that do not require the collaboration of the adjoining Arab states are now being carried out by the Israeli Government. The prestressed-concrete sections of the main 108-inch pipeline that will carry upper Jordan water down as far as the Negev are now being fabri-

cated and set in place in a great trench, and the tunnels to carry it through intervening hills are under construction.

Beyond this major undertaking the country is conserving for use and reuse such minor flows of water as are represented by the sewage of its cities and the runoff of intermittent streams along the coast. In the southern Negev, where the annual rainfall is less than six inches, the Soil Conservation Service is adopting the methods of ancient Nabataeans to impound the waters of flash floods for the irrigation of forage crops.

Prospects for the future have recently been brightened by progress in the desalting of sea water. A new method developed in the laboratories of the Government is about to be tested in two pilot plants, each with an output of 250,000 gallons per day. Success in this undertaking would be a major victory not only for Israel but also for all the other arid-land countries of the world.

On the anvil of adversity the State and people of Israel have been hammering out solutions to problems that other nations must sooner or later face up to. There are no more continents left to explore or to exploit. The best lands of the earth are occupied and in use. All of them, to a greater or lesser extent, need the same measures of reclamation and conservation that have succeeded so well in Israel. The frontiers of today are the lands under our feet.

15

HYBRID CORN

PAUL C. MANGELSDORF
August 1951

HYBRID CORN, a man-made product developed during the past 25 years, may prove to be the most far-reaching contribution in applied biology of this century. With its accompanying improvements in farming methods, it has revolutionized the agriculture of the American Corn Belt. Because of it U. S. farmers are growing more corn on fewer acres than ever before in this country's history. The new abundance of food brought by hybrid corn played a significant role in World War II and in the rehabilitation of Europe after the war. Now this product, spreading to Italy, to Mexico and to other countries where corn is an important crop, promises to become a factor of considerable consequence in solving the world food problem.

What is hybrid corn and how has it made possible these substantial contributions to the world's food resources?

In a broad sense all corn is hybrid, for this plant is a cross-pollinated species in which hybridization between individual plants, between varieties and between races occurs constantly. Such natural, more or less accidental hybridization has played a major role in corn's evolution under domestication. But the hybrid corn with which we shall deal here is a planned exploitation of this natural tendency on a scale far beyond that possible in nature.

The biological basis of hybrid corn is a genetic phenomenon known as "hybrid vigor." It means simply that crossed animals or plants have greater vigor or capacity for growth than those produced by inbreeding. This fact has been known since Biblical times. The ancient Near Eastern peoples who mated the horse and the ass to produce a sterile hybrid, the mule, were creating and utilizing hybrid vigor. The mule is an excellent example of the practical advantages that often follow crossing. This animal, said to be "without pride of ancestry or hope of posterity," has greater endurance than either of its parents; it is usually longer-lived than the horse, less subject to diseases and injury and more efficient in the use of food. Hybrid corn resembles the mule (indeed, it used to be called "mule corn") in being more useful to man than either of its parents.

Early Experiments

The idea of crossing varieties of corn is as old as some of the early American Indian tribes, who regularly planted different kinds of corn close together to promote hybridization and increase yields. Cotton Mather, of witch-hunting fame, published in 1716 observations on the natural crossing of corn varieties, and James Logan, onetime Governor of Pennsylvania, in 1735 conducted experiments which demonstrated natural crossing between corn plants.

But it was Charles Darwin who made the important studies of hybrid vigor in plants which open the story of modern hybrid corn. He investigated the effects of self-pollination and cross-pollination in plants, including corn as one of his subjects. His were the first controlled experiments in which crossed and self-bred individuals were compared under identical environmental conditions. He was the first to see that it was the crossing between unrelated varieties of a plant, not the mere act of crossing itself, that produced hybrid vigor, for he found that when separate flowers on the same plant or different plants of the same strain were crossed, their progeny did not possess such vigor. He concluded, quite correctly, that the phenomenon occurred only when diverse heredities were united. These researches, together with his theory of evolution, inspired the studies on heredity which eventually led to the discovery of the principles underlying the production of hybrid corn.

Darwin's experiments were known, even before their publication, to the American botanist Asa Gray, with whom Darwin was in more or less constant communication. One of Gray's students, William Beal, became, like Gray, an admirer and follower of Darwin. At Michigan State College Beal undertook the first controlled experiments aimed at the improvement of corn through the utilization of hybrid vigor. He selected some of the varieties of flint and dent corn then commonly grown and planted them together in a field isolated from other corn. He removed the tassels—the pollen-bearing male flower clusters—from one variety before the pollen was shed. The female flowers of these emasculated plants then had to receive their pollen from the tassels of another variety. The seed borne on the detasseled plants, being a crossed breed, produced only hybrid plants when planted the following season.

The technique Beal invented for crossing corn—planting two kinds in the same field and removing the tassels of one—proved highly successful and is still essentially the method employed today in producing hybrid seed corn. But as a device for increasing corn yield his operation of crossing two unselected varieties, each of mixed inheritance, was ineffective: the gain in yield was seldom large enough to justify the time and care spent in crossing the plants. The missing requirement—the basic principle that made hybrid corn practicable—was discovered by George H. Shull of the Carnegie Institution.

His discovery was an unexpected by-product of theoretical studies on inheritance which he had begun in 1905. Shull's contribution grew from certain earlier studies made by two great scientists: Darwin's cousin Francis Galton and the Danish botanist Wilhelm Ludwig Johannsen. Galton had recognized that the result of the combination of parental heredity could take two forms: an "alternative" inheritance, such as the coat color of basset hounds, which came from one parent or the other but was not a mixture of both, and a "blended" inheritance, such as in human stature. He observed that children of very tall parents are shorter than their parents, on the average, while children of very short parents tend to be taller. These observations led Galton to the formulation of his "law of regression," which holds that the progeny of parents above

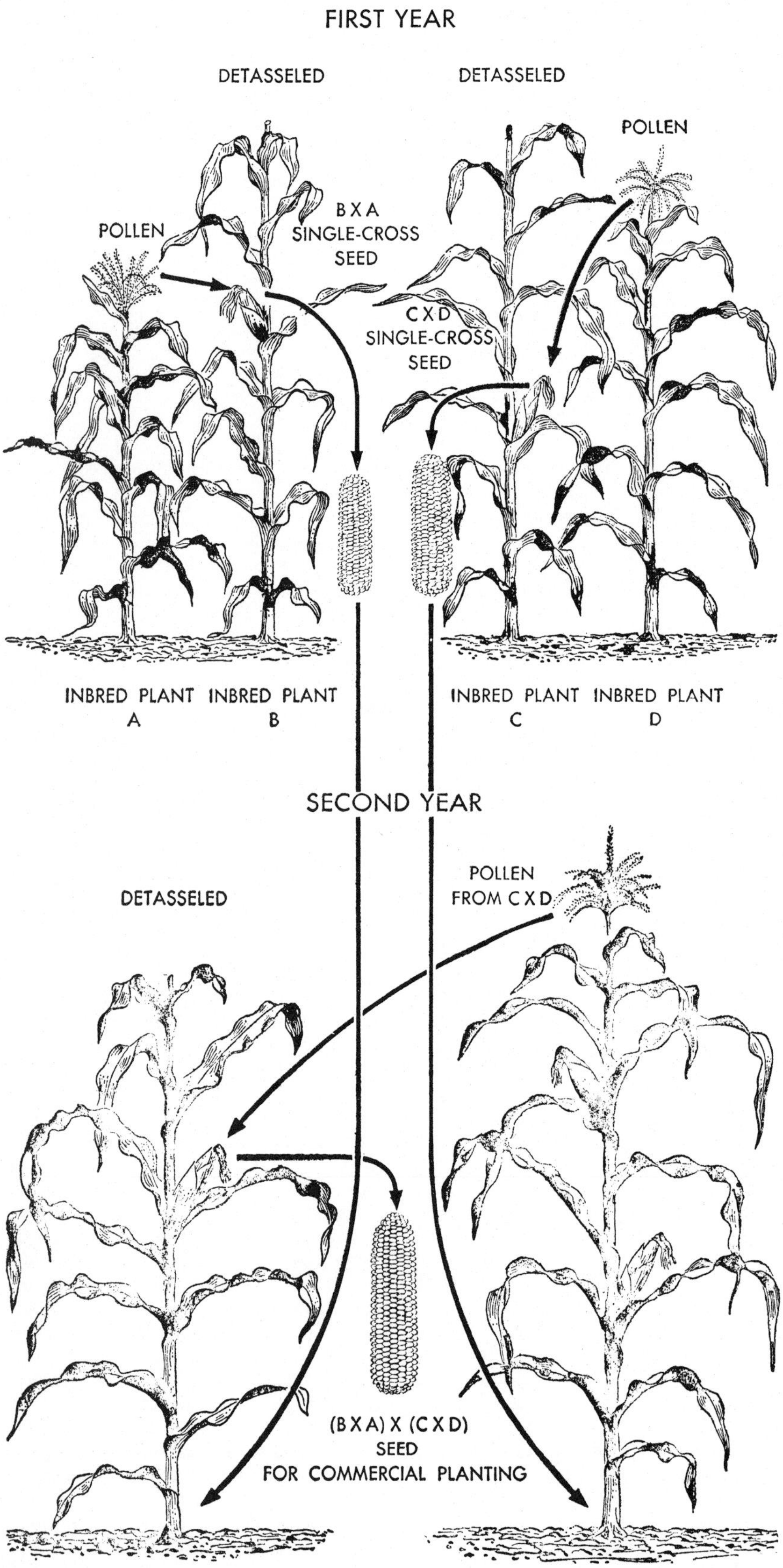

FIRST YEAR

DETASSELED   DETASSELED

POLLEN

POLLEN

B X A
SINGLE-CROSS
SEED

C X D
SINGLE-CROSS
SEED

INBRED PLANT   INBRED PLANT   INBRED PLANT   INBRED PLANT
A   B   C   D

SECOND YEAR

DETASSELED

POLLEN
FROM C X D

(B X A) X (C X D)
SEED
FOR COMMERCIAL PLANTING

SINGLE-CROSS PLANT (B X A)   SINGLE-CROSS PLANT (C X D)

DOUBLE CROSS made the experimental hybridization of corn a practical reality. First two pairs of inbred corn plants are crossed (*top*);
then the process is repeated with their hybrid descendants (*bottom*). The
second cross greatly multiplies the number of seeds produced by the first.

or below the average in any given characteristic tend to regress toward the
average.

This regression is seldom complete,
however, and Johannsen saw in that
circumstance an opportunity for controlling heredity through the selection in
successive generations of extreme variations. He tested the possibility by trying
to breed unusually large and unusually
small beans by selection. He found that,
although selection apparently was effective in the first generation, it had no
measurable effect whatever in later generations. Johannsen concluded that in
self-fertilized plants such as the bean the
progeny of a single plant represent a
"pure line" in which all individuals are
genetically identical and in which any
residual variation is environmental in
origin. He postulated that an unselected
race such as the ordinary garden bean
with which he started his experiments
was a mixture of pure lines differing
among themselves in many characteristics but each one genetically uniform.
Johannsen's pure-line theory has been
widely applied to the improvement of
cereals and other self-fertilized plants.
Many of the varieties of wheat, oats,
barley, rice, sorghum and flax grown
today are the result of sorting out the
pure lines in mixed agricultural races
and identifying and multiplying the
superior ones.

Inbreeding

Shull's contribution was to apply the
pure-line theory to corn, with spectacular, though unpremeditated, results. He
started with the objective of analyzing
the inheritance of quantitative or "blend-
ing" characteristics, and he chose the
number of rows of kernels in an ear of
corn as an inherited quantitative characteristic suitable for study. Through
self-pollination he developed a number
of inbred lines of corn with various
numbers of rows of kernels. These lines,
as a consequence of inbreeding, declined
in vigor and productiveness and at the
same time each became quite uniform.
Shull concluded correctly that he had
isolated pure lines of corn similar to
those in beans described by Johannsen.
Then, as the first step in studying
the inheritance of kernel-row number,
he crossed these pure lines. The results
were surprising and highly significant.
The hybrids between two pure lines
were quite uniform, like their inbred
parents, but unlike their parents they
were vigorous and productive. Some
were definitely superior to the original
open-pollinated variety from which they
had been derived. Inbreeding had isolated, from a single heterogeneous species, the diverse germinal entities whose
union Darwin had earlier postulated as
the cause of hybrid vigor.

Shull recognized at once that inbreed-

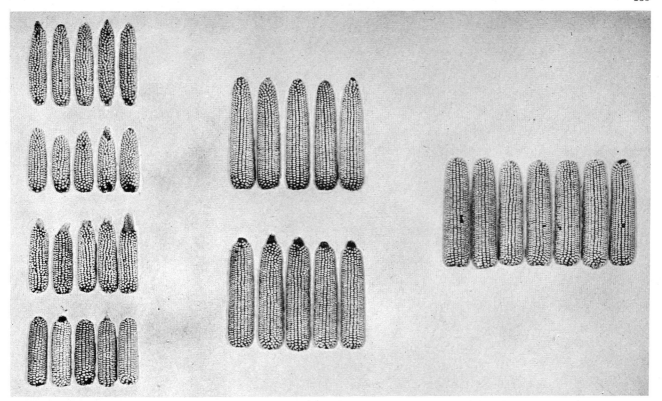

EARS of corn illustrate the effects of the double cross. At the left are the ears borne by the plants of four inbred strains. If the strains are divided into pairs, and one plant in each pair is allowed to pollinate the other, the latter bears ears like those shown in center. If a plant grown from a seed in one of the two center groups pollinates a plant grown from a seed in the other, the latter bears ears such as those on the right.

ing followed by crossing offered an entirely new method of improving the yield of corn. In two papers published in 1908 and 1909 he reported his results and outlined a method of corn breeding based upon his discoveries. He proposed the isolation of inbred strains as a first step and the crossing of two such inbred strains as a second. Only the first-generation cross was to be used for seed for crop production, because hybrid vigor is always at its maximum in the first generation. Shull's idea of growing otherwise useless inbred strains of corn solely for later crossing was revolutionary as a method of corn breeding, but it eventually won acceptance and is now the basis that underlies almost the entire hybrid seed-corn enterprise.

However, Shull's suggestion for the second step—the crossing of two weak inbred strains, known as a single cross—proved impractical as a method of seed production. Because the inbred strains are relatively unproductive, hybrid seed obtained in this way is too expensive except for certain special purposes.

The Double Cross

One further major development was needed to make hybrid corn practicable and the great boon to agriculture that it has become. This contribution came from the Connecticut Agricultural Experiment Station by the end of the second decade of this century. The story begins in 1906, when Edward M. East arrived there from the University of Illinois, where he had participated in corn-breeding experiments with some of Beal's former students. At the Connecticut Station, East began a series of studies of the effects of corn inbreeding and crossbreeding which were to continue to this day and yield a great deal of information about corn, including the effects of selection on its chemistry. It was East who called attention to the need for developing a more practical method for producing hybrid seed. It remained for Donald Jones, one of East's students, who assumed charge of the Connecticut experiments in 1915, to invent a method which solved the problem.

Jones' solution was simply to use seed from a double cross instead of a single cross. The double cross, which combines four inbred strains, is a hybrid of two single crosses. For example, two inbred strains, A and B, are combined to produce the single cross A×B. Two additional inbred strains, C and D, are combined to produce a second single cross C×D. All four strains are now brought together in the double cross (A×B) × (C×D). At first glance it may seem paradoxical to solve the problem of hybrid seed production by making three crosses instead of one. But the double cross is actually an ingenious device for making a small amount of scarce single-crossed seed go a long way. Whereas single-crossed seed is pro-

duced from undersized ears borne on stunted inbred plants, double-crossed seed is produced on normal-sized ears borne on vigorous single-cross plants. A few bushels of single-crossed seed can be converted in one generation to several thousand bushels of double-crossed seed. The difference in cost of the two kinds of seed is reflected in the units in which they are sold: double-crossed seed is priced by the bushel, single-crossed seed by the thousand seeds. Double-cross hybrids are never as uniform as single crosses, but they may be just as productive or more so.

Jones made a second important contribution to the development of hybrid corn by presenting a genetic interpretation of hybrid vigor. Shull and East had suggested that hybrid vigor was due to some physiological stimulation resulting from hybridity itself. Shull was quite certain that something more than gene action was involved. He thought that part of the stimulation might be derived from the interaction between the male nucleus and the egg cytoplasm. Jones proposed the theory that hybrid vigor is the product of bringing together in the hybrid the favorable genes of both parents. These are usually partly dominant. Thus if one inbred strain has the genes *AA BB cc dd* (to use a greatly oversimplified example), and the other has the genes *aa bb CC DD*, the first-generation hybrid has the genetic constitution, *Aa Bb Cc Dd*. Since the genes

MANY STRAINS OF CORN are tested for their value in hybridization. This photograph shows a field of the De Kalb Agricultural Association of De Kalb, Ill. On it are a number of strains illustrating the evolution of corn.

A, B, C and D are assumed not only to have favorable effects but to be partially dominant in their action, the hybrid contains the best genes of both parents and is correspondingly better than either parent. Jones' theory differs from a similar earlier theory in assuming that the genes involved are so numerous that several are borne on the same chromosome and thus tend to be inherited in groups. This explains why vigor is at its maximum in the first generation after crossing, and why it is impossible through selection in later generations to incorporate all of the favorable genes into a new variety as good or better than the first-generation hybrid. The ideal combination AA DD CC EE, which combines all of the favorable genes, is impossible to attain because of chromosomal linkage. For example, the genes B and c may be borne at adjacent loci on the same chromosome and thus be inseparably joined in their inheritance. Although Jones' theory is not universally accepted, and it now seems probable that hybrid vigor involves still other genetic mechanisms, it nevertheless gave great stimulus to practical hybrid corn breeding.

How It Is Produced

Historically, then, hybrid corn was transformed from Shull's magnificent design to the practical reality it now is when Jones' method of seed production made it feasible and his theory of hybrid vigor made it plausible. This combination proved irresistible to even the most conservative agronomists. Soon after 1917 hybrid corn-breeding programs were initiated in many states. By 1933 hybrid corn was in commercial production on a substantial scale, and the U. S. Department of Agriculture had begun to gather statistics on it. By 1950 more than three fourths of the total corn acreage of the U. S., some 65 million acres, was in hybrid corn.

This immense achievement stems from the work of many corn breeders, variously associated with the U. S. Department of Agriculture, state experiment stations and private industry. Among the pioneers in the breeding of corn were Henry A. Wallace, Herbert K. Hayes and Frederick D. Richey.

Hybrid corn is usually produced now by a process that involves three principal steps. To understand them we must consider briefly how corn produces progeny. The corn plant is unique among the major cereals in bearing its male and female flower clusters separately on the same plant. One cluster, the ear, bears only female flowers—several hundred or more enclosed in husks, each with its silk to receive the male pollen. The other, the tassel, bears only male flowers, usually more than a thousand in number. Each male flower contains three anthers, or pollen sacs, and each anther contains about 2,500 pollen grains. A single corn plant sheds several million pollen grains during its flowering period. These are so small and light and so easily carried by the wind that they seldom fall upon the silks of the same plant. As a consequence, under natural conditions cross-pollination is the rule. In experimental or seed-production plots special arrangements are made to control pollination. Experimental pollinations are usually made under bags. The young ears bearing the female flowers are covered with glassine or parchment bags before the silks have appeared. At the same time or a few days later the tassels also are bagged, for the collection of pollen. A single pollination produces an ear bearing several hundred seeds. A single bagged tassel produces enough pollen to pollinate several hundred ears.

The first step in the production of hybrid corn is the isolation of inbred strains. This is still accomplished, as in Shull's and East's experiments, by self-pollination. Hundreds of thousands of self-pollinations in corn are made each year, and tons of paper bags are consumed in the process. The manufacture of special corn-pollinating bags has become a recognized minor industry.

Self-pollination is a form of inbreeding approximately three times as intensive in its effects as matings between brothers and sisters in animals. The same plant is literally both the father and the mother of the offspring. Some plants—wheat, rice, barley and oats, for example—are naturally self-pollinated and suffer no deleterious effects from the process. But corn, a naturally cross-

LONG ROWS OF TASSELS stretch across a single-cross field of the Pioneer Hi-Bred Corn Company of Des Moines, Iowa. The tassels have been removed from two out of three rows; double-cross tassels are farther apart.

pollinated plant, responds to inbreeding with conspicuous effects. First, in the early generations many inherited abnormalities appear—defective seeds, dwarfs, albinos, stripes and a host of other chlorophyll deficiencies. These abnormalities were once supposesd to be the degenerative products of the "unnatural" process of inbreeding, but it is now known that inbreeding merely brings to light deleterious characters already present, which have previously remained hidden because they are recessive traits. Inbreeding actually helps the corn breeder, for it reveals hidden defects and allows the breeder to remove them permanently from his stocks.

After five or six generations of inbreeding the inbred strains have become remarkably uniform, much more uniform than any variety of corn occurring naturally. All the plants of a single strain are genetically identical, or almost so; and their genetic uniformity is reflected in a remarkable uniformity in all perceptible characteristics, physical and physiological. But even the best of these uniform strains yield no more than half as much as the open-pollinated varieties from which they were derived, and many yield much less. Their only value is as potential parents of productive hybrids.

Built-In Characteristics

Inbreeding accompanied by selection has given the corn breeder a remarkable degree of control over corn's heredity. Much of the breeding work today is aimed not only at greater yields but also at improvements in other characteristics. Almost all the corn now grown in the Corn Belt has been bred to possess stiff stalks that remain upright far into the fall—an important quality for mechanical harvesting. Some breeders, shaping the corn to the machines, are developing hybrids bearing two or three small ears instead of a single large one. Resistance to drought was recognized as an important characteristic during the hot dry summers of the 1930s and has been incorporated into many hybrids. Hybrid corn has also been bred for resistance to various diseases. Through selection corn varieties can even be developed to withstand the depredations of insects. Some inbred strains of corn are quite resistant to chinch-bug injury. Others are either unattractive to root worms or survive their assaults. The Southern corn breeder uses corn with long tight husks, which protect the ears against the inroads of ear worms and weevils. Corn breeders in Argentina claim to have isolated lines that contain a bitter substance rendering the foliage unattractive to grasshoppers. This same corn has been used in the U. S. in an attempt to develop new strains possibly resistant to the European corn borer. Strains resistant to corn-borer damage are frequently also unpalatable to aphids.

After inbreeding, the second step in the production of hybrid corn is the testing of the inbred lines in various crossing combinations to determine their hybrid performance. Usually the lines are first screened by crossing all to a common parent—an open-pollinated variety. This comparison allows the corn breeder to eliminate many of the poorer strains. The more promising ones are then tested further in single or double crosses. Of each hundred lines isolated, usually not more than one or two prove satisfactory for use in hybrids.

The final step in producing hybrid seed is to combine the selected strains into commercial hybrids. In sweet corn, especially for canning, where uniformity in the size and shape of the ears is a more important consideration than the cost of the seed, the product is usually a single cross. In field corn the cost of the seed is paramount, so all the seed produced for use is double-crossed. A given amount of land and labor will yield two to three times as much double-crossed seed as single-crossed.

Because the second-generation progeny of a hybrid decline markedly in yield and uniformity disappears, only one crop of corn is grown from the crossed seed. Hence the farmer must buy new hybrid seed each season. The production of hybrid seed corn has become a huge and highly specialized enterprise comparable to the pharmaceutical industry. Hundreds of different hybrids, adapted to a wide variety of soils and climates, are produced. Like

TASSELS ARE PULLED by mechanized hand labor near Vincennes, Ind. Here there are two rows of tassel plants and four rows of plants from which the tassels are removed by workers on a platform moved by tractor.

vaccines and serums, they cannot be identified by their appearance. It is their inherent genetic qualities that distinguish hybrids from one another, and farmers have learned to buy hybrid seed on the basis of these qualities.

The almost universal use of hybrid corn in the U. S., and the prospective wide adoption of it in other parts of the world, is not without its dangers. Chief among these is that farmers as a rule are no longer growing the open-pollinated varieties. These varieties, from which all inbred strains are ultimately derived, may therefore become extinct. Already more than 99 per cent of the corn acreage in several of the Corn Belt states is in hybrid corn; in Iowa it is 100 per cent hybrid. The loss of the original source of breeding material would mean not only that improvement of the present strains would be restricted but that new types of hybrid corn could not be developed to cope with new diseases or insect pests suddenly become rampant. Our corn would also lose the ability to adapt to climatic changes. Open-pollinated varieties of corn, in which cross-pollination is the rule, are admirably contrived for maintaining genetic plasticity and would be capable of surviving rather drastic changes in the environment. Hybrid corn, a small, highly selected sample of the original genetic diversity, has lost this capability.

The U. S. Department of Agriculture, recognizing the danger, has taken steps to maintain the open-pollinated varieties of the Corn Belt. It is also important, however, to preserve the indigenous corn varieties of other parts of the U. S. and of the countries of Latin America. Many of the U. S. varieties had their origin in Mexico, and Mexican corn in turn has ancient affinities with the corn of Central and South America. The in-

digenous varieties of the countries to our south may one day become of critical importance as sources of new genes to improve, or perhaps even to save, the corn of the U. S. The National Research Council is therefore planning, in cooperation with the Department of Agriculture, the State Department and other agencies, to collect and preserve the native corn varieties in the principal corn-growing countries of this hemisphere.

Future Development

What does the future hold for hybrid corn? To a large extent this will hinge on basic research in corn genetics. Unfortunately new discoveries have not kept pace with practical utilization. Corn breeders, like applied scientists in other fields, have been spending the accumulated capital of theoretical research of the past without taking adequate steps to create new capital. Still unsolved, for example, is the problem of the genetic basis of hybrid vigor, which is clearly of more than academic interest to the practical corn breeder.

Some advances can still be made by the application of present knowledge. Already a trend has begun toward developing highly specialized types of corn for particular purposes. There are special white corn varieties (lacking the pigment carotene) which are used for the manufacture of hominy; a "waxy" corn containing large amounts of the carbohydrate amylopectin has been developed for industrial purposes, including the making of tapioca; for feeding meat-animals breeders have produced a corn with a high protein content. It is possible that corn may be bred with a higher content of the pellagra-preventing vitamins of the B complex, especially niacin,

in which corn is now notoriously deficient.

The methods of production no doubt also will be improved. Two techniques for creating uniform strains without prolonged inbreeding are under trial. One is now being tested by Sherret Chase at Iowa State College. This method involves the use of "haploid" plants, which contain only half of the normal number of chromosomes. Such plants occur spontaneously. Haploid plants are weak, often sterile and of no value in themselves. But their chromosomes can be doubled by treatment with the alkaloid colchicine, or they may double spontaneously. When this happens, they produce offspring containing the normal number of chromosomes. Since all the chromosomes come from a single original parental germ cell, plants derived in this way are completely pure for all of their genes and are even more uniform than strains resulting from inbreeding.

A second short-cut method for obtaining uniform strains has been suggested by Charles R. Burnham of the University of Minnesota. By treating seed with X-rays Burnham is attempting to produce an artificial stock of corn in which the chromosomes are broken and so "scrambled" that they will no longer form normal pairs with the chromosomes in normal plants when hybridized with them. Such hybrids, when self-pollinated, should produce three kinds of plants, of which one should have only normal chromosomes and be pure for all of its genes. Plants of the latter kind would be the equivalent of inbred strains.

The operation of detasseling as a prelude to crossing is also destined to be simplified. Detasseling has been called the "peskiest and most expensive" part of producing hybrid seed corn. Each summer the seed industry must find and train thousands of temporary workers, many of them high-school students, to perform this essential task. One firm alone employs more than 20,000 laborers during the detasseling season, and it has been estimated that on the peak day of the season some 125,000 persons in the U. S. are engaged in removing tassels from corn plants. Many attempts have been made to simplify this operation or eliminate it entirely, but until recently none was notably successful. Now what promises to be a partial solution to the problem has been discovered. It involves a certain form of sterility in corn which prevents the tassels from shedding pollen but which is transmitted only through the seeds. Marcus Rhoades of the University of Illinois has shown that this kind of sterility is inherited not through the chromosomes but through the cytoplasm of the germ cells. Jones and I have found that it can easily be incorporated into any inbred strain of corn by crossing, and that it is an excellent substitute for detasseling. A sterile inbred crossed to a fertile inbred produces a sterile single cross. A sterile single cross grown in a crossing field requires no detasseling. The resulting double cross also is sterile; that is, it produces no pollen. But it can be pollinated by planting it with a certain proportion of a comparable fertile double cross. Another method of obtaining a crop from it is to prevent the double cross from being sterile by incorporating in it an inbred strain carrying fertility-restoring genes. This scheme, which has proved completely successful in Jones' experimental cultures, is the last word in the biological manipulation of the corn plant. It employs hereditary factors in the cytoplasm to make corn sterile when sterility is a distinct asset, and uses hereditary factors on the chromosomes to make it fertile when fertility is essential. Hybrid seed produced in this way is being grown on a commercial scale for the first time in 1951.

Hybrid corn weil illustrates the importance of the free interplay of theory and practice. The practical motive of improving corn has played its part, but the development of hybrid corn is due in even greater measure to fundamental research aimed only at increasing theoretical knowledge in genetics. Progress of the kind represented by this development is most likely to occur in a free society where truth is sought for its own sake and where there is no undue emphasis on utilitarian aspects. In the case of hybrid corn breeders actually had to go back before they went forward: the first step, inbreeding, led not to immediate improvement but to a drastic reduction in yield. To avoid having to defend this paradoxical procedure of "advancing backwards," corn breeders sometimes took the precaution of plant-

TASSEL IS REMOVED from "female" plant. Detasseling is one of the biggest jobs in growing hybrid seed. One firm employs 20,000 detasselers a season.

EAR IS BAGGED in an experimental plot. By bagging both the ear and the tassel on a small scale, experimenters can completely control pollination.

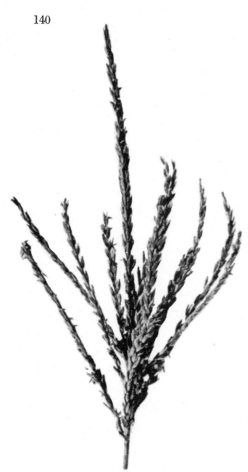

FERTILE TASSEL possesses many anthers, the pollen-bearing organs.

STERILE TASSEL has only aborted anthers, therefore sheds no pollen.

ing their experimental plots of stunted inbred corn in out-of-the-way places where the public was not likely to see them.

Impact on Food Supply

Hybrid corn's greatest significance lies in the contributions which it and similar developments in applied genetics can make to the world food supply. What hybrid corn has already accomplished toward this end is illustrated by two dramatic examples. During three war years, 1942 to 1944, the American farmer, though afflicted with an acute labor shortage and unfavorable weather, produced 90 per cent as much corn as he had during the previous four years of peace, themselves years of unprecedented production. In other words, hybrid corn enabled him to add a 20 per cent increase to the previous gains. Thanks to hybrid corn, the U. S. suffered no real food shortages at home, was able to ship vast quantities of food abroad to her Allies, and still had enough surplus grain to use large quantities in the manufacture of alcohol, synthetic rubber, explosives and other materials of war.

At the end of the war the American food surplus served a more peaceful but no less important purpose. In the year ending June 30, 1947, the U. S. shipped to hungry and war-torn Europe 18 million long-tons of food. Very little of this was corn, but the food actually sent represented, in terms of calories, the equivalent of 720 million bushels of corn. In the same year, through the use of hybrid corn, the corn crop of the U. S. had been increased by approximately 800 million bushels. That is, the U. S. gain in this one crop was sufficient to meet Europe's food deficit during the first postwar years, with food to spare.

Hybrid corn has proved to be a catalyst affecting the entire agricultural economy wherever it has touched it. Even the most skeptical farmers, once they have proved to their own satisfaction the superiority of hybrid corn, turn to the experiment stations for other innovations growing out of agricultural research. The higher cost of hybrid seed is an inducement to strive for maximum yields, and in the U. S. this has led to the adoption of improved agricultural practices, including the use of fertilizers, crop rotations and the growing of soil-improving crops of soybeans and other leguminous plants that gather soil-enriching nitrogen from the atmosphere. The result of all this is that the increases in corn yields obtained by American farmers on their own farms have been much larger than in experiment stations. Whereas hybrid corn grown in controlled experiments usually yields about 20 to 30 per cent more than the original open-pollinated corn from which it derives, the average farm yield of corn per acre in

the U. S. has increased by about 50 per cent: from about 22 bushels in the early 1930s, when hybrid corn first began to be used commercially, to approximately 33 bushels in the late 1940s, when it occupied some 75 per cent of the total corn acreage. Under favorable conditions, yields of 100 bushels per acre for hybrid corn are common, and yields exceeding 200 bushels are regularly reported. This substantial increase can be attributed to the use of fertilizers and other soil-improvement practices as well as hybrid corn.

The success of hybrid corn in the U. S. promises to be repeated in other parts of the world where corn is an important plant. One of the first countries to benefit is Italy, which fortunately has been able to use hybrids developed in the U. S. Corn hybrids are usually so well "tailored" to a particular environment that it is seldom possible to move them successfully from one country, or even from one region, to another. Italy has proved to be an exception to this rule and is now importing hybrid seed corn from the U. S. on a substantial scale—enough to plant approximately a million acres in 1950.

In the countries of Latin America, in many of which corn is a basic food plant, new hybrids especially adapted to local conditions are being developed. Corn-breeding programs aimed at this objective are in progress in Mexico, Guatemala, El Salvador, Costa Rica, Cuba, Colombia, Venezuela, Brazil, Uruguay, Argentina, Peru and Chile. The corn-breeding program in Mexico, a cooperative project of the Mexican Government and the Rockefeller Foundation, has been particularly successful. Begun in 1943, it has already made itself felt in the Mexican economy; in 1948, for the first time since the Revolution of 1911, Mexico produced enough corn to feed her own population.

To Mexico hybrid corn is perhaps even more important than to the U. S. In the U. S. three fourths of all corn is fed to livestock and is transformed into meat, milk, eggs and other animal products before reaching the ultimate consumer. In Mexico corn is used directly; it is literally the staff of life of millions of people—the daily bread, which, eaten 365 days a year, fuels most of the human metabolism. Corn has an almost sacred significance to the Mexican farmer, as it had to his ancestors for centuries past. It turns out, however, that the Mexican farmer, for all of his inherent conservatism, is, like his American counterpart, willing to try new kinds of corn.

What has been done in corn to utilize the phenomenon of hybrid vigor can be done in any crop plant that lends itself to mass hybridization. Plants of the gourd family are especially easy to hybridize. Like corn, they bear male and female flowers separately on the same plant. They are therefore easily self-pollinated to produce inbred strains and readily emasculated to effect crossing. Hybrid forms of cucumbers, squashes and watermelons are now grown. Like hybrid corn, they are characterized by vigor, productiveness and uniformity.

Plants in which both the male and female elements occur in the same flower present greater difficulties. In some, like the tomato, whose flower parts are relatively large and whose fruit contains a large number of seeds, hand pollinations to produce hybrid seed are feasible. In other species, such as onions and sugar beets, whose flowers are much too small and delicate to permit emasculation on a commercial scale, forms of cytoplasmic pollen sterility, similar in their effect to that described above for corn, have been used for some years. Since onions and sugar beets are grown for their vegetative parts, the problem of restoring fertility in the final hybrid is not involved. Other crop plants in which hybrid vigor is either being used or tested are alfalfa, barley, rye and sorghum.

Hybridized Animals

Work on the development of hybrid vigor has also been extended to domestic animals. The production of hybrid chickens has already become an enterprise second in importance only to the hybrid seed-corn industry. Hybrid pigs are coming into common use, and hybrid sheep and cattle are well along in the experimental stage. In farm animals the problem of crossing is simple, because the animals are bisexual and can reproduce only by cross-fertilization. But the problem of producing inbred strains is more difficult than in plants. Inbreeding by matings between brothers and sisters—the most intensive form possible in bisexual animals—is only one third as effective as self-pollinations in plants. Since individual animals are more valuable than individual plants, inbreeding on the vast scale on which it is practiced in plants is not yet feasible. The results so far obtained, however, have been very promising. Hybrid chickens grow faster and lay more eggs. Hybrid pigs make more pork with less feed. Hybrid cattle produce more beef in less time. The animal breeder, like the corn breeder, has found hybrid vigor a powerful force to be harnessed in raising the physiological efficiency of organisms.

The time is rapidly approaching when the majority of our cultivated plants and domestic animals will be hybrid forms. Hybrid corn has shown the way. Man has only begun to exploit the rich "gifts of hybridity."

FERTILE TASSEL FIELD in Connecticut has had tassels removed from most of its rows. The tasseled rows appear in center and at left and right.

STERILE TASSEL FIELD in Connecticut has tassels on every row. Plants bearing fertile tassels are planted only in the rows where they are needed.

8

16

HYBRID WHEAT

BYRD C. CURTIS AND DAVID R. JOHNSTON
May 1969

So far in this century the production of two major grain crops, corn and sorghum, has been revolutionized by the technique of mass hybridization: the crossing of two dissimilar inbred lines or varieties to obtain offspring with more desirable qualities than those possessed by either of the parent lines. It now appears that another important grain, wheat, is on the verge of a similar revolution. The problems associated with the production of hybrid wheat on a commercially feasible scale have been particularly difficult, but enough progress has been made in the past few years to predict with some assurance that the eventual widespread introduction of hybrid wheat varieties will have a far greater economic and nutritional impact than the introduction of any other hybrid crop grown in the world today.

The basis of all such attempts at genetic manipulation is the phenomenon of hybrid vigor, the tendency for the offspring of crossed varieties to have greater vitality than the offspring of inbred varieties. Hybrid vigor can be manifested in a number of ways: increased yield, greater resistance to disease or insects or harsh climate, a shorter growing season and better milling or baking qualities. In the case of wheat the primary benefit being sought is increased yield.

It turns out that the extra vigor of hybrids is expressed at a maximum in the first generation after the cross; later generations show a drastic reduction in vigor. Hence the object of any hybridi-

zation program is to perfect a technique for producing enough hybrid seed to grow first-generation plants on a large scale.

This was easy to accomplish with corn, because in a broad sense all corn is hybrid. Corn is a cross-pollinated species in which the male sex parts (in the tassel) and the female sex parts (in the ear) are located in quite separate parts of the same plant. Removing the tassel by hand makes the plant female and therefore incapable of self-fertilization; thus all the seed produced on the ear will be hybrid since it must be fertilized by pollen from other plants. Enough hybrid seed was produced by this method to plant the entire corn acreage of the U.S. within the first 20 years after the first hybrid varieties were introduced in the 1930's.

In the early 1950's a new technique for producing hybrid corn seed was developed. It involves a sophisticated genetic procedure for inducing male sterility in a generation of corn plants; these plants are then crossed with a variety that is capable of restoring full fertility to the first-generation offspring of the cross. This approach eliminated the need for the laborious hand-detasseling operation and is currently employed in the production of a large percentage of the world's crop of hybrid corn. A similar technique has proved successful for producing hybrid sorghum; in only a few years it has resulted in the hybridization of the entire sorghum crop of the U.S. It

is basically a variation of this general approach that has been applied to the hybridization of wheat.

Unlike corn, wheat is almost 100 percent self-pollinated. Both the male and the female sex parts—the male stamens containing the pollen and the female pistil containing the egg—are located in the same floret, or flower [*see illustration on next page*]. Normally the anthers (the elongated pollen-bearing portions of the stamens) supply pollen to the stigma (the feathery portion of the pistil) before the floret opens enough to allow the entrance of pollen from other plants. To obtain a single hybrid seed it is necessary to remove the three anthers in a floret with small forceps and later apply pollen from another plant to the stigma by hand. To ensure success all these operations must be timed precisely and executed with great care. Obviously commercial quantities of hybrid seed cannot be produced in this manner.

Here it should be noted that many varieties of wheat already under cultivation are often referred to incorrectly as hybrid wheat. This misnomer arises from the fact that it is possible to derive improved varieties from a handmade hybrid several generations removed. The best descendants of the original hybrid are selected and inbred for five or six successive self-pollinated generations until a "pure line" or "true breeding" plant is obtained. Such a plant, combining the good traits of the original parents, will

144

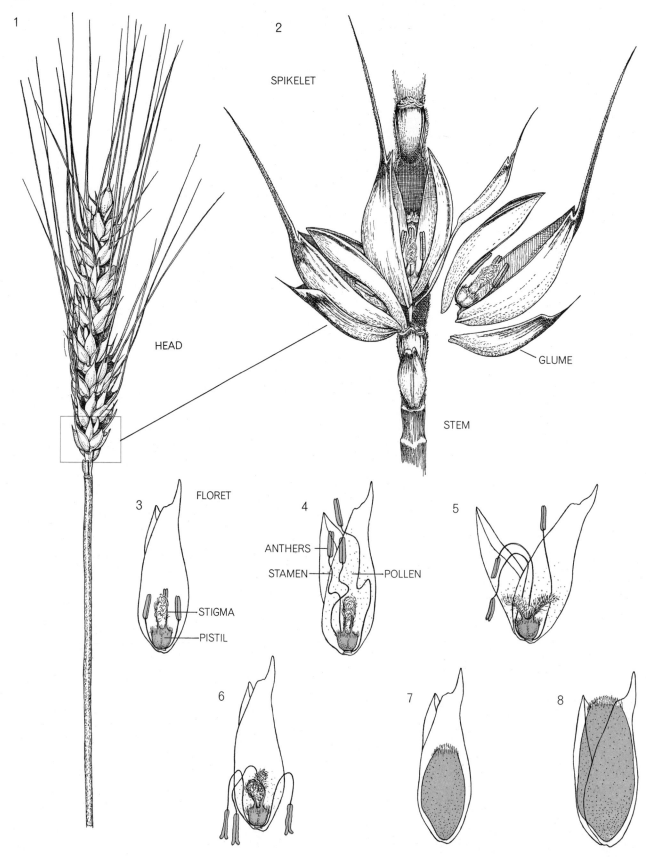

1

HEAD

2

SPIKELET

GLUME

STEM

3   FLORET

STIGMA

PISTIL

4

ANTHERS

STAMEN   POLLEN

5

6

7

8

SELF-FERTILIZATION of an individual floret, or flower, of wheat is represented in this series of drawings. Drawing *1* shows a typical mature wheat spike; drawing *2* shows how the wheat head is composed of alternating rows of spikelets, each of which contains several florets (in this case three are shown). Drawings *3* through *6* show how the anthers (the elongated pollen-bearing portions of the male stamens) normally supply pollen on the stigma (the feathery portion of the female pistil) before the floret opens enough to allow the entrance of pollen from other plants. Drawings *7* and *8* show how the fertilized egg develops into the seed, or grain, of wheat. Threshing separates the full-grown grains from the rest of the plant, which is referred to collectively as chaff.

produce similar plants in all subsequent generations, provided that it is self-pollinated and that mutation does not take place. Most modern wheat varieties were obtained by some such procedure, but they are clearly not true hybrids.

The development of a technique for producing true hybrid wheat seed on a large scale began in the early 1950's, at about the same time that the male-sterile mechanism for producing hybrid corn was being perfected. The Japanese investigator H. Kihara reported in 1951 that he had succeeded in inducing "cytoplasmic male sterility" in wheat; this type of male sterility sometimes results when the nucleus of one cell interacts with the cytoplasm of an unrelated cell. Kihara had transferred the nucleus of a common bread wheat (*Triticum aestivum* subspecies *vulgare*) into the cytoplasm of a wild relative of wheat called goat grass (*Aegilops caudata*) and had found that the progeny were female-fertile but male-sterile. In 1953 H. Fukasawa of Japan obtained similar results with a different wild species (*Aegilops ovata*) as the female parent in a cross with durum wheat (*Triticum durum*), a species used primarily for making macaroni. Later Kihara developed still another male-sterile variety by crossing emmer wheat (*Triticum dicoccum*) with a species known only as *Triticum timopheevi*.

None of these developments proved successful in providing a commercially useful male-sterile mechanism for hybrid wheat, owing to adverse side effects produced by the particular lines that were crossed. The findings did, however, stimulate later and more successful experiments. In 1961 J. A. Wilson and W. M. Ross, working at the Kansas Agricultural Experiment Station, obtained stable cytoplasmic-male-sterile bread wheats by crossing *T. timopheevi* as the female or seed parent with a variety of *T. aestivum* called Bison wheat. Repeated backcrossing of the Bison variety with the male-sterile progeny plants resulted in stable, cytoplasmic-male-sterile Bison lines. These lines were widely distributed internationally and are the source of many of the male-sterile wheat varieties currently available.

For the cytoplasmic-male-sterile system to be useful in producing hybrid wheat a corresponding fertility-restoring mechanism must be found. A male-sterile Bison line, when pollinated by normal fertile Bison wheat or other normal, nonrestoring varieties, will produce offspring that are also male-sterile. This

EMASCULATION of a wheat floret is accomplished by removing the three anthers in the floret with small forceps. Pollen from another plant can later be applied to the stigma by hand in order to obtain a single seed of hybrid wheat. Such a procedure is obviously not commercially feasible as a method of producing large quantities of hybrid wheat seed.

MALE-STERILE SEED on the wheat head at center was obtained by hand-fertilizing the florets of an emasculated plant called *Triticum timopheevi* with pollen from a common bread-wheat variety of *Triticum aestivum*. Normal spikes of the female and male parents are shown at left and right respectively. The "cytoplasmic male sterility" of the crossed variety results from a little-understood interaction of the chromosomal genes in the nuclei of the male parent's cells with an unknown heredity factor in the cytoplasm of the female parent's cells. Repeated backcrossing of fully fertile plants of the common bread-wheat variety with the male-sterile progeny plants results in a stable cytoplasmic-male-sterile line.

variety called Primepi wheat by E. Oehler and M. Ingold of France. Much research is currently under way to collect enough of the restorer genes into single agronomically desirable varieties in order to make hybrids completely fertile when crossed with agronomically sound male-sterile varieties. One private seed company has announced success in this effort and has distributed several varieties of hybrid wheat seed for test planting by farmers.

The cytoplasm-nucleus reaction that leads to male sterility is not a well-understood process. In fertile plants there is apparently a good balance between the chromosomal genes and an unknown heredity mechanism carried in the cytoplasm. In cytoplasmic-male-sterile plants this delicate balance is undoubtedly upset in some way, giving rise to deformed anthers and empty or sterile pollen grains [*see illustration on page 148*]. In contrast to normal anthers the anthers of male-sterile plants are more slender and tend to curl at the base into the shape of an arrowhead. Such anthers produce little pollen, and what little is produced cannot effect fertilization. Apparently the induced imbalance has no effect on the pistil. Normal pollen applied to the stigma will function properly, and a seed will be produced. In fact, there is scant evidence that male-sterile plants are morphologically different from normal plants except for defective anthers and pollen.

Cytoplasmic-male-sterile lines can be developed, starting with a stable line such as Bison wheat or Wichita wheat, by means of the backcross method [*see top illustration on page 149*]. The final progeny will be identical with the recurrent parent in most characteristics except that it will be male-sterile instead of fully fertile. In order to prevent contamination by foreign pollen and to ensure that only pollen from the recurrent parent effects fertilization, some stratagem must be found for isolating the individual plant, such as placing a small plastic bag over its head.

A few wheat varieties have proved difficult to sterilize. These plants, which usually have a restorer gene that prevents sterilization, include several important commercial varieties. Just why these varieties have such genes is not known. Perhaps they arose by mutation or are carry-over genes from natural crosses to *T. timopheevi* in past generations.

Male-sterile varieties are maintained and increased by growing the male-ster-

outcome is of course an important and integral step in maintaining and increasing seed of the male-sterile variety, but seed used by the farmer to plant his hybrid crop must have, in the resulting plants, the capacity for male fertility. This is achieved by pollinating the male-sterile plants with a variety that restores fertility. Such varieties have in them dominant restorer genes capable of overcoming the cytoplasm-nucleus reaction that causes male sterility. Finding highly effective dominant gene systems for fertility-restoration has proved difficult indeed, but a few such systems that restore fertility well under certain environments have been reported.

In 1960 Wilson and Ross announced the discovery of partial fertility-restoring factors in bread wheat for one of the cytoplasms on which Fukasawa had worked. Then in February, 1962, Wilson suggested that restorer genes must exist in *T. timopheevi*, since it carried the sterile cytoplasm; otherwise *T. timopheevi* would be male-sterile and unable

to reproduce itself. Several months later John W. Schmidt, V. A. Johnson and S. D. Mann of the Nebraska Agricultural Experiment Station demonstrated that a bread wheat derived from *T. timopheevi* was effective in restoring reasonable fertility to male-sterile Bison wheat. Shortly thereafter Wilson, working independently, reported similar results. Subsequent studies by these same investigators and others have shown that the original restorer sources were not completely effective in restoring fertility to the Bison plants and other sterile wheats in all environments.

In the years since the discovery of these partial fertility-restoring factors it has become obvious that restorer genes and modifiers of restorer genes are distributed among several of the world's existing wheat varieties. For example, Ronald W. Livers of the Kansas Agricultural Experiment Station has reported a number of common varieties that carry genes for partial restoration. An important restorer gene was also found in a

ile, or *A*-line, plants in "drill strips" in the field. These are situated between drill strips of the normal fertile, or *B*-line, plants. The male-sterile plants are pollinated by windblown pollen from the fertile plants. Seed from the male-sterile plants is harvested (by combine) separately from seed from the fertile plants; after a sufficient increase is attained the male-sterile seed is ready to be planted as the female next to a restorer line as the male for hybrid-seed production.

Developing additional restorer, or *R*-line, plants from a given source is much more difficult and laborious than developing additional male-sterile lines. For example, in order to produce a restorer line of the Scout variety one must select and backcross the dominant restorer plants in later generations to the parent Scout line. After several repetitions of the cycle the characteristics of the Scout plants are combined with the

dominant restorer genes, resulting in a Scout restorer variety [*see bottom illustration on page 149*].

The same general procedure can be used to produce a new restorer variety with the desirable characteristics of the Scout line combined with the desirable characteristics of the original restorer source. Here, as before, the second-generation restorer plants would be selected for producing the third and succeeding generations. In each generation, however, the best agronomic plants would be selected for continuation. Thus after the fifth generation it should be possible to select plants approaching the pure-line status.

An important requirement for any breeding scheme designed to produce restorer lines is that the selected plants of each generation be crossed to a cytoplasmic-male-sterile variety to determine if the selected plants actually carry the restorer genes. The first-generation

plants from these test crosses will be fully restored to male fertility if the selected plants carry the necessary fertility-restoring genes.

After a number of male-sterile and restorer lines are produced, various combinations of hybrids are made and tested under field conditions to determine which hybrids display the greatest hybrid vigor for a given production area. Once this is established the hybrid seed can be produced on a large scale for sale to farmers. Such seed will produce first-generation hybrid plants that one hopes will possess the desired amount of hybrid vigor. Obviously hybrid vigor must be manifested or the farmer gains nothing from planting the hybrid.

The amount of hybrid vigor to be gained from hybrid wheat in large-scale field plantings has not yet been established, owing to the unavailability of sufficient seed stocks for such plantings. Many reports on the gain in hybrid vigor

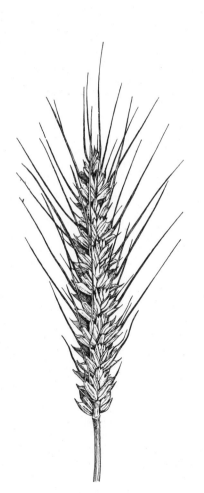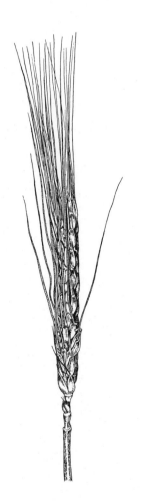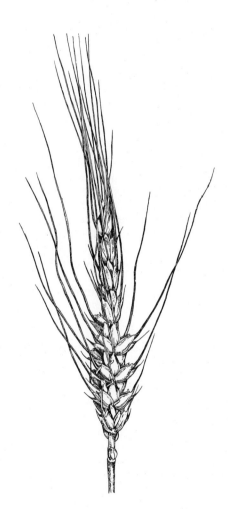

THREE MATURE SPIKES of Bison wheat are compared in this illustration. At left is a cytoplasmic-male-sterile spike open and ready for cross-pollination. At center is a fully fertile spike undergoing self-pollination. At right is a partially fertile spike derived from a cytoplasmic-male-sterile line; its fertility has been partially restored by crossing its male-sterile parent as the female with a variety of bread wheat derived from *T. timopheevi* as the male. The discovery of the fertility-restoring factors in bread wheat in the early 1960's stimulated a major expansion in the research effort devoted to the development of new hybrid wheat varieties.

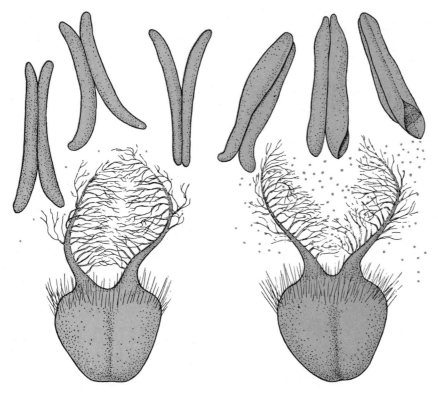

SEX PARTS from a male-sterile plant of Bison wheat (*left*) are compared with the sex parts from a normal plant of Wichita wheat (*right*). The anthers of the male-sterile plant (*top left*) are more slender than the normal anthers (*top right*) and tend to curl at the base into the shape of an arrowhead. The little pollen produced by the anthers of the male-sterile wheat plant is sterile and cannot effect fertilization. The pistil of the male-sterile plant (*bottom left*) is morphologically identical with that of the normal plant (*bottom right*).

from small experimental plantings in various parts of the world are in the literature on wheat. One can surmise from these reports that a 20 to 30 percent increase in yield should not be difficult to obtain from large-scale field plantings. Data from such plantings should be available in two to four years; it will then be possible to ascertain the degree of hybrid vigor to be realized from wheat.

So far we have discussed only the mechanics of producing hybrid wheat plants experimentally. No mention has been made of the difficulties usually encountered in increasing the male-sterile lines or in growing hybrid seed under field conditions.

The major difficulty in producing economically feasible hybrids, other than developing adequate restorer lines, has been the failure to obtain consistently good cross-pollination in the field. Cross-pollination is affected by many factors, including the synchronization of the flowering times of the male-sterile plants and the pollinator plants. The vagaries of the weather also have a strong bearing on cross-pollination. The problem of

providing receptive female florets at the time of maximum pollen dispersal depends on the relative maturity of the two parent varieties. Concurrent plantings of male-sterile and pollinator varieties, with the pollinator plants reaching the "heading" stage one to three days later than the male-sterile plants, will usually provide adequate synchronization.

Under conditions of cool weather and an adequate supply of moisture, the female sex parts will remain receptive for as long as eight to 10 days. Maximum receptivity is attained three to five days after the wheat heads are fully formed. If the pollinator variety happens to mature earlier than the male-sterile one, maximum pollen dispersal may precede maximum floret receptivity, and the result will be a reduced yield of seed. Solutions to such problems may be found in adjusting the planting times or in varying the seeding rates. Such activities will, however, add considerably to production costs and thereby increase the cost of hybrid seed. Extra seed costs to the farmer reduce the value he gains from hybrid vigor.

Wheat pollen is short-lived and is adversely affected by hot and dry con-

ditions. Richard E. Watkins of Colorado State University has found under laboratory conditions that the life-span of pollen from typical wheat varieties is less than five minutes after anthesis (the opening of the anther) at a temperature of 95 degrees Fahrenheit and a relative humidity of 20 percent. At 65 degrees F. and 80 percent relative humidity some pollen remained alive for as long as an hour, with more than 60 percent still viable 20 minutes after anthesis. In addition to affecting the viability of pollen, high temperatures and low humidity reduce the pollen load in the air, since the wilting of the plant parts impedes the release of pollen.

In general the best cross-pollination has been achieved at cool (but not too cool) temperatures and medium humidity. Too much rainfall or fog also impedes pollen dispersal. Under these conditions pollen is either washed down or made so soggy that it fails to become airborne. Such weather conditions may also cause "lodging," or bending of the entire wheat plant toward the ground, another situation that could prevent cross-pollination. Gentle to strong breezes are necessary to move pollen from one plant to another, but too much wind may result in the loss of pollen. Wheat pollen is heavier than air, but it is easily borne aloft by air movement. Some workers have envisioned the use of wind machines to enhance cross-pollination, but no reports of the successful application of this technique are at hand.

Cross-pollination of more than 90 percent of the male-sterile plants has been achieved under field conditions, but much lower percentages are usually the case. It is believed that at least 70 percent cross-pollination is needed to keep the cost of hybrid wheat seed at an acceptable level. Efforts are under way to enhance cross-pollination by incorporating (through breeding) larger and more prominent anther types into pollinators and wider-opening florets with larger stigmas into male-sterile lines.

What will the advent of hybrid wheat mean to farmers, millers, bakers and the ultimate consumers of wheat products? To the farmer hybrid wheat should provide greater returns per acre of land, resulting not only from hybrid vigor but also from the more intensive and efficient management practices that seem to accompany the introduction of a hybrid crop. When hybrid corn was introduced, improved management accounted for as much or more of the increase in yield as hybrid vigor did. The production prob-

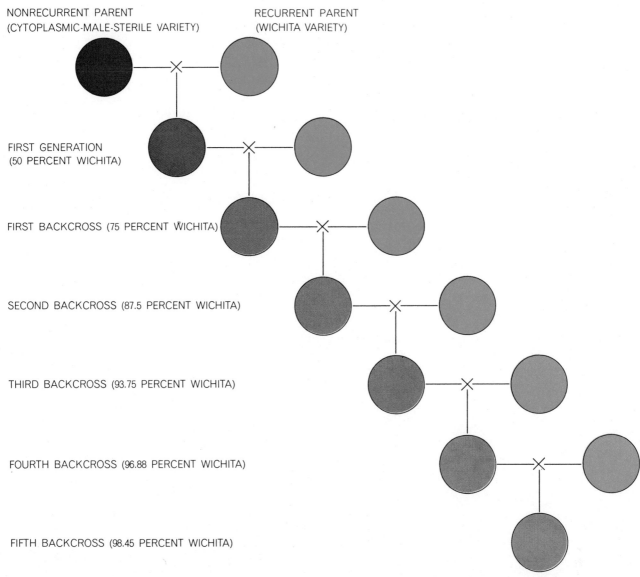

NONRECURRENT PARENT
(CYTOPLASMIC-MALE-STERILE VARIETY)

RECURRENT PARENT
(WICHITA VARIETY)

FIRST GENERATION
(50 PERCENT WICHITA)

FIRST BACKCROSS (75 PERCENT WICHITA)

SECOND BACKCROSS (87.5 PERCENT WICHITA)

THIRD BACKCROSS (93.75 PERCENT WICHITA)

FOURTH BACKCROSS (96.88 PERCENT WICHITA)

FIFTH BACKCROSS (98.45 PERCENT WICHITA)

BACKCROSS METHOD is used to develop and increase a cyto-plasmic-male-sterile line of Wichita wheat. The final progeny will be identical with the recurrent Wichita parent in most character-istics except that it will be male-sterile instead of fully fertile. Theoretically the genetic content of the nonrecurrent male-sterile parent will be reduced by half for each generation of backcrossing.

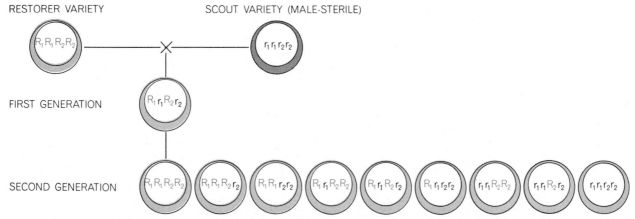

RESTORER VARIETY

SCOUT VARIETY (MALE-STERILE)

FIRST GENERATION

SECOND GENERATION

RESTORER LINE, consisting of wheat plants of the Scout variety, is developed by selecting and backcrossing the dominant restorer plants of later generations to the parent Scout line. After several repetitions of the cycle the characteristics of the Scout plants are combined with the dominant restorer genes, resulting in a Scout re-storer variety. In this schematic representation of one such cycle the dominant restorer genes, located in the nuclei of plants derived from the original restorer plants, are denoted by R_1 and R_2; the recessive, or nonrestorer, genes are denoted by r_1 and r_2. The cyto-plasm from the male-sterile variety is colored; the normal cyto-plasm from the Scout variety is gray. The completely dominant restorer plant in the second generation is designated $R_1R_1R_2R_2$.

lems associated with hybrid wheat should be no greater than those encountered with high-yield pure-line wheat varieties. Hybrids must of course be acceptable in all the agronomic characteristics to which the wheat farmer is accustomed.

One specific trait that hybrid wheats must possess is good resistance to lodging, particularly in the high-yield areas. Increased grain yields of 25 to 50 percent impose a proportionately increased weight load on the wheat straw and may result in severe lodging. Heavily lodged plants result in high harvest losses in grain, manpower and harvesting-machine time. One way of increasing resistance to lodging is to breed hybrids with shorter and stiffer stems. Most breeding programs aimed at developing hybrid wheat are taking this approach to attain resistance to lodging. Sources for these hybrids include the highly productive semidwarf Mexican wheat varieties and semidwarf wheat varieties from the northwestern U.S.

Hybrids with shorter stems may not be necessary in some of the important wheat areas that normally do not support the development of tall straw. Much of the high plains of the U.S. is such an area. Irrigation is on the increase in some of the high plains areas, however, and there are indications that better lodging resistance will be needed.

Resistance to disease is an important attribute for stabilizing wheat production in many parts of the world. Hybrid wheat should offer more flexibility than pure-line varieties in the control of diseases, particularly parasitic diseases such as the rusts. When a pure-line variety that is resistant to the prevalent rust fungi is released, new rust species usually arise to attack the new variety. The same pattern will probably hold true for hybrids, except that other hybrids, resistant to the new rusts, can be easily substituted for the old hybrid. The reason is that farmers must obtain new seed for each crop from the producer of the original hybrid. Farmers growing pure-

line varieties would be more likely to plant subsequent crops from the seed of preceding crops and thus perpetuate the rust-susceptible varieties. The producers of hybrid wheat have a great responsibility to keep abreast of the rust situation and to develop resistant hybrids to combat the disease.

A large number of wheat varieties are required to fit the varied ecological conditions of the areas where wheat is grown. This suggests that many different hybrids will also be required to fit many ecological niches. There is some hope, however, that one of the benefits of hybrid vigor will be increased adaptability. Some experiments have indicated that the root system of a wheat plant may be improved to allow the plant to perform better under drought conditions. Hybrids so designed will help to stabilize production in areas of the world where rainfall is highly variable from season to season.

To millers, bakers and consumers the quality of wheat produced by hybrids

IN THE FIELD hybrid wheat seed is produced by growing alternating "drill strips" of the male-sterile line and the fertility-restoring line. Cross-pollination of the male-sterile plants is effected by wind-borne pollen from the restorer plants. The hybrid seed from the

is of great importance. Approximately three-fourths of the wheat produced in the world is destined for human consumption. This being so, a close relation has developed between breeders and cereal chemists seeking to maintain or improve quality as new varieties are developed.

The quality of a particular variety of wheat is defined in terms of its ultimate use. Some flours require a high protein content; others do not. Some call for the ability to absorb more water than others. The large number of wheat products results in an equally large number of flour specifications. Most of the wheat produced in North America is classed as bread wheat. These wheats are generally high in protein content and have good water-absorption and gas-retention properties. The best of them are blended with poorer wheats to improve the flour quality in the making of bread.

Wheat varieties differ in their ability to confer favorable characteristics on their progeny. Studies have shown that the protein content of first-generation hybrids can be higher than that of the superior parent. This is somewhat at variance with regular varieties, in which high yield has been associated with low protein content. Conversely, some hybrids have been lower in protein (and in yield) than the inferior parent. Other crosses have produced hybrids whose protein content is intermediate between the protein content of the parents. Similar results have been obtained for water-absorption and gas-retention properties. There is still very little information on the quality characteristics of hybrid wheat, but the information available indicates that with the proper selection of parents hybrids of any desired quality can be obtained.

Research on the hybridization of wheat has been expanded on all fronts in the past decade. Much of the expansion was triggered by the discovery of the male-sterility and fertility-restoring system. Until the early 1960's wheat research was centered in the land-grant institutions and supported primarily by tax funds, both state and Federal. Credit for the discovery of the mechanism to produce hybrid wheat belongs to those institutions. The prospect of an extremely large and continuing market for hybrid wheat seed has prompted several private seed companies, most of them experienced in breeding other crops, to initiate research programs aimed at the development of hybrid wheat varieties. The combined efforts of public and private breeders have already produced dramatic results, but there is a clear need for continued research in many areas. The value of hybrid wheat can only be maintained and improved by further development of improved parental lines. These are the backbone of a successful hybridization program. The development of high-yielding, strong-strawed, disease-resistant parental lines producing grain of good milling and baking quality will ensure the success of hybrid wheat for the future.

male-sterile strips is harvested (by combine) separately from the inbred seed from the restorer strips. The photograph on these two pages shows a typical drill-strip wheat field that is located at the Colorado State University Agronomy Farm in Fort Collins, Colo.

17

CHEMICAL FERTILIZERS

CHRISTOPHER J. PRATT
June 1965

Whatever estimate one accepts of the increase of the human population in the finite future, or whatever estimate of how long it will take to bring this increase under control, it is clear that the present rate of increase is alarmingly high. Three centuries ago the number of people in the world was probably about 500 million; now it is more than three billion, and if the current rate of increase holds, it will be six billion by the end of the century and millennium. In some underdeveloped areas, where the rate is highest, the Malthusian prediction that population would eventually outrun food supplies seems close to reality.

Clearly mankind faces a formidable problem in making certain that future populations have enough to eat. Doubtless a partial solution lies in improved technology, which has already done so much to keep the food supply abreast of population, and in the spread of existing technology from the developed to the underdeveloped countries. It should also be possible to bring some new areas under cultivation or grazing, but the opportunities in that direction appear to be limited. Even though only about 2.4 billion acres, or approximately 7 percent of the earth's land area, are used for crop production in any one year, most of the unused land is too dry or too cold for agriculture or is in some other way unsuitable. Neither extensive clearing of forests nor large-scale cultivation of tropical lands offers as much promise as one might think, because much of the soil in such regions is lateritic and turns hard as the result of an oxidizing effect when it is put to the plow [see "Lateritic Soils," by Mary McNeil; SCIENTIFIC AMERICAN Offprint 870].

With huge amounts of capital and carefully planned projects it would be possible to create much new cropland by vast undertakings of irrigation, drainage and other kinds of reclamation. Even if such projects were launched, however, they would take decades to complete. It seems more feasible to look to shorter-range ventures, particularly in those developing areas where famine is an imminent threat.

Of all the short-range factors capable of increasing agricultural production readily—factors including pesticides, improved plant varieties and mechanization—the largest yields and the most substantial returns on invested capital come from chemical fertilizers. The application of these substances to underfertilized soils can have dramatic results. In a typical situation the ratio of the extra weight of grain produced per unit weight of nutrients applied can be as high as 10 to 1. To put it another way, an investment of this kind alone can quickly produce increases in crop yields of 100 to 200 percent.

Today some 30 million tons of the so-called primary nutrients—nitrogen, phosphorus and potassium—are annually supplied to world agriculture by chemical fertilizers. This amount is hardly adequate, for reasons I shall discuss. Moreover, crop yields diminish in proportion to the amount of fertilizer applied. Therefore it can be estimated that a population of six billion in the year 2000 will require at least 120 million tons of primary nutrients. An increase of 90 million tons of nutrients for three billion more people means that 60 pounds of primary nutrients will be needed to help sustain each additional person for a year. This is equivalent to about one 100-pound bag of modern high-analysis chemical fertilizer.

Stated in such a way, the amount of effort required to supply the additional fertilizer may seem modest. Actually

the expansion of capacity required is enormous; achieving it may well become a major preoccupation of technology. Fortunately processes for manufacturing the needed substances are already well established on a large scale and are capable of rapid expansion, provided

CHEMICAL FERTILIZER is meticulously placed on a field in Oklahoma by a spreader

that enough capital is made available and the necessary priorities are given. Considering all these factors, it is appropriate to review briefly the fertilizer situation: how plants utilize nutrients, how chemical fertilizers came into use, how they are manufactured, how they are best applied and how the increasing demand for them can be met by chemical technology.

Plants and Nutrients

A growing plant requires most or all of 16 nutrients, nine in large amounts and seven in small. The former are sometimes called macronutrients, the latter micronutrients. Most plants obtain three of the macronutrients—carbon, hydrogen and oxygen—from the air and all the other nutrients from the soil. (A few species, such as clover, are able to fill their nitrogen needs from the air.) The primary soil macronutrients—nitrogen, phosphorus and potassium—are the N, P and K often seen on bags of fertilizer; they are also the substances represented by the set of three figures, such as 10-12-8, that normally designates the nutrient content of a fertilizer. Usually these figures respectively denote the percentage in the fertilizer of total nitrogen (N), of phosphorus pentoxide (P_2O_5, often called phosphoric acid or phosphate) in a form available for use by plants and of water-soluble potassium oxide (K_2O, usually called potash).

The three other soil macronutrients—calcium, magnesium and sulfur—are often called secondary. Agricultural lime, limestone and dolomite, which are used to correct soil acidity, also serve as sources of calcium and magnesium. Sulfur deficiencies can be remedied by certain commercial fertilizers. The seven micronutrients, which are sometimes added in traces to fertilizers providing one or more of the primary nutrients, are boron, copper, iron, manganese, zinc, molybdenum and chlorine.

The growth of plants is a highly complicated process that is far from fully understood. For the purposes of this article it is enough to say that the usual path of mineral nutrients from the soil to the plant is from the solid particles of soil to the water in the soil and thence into the root. The actual transfer of nutrients from soil to root involves the movement of mineral ions. These ions are contained mostly in the soil water, but some of them are adsorbed on solid soil particles.

It follows that nutrients must be in ionic form or capable of transformation to ionic form by soil processes if they are to be of any value to the plant. Hence it is not necessarily a lack of minerals in a soil that causes plants to show signs of nutrient deficiency; the problem can also be that the nutrients are not in a form readily available to the plant. For example, it is quite possible for crops to starve in soils that are amply supplied with phosphorus and potassium if these nutrients are insoluble in water or plant juices. Essentially what the chemical fertilizer industry

34 feet wide. The spreader is of the "drill" type, meaning that it lays fertilizer in precise rows instead of broadcasting it generally over the field as would be done by other types of spreaders. Careful placing is often important for economic or nutritional reasons.

does, in addition to converting inert nitrogen from the air into soluble salts, is employ processes to "open" the molecules containing the vital nutrients so that these molecules form soluble salts that plants can assimilate readily.

One can best grasp the need for mineral nutrients in agriculture by taking account of the nutrients that are removed from the soil by cropping and grazing. A ton of wheat grain is equivalent to about 40 pounds of nitrogen, eight pounds of phosphorus and nine pounds of potassium. If the straw, husks, roots and other agricultural wastes of such a crop are not returned to the soil, they represent additional large losses of nutrients. A ton of fat cattle corresponds to a depletion of about 54 pounds of nitrogen, 15 pounds of phosphorus, three pounds of potassium and 26 pounds of calcium. Such rates of removal will quickly exhaust a typical soil unless the losses are made up by regular additions of suitable fertilizer.

Equivalent additions of fertilizer, however, are not really enough. There are other factors to be taken into account, and they explain why the present consumption of fertilizers is barely adequate. Nutrients are leached from soils by the flow of water; moreover, they are fixed in forms not readily available to plants. As a result of such losses the proportion of soil nitrogen and phosphorus utilized by a crop is rarely more than 75 percent. In some instances the utilization of phosphorus is as low as 10 percent.

Even allowing for losses, the increased crop value resulting from the proper application of fertilizer can be substantial. On a poor soil the gain can approach 10 times the cost of the material applied. Where the soil is good and the crop yields are high the gain from fertilizing is more likely to be three to five times the cost of the fertilizer. Because of this diminishing return there eventually comes a point at which the additional yield no longer

justifies the cost of the corresponding extra fertilizer. There is also an agronomic reason for avoiding the overapplication of fertilizer: ultimately a point can be reached at which the high concentration of nutrient salts in the soil can damage the plants.

The Evolution of Fertilizers

Long before men began to write history they knew about the effects of organic, or natural, fertilizers on growing plants. The effects must surely have been evident in the relatively lush growth induced by animal droppings and carcasses. Eventually farmers began to collect dung and apply it to crops. The first English settlers in North America reported that the Indians substantially increased their yield of maize by burying a fish with each seed they planted. In medieval times farmers in Europe had commonly undertaken to grow nitrogen-converting legumes such as clover and to rotate crops in order to

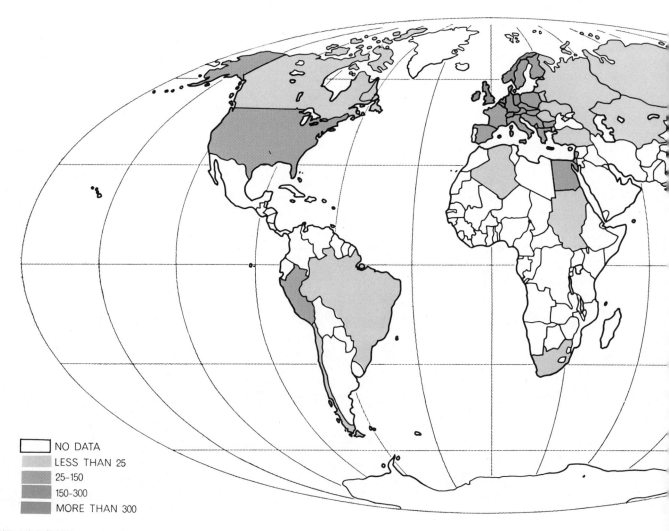

NO DATA
LESS THAN 25
25–150
150–300
MORE THAN 300

USE OF CHEMICAL FERTILIZERS is shown according to data assembled by the Food and Agriculture Organization of the United Nations. Figures give average consumption of fertilizer in metric tons per 1,000 hectares of arable land, defined as land planted to

maintain soil fertility. By the early 19th century the use of farm manure, blood, bones, animal wastes and Peruvian guano became widespread, particularly in England, where the Industrial Revolution had brought about a rapid expansion of population and a simultaneous movement of workers from the land to the manufacturing towns. For a time it appeared that the limited supply of organic fertilizer would be insufficient to meet the rising demand for food in the industrializing countries; it is said that even human bones from the battlefields of Europe were recovered, crushed and used as plant foods.

Although the use of organic fertilizers was well established by the 19th century, the basic reasons for their effectiveness were not understood. This lack of knowledge hampered the discovery of alternative substances that could relieve the pressure on the limited supply of organic matter. Another obstacle was the passionate belief held by many that organic materials had special fertiliz-

ing properties not shared by inorganic substances. Even after the Swiss chemist Nicolas de Saussure demonstrated in 1804 that plants can grow luxuriantly on carbon and oxygen from the air and mineral nutrients from the soil, strong feelings about organic fertilizers persisted. (Today the view is still sometimes expressed that organic fertilizers possess inexplicable virtues unrelated to their content of primary nutrient. Such materials—manure, sewage sludge, compost and the like—are indeed valuable as conditioners of soil and as minor contributors of plant nutrients, but there is not nearly enough organic material to meet present needs, let alone those of the future.)

Gradually the advances of chemistry revealed the processes of plant nutrition and pointed the way toward the substitution of chemical fertilizers for organic fertilizers. In some instances the process was very slow. Nitrogen, for example, was recognized as an important plant nutrient in manures and other organic matter long before anybody understood the complex cycle by which unreactive atmospheric nitrogen is converted by legumes and soil bacteria into ammonia and soluble nitrogen salts. By the time the process was understood, early in this century, conditions were ripe for a rapid evolution of industrial replacements for organic nitrogen in fertilizer. For one thing, ammonia in the form of ammonium sulfate had become available as a by-product of coal-gas works. For another, mine operators in Chile had begun large-scale production and export of sodium nitrate for use in explosives and other chemicals. As a result of their availability these salts rapidly overtook organic nitrogen as an ingredient of fertilizer. The speed of the transformation is indicated by the fact that the proportion of organic nitrogen materials in fertilizers used in the U.S. fell from 91 percent in 1900 to 40 percent in 1913.

Chilean nitrate was not to hold its position for long. A prolonged effort to synthesize ammonia by combining nitrogen with hydrogen succeeded at last in 1910, when the German chemist Fritz Haber found that the reaction would proceed at high pressure (at least 3,000 pounds per square inch) and in the presence of osmium as a catalyst. The achievement gave rise to a revolution in chemical fertilizer technology. In 1913 Haber and Karl Bosch, having worked out many difficult engineering problems, designed a commercial plant that soon produced 20 tons of ammonia a day. The requirements of the two world wars

made ammonia available on a large scale, together with such derivatives as ammonium nitrate and urea. These compounds in time largely replaced Chilean nitrate as a source of nitrogen and also reduced the proportion of fertilizers containing organic nitrogen to a few percent of total fertilizer consumption.

Phosphorus moved from the organic to the chemical stage in fertilizer sooner than nitrogen but by a similarly slow process. The first association of phosphorus with bones was made by the Swedish mineralogist and chemist Johan Gottlieb Gahn in 1769. It took until 1840, however, for chemistry to advance to the stage where it was possible to recognize that phosphorus was the key ingredient in the bone manure that had come into wide use. In that same year the great German chemist Justus von Liebig, who is regarded by many scholars as the founder of agricultural chemistry, put forward the thesis that the action of sulfuric acid on bones would make the phosphorus in the bones more readily available to plants.

This idea was promptly developed in England, where the need for additional sources of fertilizer was acute. In 1842 John Bennet Lawes, a wealthy farmer and industrialist who spent many years conducting agricultural experiments on his estate at Rothamsted, obtained a patent covering the treatment of bones and bone ash with sulfuric acid to make an improved phosphorus-containing fertilizer. Significantly he included "other phosphoritic substances" in his patent, indicating that he foresaw the role of minerals as sources of phosphate. Within 20 years the production in Britain of "chemical manures" made from sulfuric acid, local coprolites (fossil manures) and various phosphatic minerals had risen to a level of 200,000 tons a year. The phosphate fertilizer industry, thus firmly established, spread rapidly to other countries. Toward the end of the 19th century slag removed during the production of iron and steel from high-phosphate ores became another major source of phosphorus for agricultural purposes in Britain and Europe, where even now several million tons of "basic slag" are used annually as a phosphate fertilizer.

As for potassium, the benefits of adding wood ashes ("pot ash") to the soil must have been recognized in ancient times. By early in the 19th century the progress of chemistry was sufficient for a start to be made in the use of potassium chloride deposits in Germany and France as sources of potassium in fertilizer. The first factory producing potash from these deposits was built in 1861. Germany and France continued to be the principal sources of potash until rather recently, when major deposits were developed in the U.S., Israel, the U.S.S.R. and Canada.

Modern Fertilizer Production

Today a farmer can buy a wide variety of chemical fertilizers. If he wants only one nutrient, he can find a fer-

tilizer that provides it; he can also find fertilizers that contain almost any combination of nitrogen, phosphorus, potassium and the micronutrients. The industry that produces them is enormous, having a worldwide output, according to a recent estimate by the Food and Agriculture Organization, of more than 33 million tons a year. I shall briefly describe the processes now involved in producing the primary nutrients.

Synthetic ammonia is firmly estab-

lished as the principal source of nitrogen in fertilizer. Ammonia synthesis remains unchanged in principle from the technique developed by Haber and Bosch. Large-scale production often presents additional problems, however, because of the need to obtain huge supplies of pure gaseous nitrogen and hydrogen at low cost. Pure nitrogen can be produced in quantity with relative ease by removing oxygen and other gases from air through liquefaction or combustion. Hy-

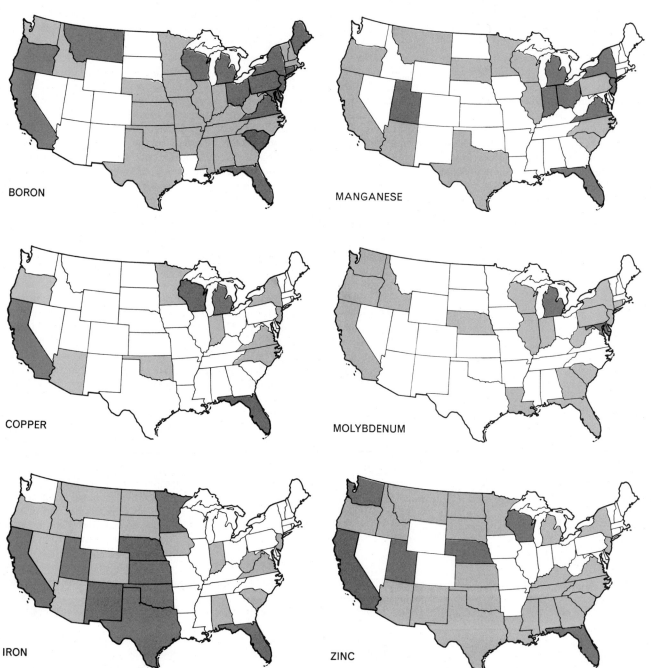

BORON

MANGANESE

COPPER

MOLYBDENUM

IRON

ZINC

NUTRIENT DEFICIENCIES appearing in various parts of the U.S. mainland are indicated. The findings, based on work done by K. C. Berger of the University of Wisconsin, pertain to several "micronutrients," meaning minerals needed by plants in small but important amounts. "Macronutrients" such as nitrogen, phosphorus and potassium are needed by plants in large amounts and usually must be supplied wherever commercial crops are grown. The colors indicate the degree of deficiency from modest (*light*) to severe (*dark*) as based on reports of the number of crops affected. The absence of color means that the state has not reported a deficiency.

drogen is another matter. Some early ammonia plants used hydrogen made by electrolysis, but the prohibitive cost led to a search for cheaper sources. Methods for producing hydrogen from solid fuels such as coal and lignite were developed in Europe. In the U.S., where natural gas is plentiful, the simpler catalytic re-forming of methane has proved an ideal way of making hydrogen. More recently the catalytic re-forming of light petroleum fractions such as naphtha with the aid of steam and the partial oxidation of heavy oil with oxygen have been widely used in countries that lack natural gas.

Although there is a strong trend, particularly in the U.S., toward injecting ammonia directly into the soil in the form of anhydrous ammonia or aqueous solutions, most agricultural ammonia is still converted into solid derivatives. Ammonium nitrate is a form popular among manufacturers, since the nitric acid needed to produce it is also made from ammonia. Similarly, large amounts of urea are produced by combining ammonia with carbon dioxide derived from oxidation of the raw material used to produce the hydrogen. Ammonium sulfate is also made on a large scale by reacting ammonia with sulfuric acid. In the Far East substantial quantities of ammonium chloride are made from ammonia and salt or hydrochloric acid. Ammonium phosphates and nitrophosphates are additional fertilizers derived from ammonia. A principal advantage of most solid forms of ammonia is the ease with which they can be transported and applied to the soil. The high nitrogen content of urea (46 percent) and ammonium nitrate (33.5 percent) make them particularly advantageous.

Most phosphate fertilizers now come from mineral deposits, chiefly those in Florida, the western U.S., North Africa and parts of the U.S.S.R. Although both igneous and sedimentary phosphate deposits exist, about 90 percent of the world's fertilizer needs are supplied from the sedimentary sources because they are more plentiful than the igneous minerals and also easier to mine and process. The origin of sedimentary phosphates has generated much speculation among geologists. Some of them believe that the minerals were precipitated from seawater after it had been saturated with phosphate and fluorine ions derived from the contact of the water with igneous rocks and gases. It is also possible that these phosphates resulted to some extent from the replacement of calcium carbonate with calcium phos-

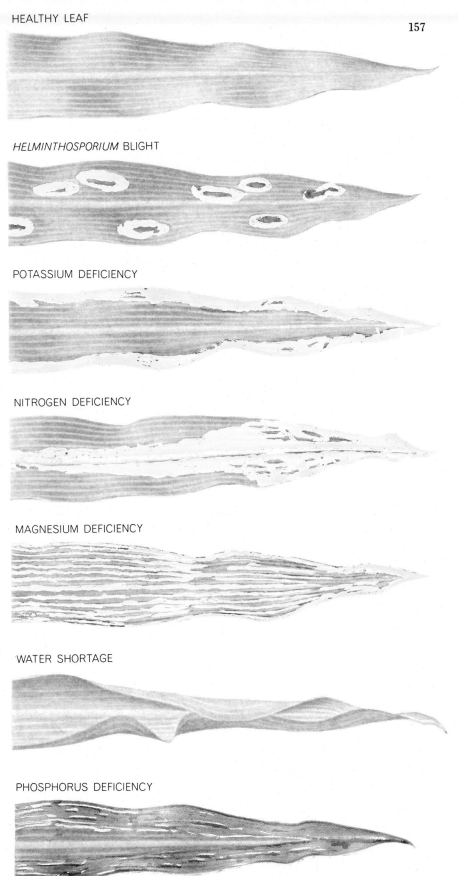

HEALTHY LEAF

HELMINTHOSPORIUM BLIGHT

POTASSIUM DEFICIENCY

NITROGEN DEFICIENCY

MAGNESIUM DEFICIENCY

WATER SHORTAGE

PHOSPHORUS DEFICIENCY

CORN-LEAF VARIATIONS directly or indirectly related to the amount of nutrients and water available to the plant are depicted. Gray represents green; the other colors are approximately as they appear in nature. *Helminthosporium* blight is a common fungus disease to which poorly nourished plants are vulnerable. Signs of potassium deficiency usually appear at the tips and along the edges of the lower leaves; of nitrogen deficiency, at the leaf tip, and of phosphorus, on young plants. Water shortage makes leaves a grayish-green.

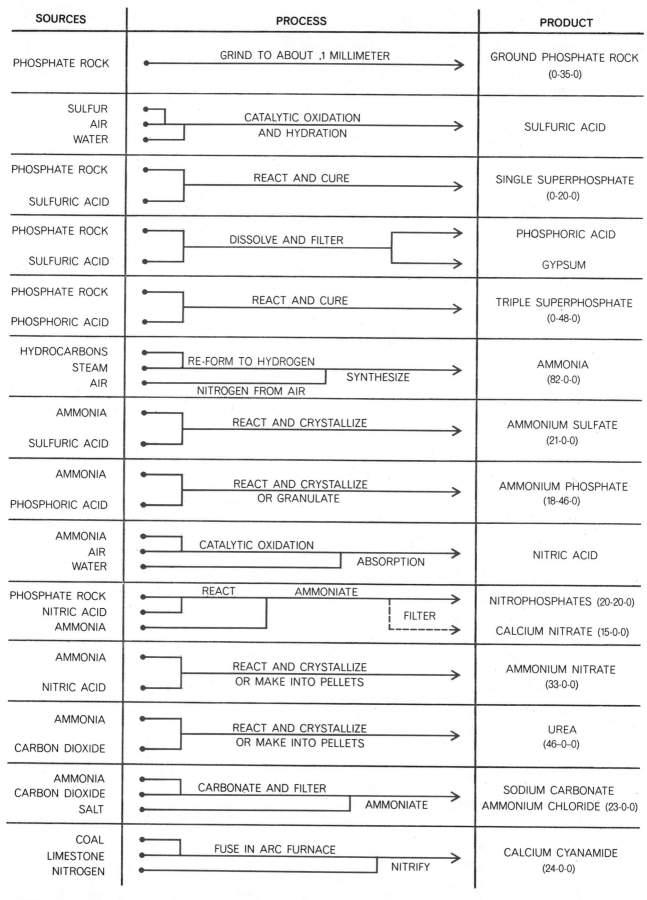

SOURCES	PROCESS	PRODUCT
PHOSPHATE ROCK	GRIND TO ABOUT .1 MILLIMETER	GROUND PHOSPHATE ROCK (0-35-0)
SULFUR / AIR / WATER	CATALYTIC OXIDATION AND HYDRATION	SULFURIC ACID
PHOSPHATE ROCK / SULFURIC ACID	REACT AND CURE	SINGLE SUPERPHOSPHATE (0-20-0)
PHOSPHATE ROCK / SULFURIC ACID	DISSOLVE AND FILTER	PHOSPHORIC ACID / GYPSUM
PHOSPHATE ROCK / PHOSPHORIC ACID	REACT AND CURE	TRIPLE SUPERPHOSPHATE (0-48-0)
HYDROCARBONS / STEAM / AIR	RE-FORM TO HYDROGEN / SYNTHESIZE / NITROGEN FROM AIR	AMMONIA (82-0-0)
AMMONIA / SULFURIC ACID	REACT AND CRYSTALLIZE	AMMONIUM SULFATE (21-0-0)
AMMONIA / PHOSPHORIC ACID	REACT AND CRYSTALLIZE OR GRANULATE	AMMONIUM PHOSPHATE (18-46-0)
AMMONIA / AIR / WATER	CATALYTIC OXIDATION / ABSORPTION	NITRIC ACID
PHOSPHATE ROCK / NITRIC ACID / AMMONIA	REACT AMMONIATE FILTER	NITROPHOSPHATES (20-20-0) / CALCIUM NITRATE (15-0-0)
AMMONIA / NITRIC ACID	REACT AND CRYSTALLIZE OR MAKE INTO PELLETS	AMMONIUM NITRATE (33-0-0)
AMMONIA / CARBON DIOXIDE	REACT AND CRYSTALLIZE OR MAKE INTO PELLETS	UREA (46–0–0)
AMMONIA / CARBON DIOXIDE / SALT	CARBONATE AND FILTER / AMMONIATE	SODIUM CARBONATE AMMONIUM CHLORIDE (23-0-0)
COAL / LIMESTONE / NITROGEN	FUSE IN ARC FURNACE / NITRIFY	CALCIUM CYANAMIDE (24-0-0)

BASIC PROCESSES used in manufacturing the major kinds of chemical fertilizers are charted. Each horizontal line shows the flow of the ingredient listed at left opposite the line. A vertical line shows a combining of ingredients. Numbers in parentheses show respectively the typical percentage of nitrogen, phosphorus and potassium materials used as fertilizer. For example, 0-35-0 means no nitrogen, 35 percent phosphorus pentoxide and no potassium oxide. Figures thus show amounts of primary nutrients.

phate in particles of the mineral aragonite on the ocean floor, a slow process that may still be taking place. Marine deposits of this nature may well become future sources of phosphate.

In any event, most of the primary deposits of sedimentary phosphate were laid down on ocean floor that subsequently became dry land. In time the weathering of such areas removed cementing substances such as calcium carbonate and magnesium carbonate, leaving extensive deposits of phosphate in the form of small pellets. Some of these deposits were later moved by surface water and redeposited elsewhere. Because of this extensive redeposition, and because pellet phosphates are insoluble in water, few minerals are found more widely scattered. By the same token, few have been formed over a longer span of time; phosphate minerals were laid down over the 400 million years from the Ordovician period to the Tertiary period and even later.

Often the phosphate pellets are covered by several feet of sand, clay or leached ore that must be taken off by scrapers or draglines before the phosphate matrix can be removed. In the extensive operations in Florida the matrix is excavated, dropped into sumps, slurried with powerful jets of water and then pumped to the processing plants. The material thus obtained may be only about 15 percent phosphate because of the large amounts of sand and clay in the matrix. Much of the sand and clay is removed by various processes to yield concentrates containing 30 to 36 percent phosphate. These concentrates are then blended and dried before further processing or shipment. Somewhat different methods are used in North Africa; there large tonnages of high-grade phosphate rock are mined by underground methods. Often they are only crushed, screened and dried before shipment.

Several types of fertilizer are made from the phosphate rock processed by the methods I have described. The simplest type consists of high-grade rock ground to particles less than .1 millimeter in size. This type is used directly on acid soils, which slowly attack the water-insoluble phosphate to make it available to plants. Next in simplicity is superphosphate, made by mixing ground phosphate rock with sulfuric acid to form a slurry that quickly hardens in a curing pile. After several weeks the hardened superphosphate is excavated and pulverized; often the powder is formed into granules. The pulverized or granulated material is marketed ei-

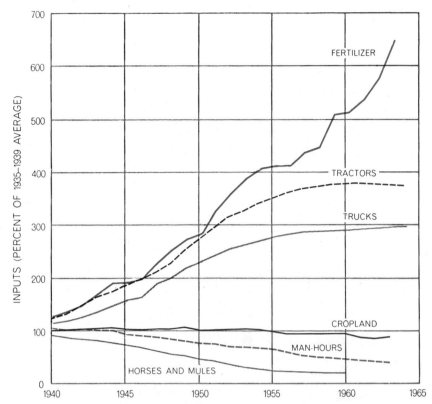

CHANGED TECHNOLOGY of U.S. agriculture over the past 30 years is reflected in a comparison of current inputs with those of 1935 through 1939. The changes are expressed as percentages of the average input in each category for the five-year base period. Concurrent with these changes of input has been a steady rise in the nation's agricultural output.

ther alone, as a phosphate fertilizer containing about 18 percent water-soluble phosphorus pentoxide, or in conjunction with other fertilizer materials. The various processing steps convert insoluble tricalcium phosphate to water-soluble monocalcium phosphate and gypsum.

Gypsum, however, is of little use in soil except when deficiencies in calcium or sulfur exist or when salinity is excessive. It also has a diluting effect on the phosphorus pentoxide content. Therefore it was a substantial advance when methods were devised for producing monocalcium phosphate without gypsum. The technique is to dissolve phosphate rock in a mixture of sulfuric and phosphoric acid to form gypsum and additional phosphoric acid, which can be separated by filtration. Thereafter the gypsum is usually discarded; the phosphoric acid is concentrated and mixed with finely ground phosphate rock to form a slurry that soon hardens into a product known as triple superphosphate. Its content of water-soluble phosphorus pentoxide is about 48 percent. Moreover, the product is cheaper to transport and to apply per unit of phosphorus pentoxide than ordinary superphosphate.

Substantial tonnages of phosphate fertilizers are also made by treating phosphate rock with nitric acid and ammonia to yield a range of materials that contain nitrogen as well as phosphorus. Potash can be added to form high-analysis fertilizers with a content of primary nutrients as high as 60 percent, as for example in a 20-20-20 grade. Another popular fertilizer is diammonium phosphate, which is made by neutralizing phosphoric acid with ammonia to yield a material containing about 20 percent nitrogen and 50 percent water-soluble phosphorus pentoxide. Potash can be added to this product to make another high-analysis mixture containing all the primary nutrients.

Potassium exists in enormous quantities in the rocks and soils of the world. Often, however, it is in the form of insoluble minerals unsuitable for agriculture. Fortunately large deposits of soluble potassium chloride are available, mostly as sylvite and sylvinite or, in conjunction with magnesium, as carnallite and langbeinite. Such deposits are often mixed with sodium chloride in the form of halite, which is toxic to many crops and must be removed.

Extensive supplies of sylvite and

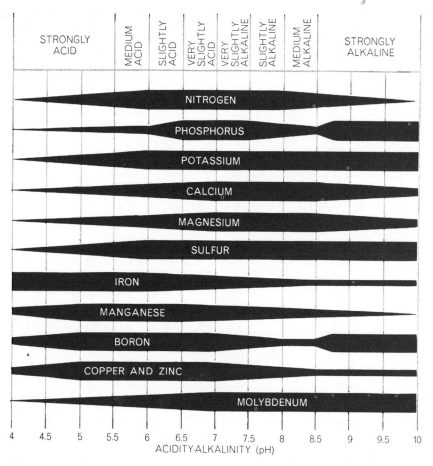

AVAILABILITY OF NUTRIENTS to plants is affected by the condition of the soil. The more soluble a nutrient is under a particular condition of soil acidity or alkalinity, the thicker is the horizontal band representing the nutrient. Solubility in turn is directly related to the availability of the nutrient in an ionic form that is assimilable by the plant.

The grower must know the condition of his soil and treat it accordingly. In most cases he must apply the bulk of the treatment before sowing, because as a rule most of the nutrient needed by a plant is taken up in the early stages of its growth. The correct nutrient balance is additionally important because a deficiency of any one plant food in the soil will reduce the effect of others, even if they are in oversupply.

A deficiency of nitrogen usually appears in plants as a yellowing of the leaves, accompanied by shriveling that proceeds upward from the lower leaves. The principal effects of nitrogen on plants include accelerated growth and increased yield of leaf, fruit and seed. Nitrogen also promotes the activity of soil bacteria. Nitrate nitrogen is quickly available to root systems, but it may therefore make the plants grow too rapidly. Moreover, nitrate is easily lost by leaching. Ammoniacal nitrogen, on the other hand, is immediately fixed in the soil by ion-exchange reactions and is released to the plants over a longer period than nitrate nitrogen. For these reasons it is sometimes the practice to inject free ammonia in anhydrous or aqueous form a few inches below the surface of a moist soil. With many crops optimum results are obtained by the proper combination of nitrate nitrogen and ammoniacal nitrogen in either solid or liquid form.

Phosphorus deficiency is often represented by purplish leaves and stems, slow growth and low yields. Phosphorus stimulates the germination of seedlings and encourages early root formation. Since these results are less evident than those induced by applications of nitrogen, many farmers, particularly in the Far East, use insufficient quantities of phosphate fertilizer.

Potassium deficiency can often be detected by a spotting or curling of lower leaves. Additional symptoms are weak stalks and stems, a condition that can cause heavy crop losses in strong winds and heavy rains. The application of potassium improves the yield of grain and seed, and it enhances the formation of starches, sugars and plant oils. It also contributes to the plant's vigor and its resistance to frost and disease.

As for deficiencies of secondary nutrients, a lack of magnesium may cause a general loss of color, weak stalks and white bands across the leaves in corn and certain other plants. A calcium deficiency may give rise to the premature death of young leaves and poor formation of seed. An inadequate supply of sulfur frequently leads to pale leaves,

carnallite were found first in Germany and later in France, the western U.S. and many other countries. Most of these deposits resulted from the evaporation of ancient seas during the Permian period (about 230 to 280 million years ago). In the Canadian province of Saskatchewan huge quantities of sylvite and carnallite were more recently found at depths of 3,000 to 4,000 feet, in the upper portion of a Devonian halite formation. Although these deposits are considerably deeper than U.S. and European potash sources, mining difficulties have now been overcome and the production of several million tons annually of Canadian potash will be of great benefit to world agriculture. Another Canadian development of growing importance is the large-scale production of potash by solution mining, which involves pumping water into the potash beds and bringing the resulting solution to the surface for evaporation and the recovery of potash in solid form.

After solid potash minerals are mined they are sometimes crushed and separated from their impurities by washing and froth-flotation, in which treatment with amine salts and air causes the sylvite particles to float away from the unwanted substances. In other cases potash is recovered by solution and crystallization. The relatively pure product is dried, treated with an amine anticaking agent and sold for agricultural purposes as muriate of potash containing 60 to 62 percent of potassium oxide. Most potash is used in conjunction with nitrogen and phosphorus compounds. Potassium sulfate and potassium nitrate are also used to a limited extent in agricultural situations where the chloride ions of potash would be harmful, as they are to tobacco.

Agronomic Considerations

It is appropriate now to consider the role of nutrients in plant growth, together with some other factors that must be taken into account in the use of fertilizers. As anyone experienced in agronomy or gardening knows, it is wasteful and sometimes even harmful to broadcast fertilizer indiscriminately.

stunted growth and immature fruit.

Typical examples of micronutrient deficiency are heart rot in vegetables and fruits as a result of a shortage of boron and stunted growth of vegetables and citrus plants resulting from insufficient manganese and molybdenum. Micronutrient deficiencies may be hard to detect and even harder to rectify, because the balance between enough of a micronutrient and a toxic oversupply can be delicate.

An important consideration in the use of fertilizers is the acidity of the soil, which considerably influences the availability of many nutrients to the plant [see illustration on opposite page]. To complicate matters, nitrogen fertilizers such as ammonia, urea ammonium nitrate and other ammonia derivatives can themselves raise the acidity of soils, by means of complex ion-exchange reactions. In most cases the acidity of a soil can be controlled by adding appropriate amounts of lime, ground limestone or other forms of calcium carbonate.

Soil tilth, or structure, is also important. For example, the richer chernozem soils found in the middle of the North American continent and in the Ukraine are in many cases well supplied with organic humus and lime salts and need only regular supplies of plant nutrients to replace those removed by agriculture and leaching. On the other hand, the podzol soils that cover the northeastern U.S., most of Britain and much of central Europe have been intensively leached by centuries of farming and exposure; they need not only liberal supplies of plant nutrients but also lime and organic humus. Desert soils may be rich in certain minerals and yet lacking in available nutrients and in the organic matter usually necessary to retain moisture and to provide good tilth. Such soils can be made productive, however, as has been amply demonstrated in Israel.

Prospective Developments

In spite of the many improvements made in chemical fertilizers during the past 50 years, several problems still confront the fertilizer industry. One major concern is achieving the controlled release of nutrients so that waste and also damage to young plants can be avoided. Methods now being tested in-

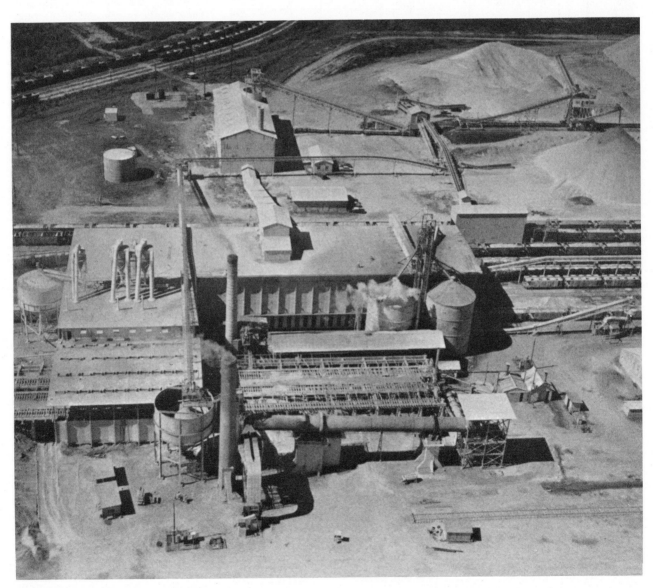

TREATMENT PLANT of the V-C Chemical Company in Florida removes organic material and some carbon dioxide from phosphate rock to provide a raw material for making phosphoric acid, which is used in the manufacture of such chemical fertilizers as triple superphosphate and diammonium phosphate. Piles of phosphate rock as brought from the mine are at top right. Horizontal tube in foreground is a calcine kiln in which rock is given thermal treatment. Treated rock is stored in tanks behind kiln until shipped.

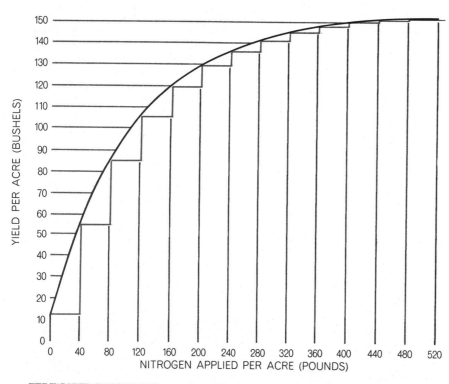

FERTILIZER ECONOMICS are indicated by the curve of yield compared with applications of fertilizer. This curve is based on a crop of irrigated corn grown in the state of Washington. With nitrogen as with other fertilizers, crop yields diminish with increasing applications of nutrients. In time additional increments of fertilizer become uneconomical.

SUBSTANCE	APPROXIMATE POUNDS PER ACRE	SUPPLIED BY
NITROGEN	310	
PHOSPHORUS	120 (PHOSPHATE) 52 (PHOSPHORUS)	1,200 POUNDS OF 25-10-20 FERTILIZER
POTASSIUM	245 (POTASH) 205 (POTASSIUM)	
CALCIUM	58	APPROXIMATELY 150 POUNDS OF AGRICULTURAL LIMESTONE
MAGNESIUM	50	APPROXIMATELY 275 POUNDS OF EPSOM SALT, OR 550 POUNDS SULFATE OF POTASH-MAGNESIA
SULFUR	33	33 POUNDS OF SULFUR
IRON	3	15 POUNDS OF IRON SULFATE
MANGANESE	.45	APPROXIMATELY 1.3 POUNDS OF MANGANESE SULFATE
BORON	.10	APPROXIMATELY 1 POUND OF BORAX
ZINC	TRACE	SMALL AMOUNT OF ZINC SULFATE
COPPER	TRACE	SMALL AMOUNT OF COPPER SULFATE OR OXIDE
MOLYBDENUM	TRACE	VERY SMALL AMOUNT OF SODIUM OR AMMONIUM MOLYBDATE
OXYGEN	10,200	AIR
CARBON	7,800	AIR
WATER	3,225 TO 4,175 TONS	29 TO 36 INCHES OF RAIN

NUTRIENTS REQUIRED to produce 150 bushels of corn are indicated. Most plants take all their nutrients from the soil except carbon, oxygen and hydrogen, obtained from the air.

clude the use of slowly decomposing inorganic materials such as magnesium ammonium phosphate and synthetic organic compounds such as formamide and oxamide. Another technique being studied is the encapsulation of fertilizer particles with sulfur or plastic. Investigators are also exploring the possibilities of producing chemical fertilizers in which a plant nutrient would be "sequestered" in molecules of the chelate type. Chelation involves a tight molecular bonding that would protect the nutrient against rapid attack. In this way the desired plant food would be released slowly and in a prescribed manner by chemical reactions in the soil. An ultimate possibility is the production of "packaged" granules, each containing a seed and whatever substances are needed during the lifetime of the plant. They would be released in the proper amounts and sequence.

A new agricultural technique already in use on a small scale is "chemical plowing." Instead of turning stubble and cover crops into the ground mechanically, the farmer kills them by spraying them with the appropriate herbicides. Eventually the dead plant materials become sources of humus and plant nutrient. Any excess of herbicide is rendered harmless by the action of soil colloids. New seeds and fertilizer are drilled directly through the dead cover material, which also gives protection against erosion, frost and drought.

Efforts are also under way to reduce the cost of transporting fertilizers and their raw materials. The approach here is to try to produce them in highly concentrated liquid or solid form. They are then appropriately diluted or combined at the point of use.

Perhaps the most vital work is the education of farmers—particularly farmers in the developing countries—in modern agricultural methods, including the use of chemical fertilizers. In addition the developing nations must establish low-cost credit plans so that impoverished farmers can buy adequate supplies of fertilizer. Similarly, credit must be extended by the developed nations to the less developed ones on an even bigger scale than at present in order to help the less developed nations obtain the materials, equipment and expert advice they need to build their own chemical fertilizer plants. Until these steps are taken to spread modern agricultural technology, the developing nations will fall far short of the contribution they could make to the intensifying problem of producing enough food for the world's growing population.

18

THIRD-GENERATION PESTICIDES

CARROLL M. WILLIAMS
July 1967

Man's efforts to control harmful insects with pesticides have encountered two intractable difficulties. The first is that the pesticides developed up to now have been too broad in their effect. They have been toxic not only to the pests at which they were aimed but also to other insects. Moreover, by persisting in the environment—and sometimes even increasing in concentration as they are passed along the food chain—they have presented a hazard to other organisms, including man. The second difficulty is that insects have shown a remarkable ability to develop resistance to pesticides.

Plainly the ideal approach would be to find agents that are highly specific in their effect, attacking only insects that are regarded as pests, and that remain effective because the insects cannot acquire resistance to them. Recent findings indicate that the possibility of achieving success along these lines is much more likely than it seemed a few years ago. The central idea embodied in these findings is that a harmful species of insect can be attacked with its own hormones.

Insects, according to the latest estimates, comprise about three million species—far more than all other animal and plant species combined. The number of individual insects alive at any one time is thought to be about a billion billion (10^{18}). Of this vast multitude 99.9 percent are from the human point of view either innocuous or downright helpful. A few are indispensable; one need think only of the role of bees in pollination.

The troublemakers are the other .1 percent, amounting to about 3,000 species. They are the agricultural pests and the vectors of human and animal disease. Those that transmit human disease are the most troublesome; they have joined with the bacteria, viruses and protozoa in what has sometimes seemed like a grand conspiracy to exterminate man, or at least to keep him in a state of perpetual ill health.

The fact that the human species is still here is an abiding mystery. Presumably the answer lies in changes in the genetic makeup of man. The example of sickle-cell anemia is instructive. The presence of sickle-shaped red blood cells in a person's blood can give rise to a serious form of anemia, but it also confers resistance to malaria. The sickle-cell trait (which does not necessarily lead to sickle-cell anemia) is appreciably more common in Negroes than in members of other populations. Investigations have suggested that the sickle cell is a genetic mutation that occurred long ago in malarial regions of Africa. Apparently attrition by malaria-carrying mosquitoes provoked countermeasures deep within the genes of primitive men.

The evolution of a genetic defense, however, takes many generations and entails many deaths. It was only in comparatively recent times that man found an alternative answer by learning to combat the insects with chemistry. He did so by inventing what can be called the first-generation pesticides: kerosene to coat the ponds, arsenate of lead to poison the pests that chew, nicotine and rotenone for the pests that suck.

Only 25 years ago did man devise the far more potent weapon that was the first of the second-generation pesticides. The weapon was dichlorodiphenyltrichloroethane, or DDT. It descended on the noxious insects like an avenging angel. On contact with it mosquitoes, flies, beetles—almost all the insects—were stricken with what might be called the "DDT's." They went into a tailspin, buzzed around upside down for an hour or so and then dropped dead.

The age-old battle with the insects appeared to have been won. We had the stuff to do them in—or so we thought. A few wise men warned that we were living in a fool's paradise and that the insects would soon become resistant to DDT, just as the bacteria had managed to develop a resistance to the challenge of sulfanilamide. That is just what happened. Within a few years the mosquitoes, lice, houseflies and other noxious insects were taking DDT in their stride. Soon they were metabolizing it, then they became addicted to it and were therefore in a position to try harder.

Fortunately the breach was plugged by the chemical industry, which had come to realize that killing insects was —in more ways than one—a formula for

164

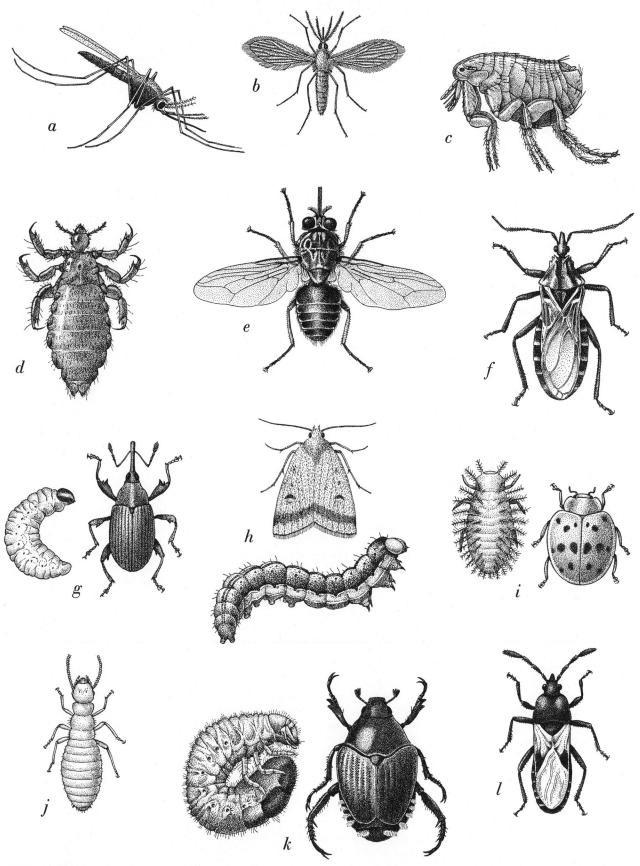

INSECT PESTS that might be controlled by third-generation pesticides include some 3,000 species, of which 12 important examples are shown here. Six (a–f) transmit diseases to human beings; the other six are agricultural pests. The disease-carriers, together with the major disease each transmits, are (a) the *Anopheles* mosquito, malaria; (b) the sand fly, leishmaniasis; (c) the rat flea, plague; (d) the body louse, typhus; (e) the tsetse fly, sleeping sickness, and (f) the kissing bug, Chagas' disease. The agricultural pests, four of which are depicted in both larval and adult form, are (g) the boll weevil; (h) the corn earworm; (i) the Mexican bean beetle; (j) the termite; (k) the Japanese beetle, and (l) the chinch bug. The species in the illustration are not drawn to the same scale.

getting along in the world. Organic chemists began a race with the insects. In most cases it was not a very long race, because the insects soon evolved an insensitivity to whatever the chemists had produced. The chemists, redoubling their efforts, synthesized a steady stream of second-generation pesticides. By 1966 the sales of such pesticides had risen to a level of $500 million a year in the U.S. alone.

Coincident with the steady rise in the output of pesticides has come a growing realization that their blunderbuss toxicity can be dangerous. The problem has attracted widespread public attention since the late Rachel Carson fervently described in *The Silent Spring* some actual and potential consequences of this toxicity. Although the attention thus aroused has resulted in a few attempts to exercise care in the application of pesticides, the problem cannot really be solved with the substances now in use.

The rapid evolution of resistance to pesticides is perhaps more critical. For example, the world's most serious disease in terms of the number of people afflicted continues to be malaria, which is transmitted by the *Anopheles* mosquito—an insect that has become completely resistant to DDT. (Meanwhile the protozoon that actually causes the disease is itself evolving strains resistant to antimalaria drugs.)

A second instance has been presented recently in Vietnam by an outbreak of plague, the dreaded disease that is conveyed from rat to man by fleas. In this case the fleas have become resistant to pesticides. Other resistant insects that are agricultural pests continue to take a heavy toll of the world's dwindling food supply from the moment the seed is planted until long after the crop is harvested. Here again we are confronted by an emergency situation that the old technology can scarcely handle.

The new approach that promises a way out of these difficulties has emerged during the past decade from basic studies of insect physiology. The prime candidate for developing third-generation pesticides is the juvenile hormone that all insects secrete at certain stages in their lives. It is one of the three internal secretions used by insects to regulate growth and metamorphosis from larva to pupa to adult. In the living insect the juvenile hormone is synthesized by the corpora allata, two tiny glands in the head. The corpora allata are also responsible for regulating the flow of the hormone into the blood.

At certain stages the hormone must be

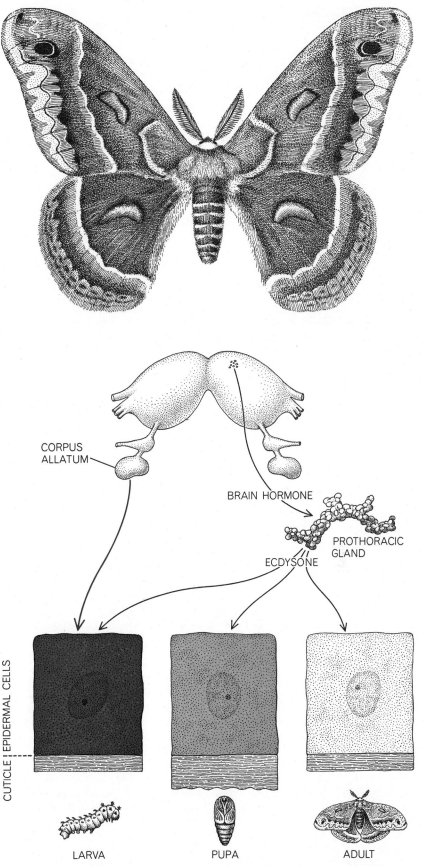

HORMONAL ACTIVITY in a Cecropia moth is outlined. Juvenile hormone (*color*) comes from the corpora allata, two small glands in the head; a second substance, brain hormone, stimulates the prothoracic glands to secrete ecdysone, which initiates the molts through which a larva passes. Juvenile hormone controls the larval forms and at later stages must be in low concentration or absent; if applied then, it deranges insect's normal development. The illustration is partly based on one by Howard A. Schneiderman and Lawrence I. Gilbert.

CHEMICAL STRUCTURES of the Cecropia juvenile hormone (*left*), isolated this year by Herbert Röller and his colleagues at the University of Wisconsin, and of a synthetic analogue (*right*) made in 1965 by W. S. Bowers and others in the U.S. Department of Agriculture show close similarity. Carbon atoms, joined to one or two hydrogen atoms, occupy each angle in the backbone of the molecules; letters show the structure at terminals and branches.

JUVENILE HORMONE ACTIVITY has been found in various substances not secreted by insects. One (*left*) is a material synthesized by M. Romanuk and his associates in Czechoslovakia. The other (*right*), isolated and identified by Bowers and his colleagues, is the "paper factor" found in the balsam fir. The paper factor has a strong juvenile hormone effect on only one family of insects, exemplified by the European bug *Pyrrhocoris apterus*.

secreted; at certain other stages it must be absent or the insect will develop abnormally [*see illustration on preceding page*]. For example, an immature larva has an absolute requirement for juvenile hormone if it is to progress through the usual larval stages. Then, in order for a mature larva to metamorphose into a sexually mature adult, the flow of hormone must stop. Still later, after the adult is fully formed, juvenile hormone must again be secreted.

The role of juvenile hormone in larval development has been established for several years. Recent studies at Harvard University by Lynn M. Riddiford and the Czechoslovakian biologist Karel Sláma have resulted in a surprising additional finding. It is that juvenile hormone must be absent from insect eggs for the eggs to undergo normal embryonic development.

The periods when the hormone must be absent are the Achilles' heel of insects. If the eggs or the insects come into contact with the hormone at these times, the hormone readily enters them and provokes a lethal derangement of further development. The result is that the eggs fail to hatch or the immature insects die without reproducing.

Juvenile hormone is an insect invention that, according to present knowledge, has no effect on other forms of life. Therefore the promise is that third-generation pesticides can zero in on in-

sects to the exclusion of other plants and animals. (Even for the insects juvenile hormone is not a toxic material in the usual sense of the word. Instead of killing, it derails the normal mechanisms of development and causes the insects to kill themselves.) A further advantage is self-evident: insects will not find it easy to evolve a resistance or an insensitivity to their own hormone without automatically committing suicide.

The potentialities of juvenile hormone as an insecticide were recognized 12 years ago in experiments performed on the first active preparation of the hormone: a golden oil extracted with ether from male Cecropia moths. Strange to say, the male Cecropia and the male of its close relative the Cynthia moth remain to this day the only insects from which one can extract the hormone. Therefore tens of thousands of the moths have been required for the experimental work with juvenile hormone; the need has been met by a small but thriving industry that rears the silkworms.

No one expected Cecropia moths to supply the tons of hormone that would be required for use as an insecticide. Obviously the hormone would have to be synthesized. That could not be done, however, until the hormone had been isolated from the golden oil and identified.

Within the past few months the difficult goals of isolating and identifying the hormone have at last been attained by a team of workers headed by Herbert Röller of the University of Wisconsin. The juvenile hormone has the empirical formula $C_{18}H_{30}O_3$, corresponding to a molecular weight of 294. It proves to be the methyl ester of the epoxide of a previously unknown fatty-acid derivative [*see upper illustration on this page*]. The apparent simplicity of the molecule is deceptive. It has two double bonds and an oxirane ring (the small triangle at lower left in the molecular diagram), and it can exist in 16 different molecular configurations. Only one of these can be the authentic hormone. With two ethyl groups ($CH_2 \cdot CH_3$) attached to carbons No. 7 and 11, the synthesis of the hormone from any known terpenoid is impossible.

The pure hormone is extraordinarily active. Tests the Wisconsin investigators have carried out with mealworms suggest that one gram of the hormone would result in the death of about a billion of these insects.

A few years before Röller and his colleagues worked out the structure of the authentic hormone, investigators at several laboratories had synthesized a number of substances with impressive juvenile hormone activity. The most potent of the materials appears to be a crude mixture that John H. Law, now at the University of Chicago, prepared by a simple one-step process in which hydrogen chloride gas was bubbled through an alcoholic solution of farnesenic acid. Without any purification this mixture was 1,000 times more active than crude Cecropia oil and fully effective in killing all kinds of insects.

One of the six active components of Law's mixture has recently been identified and synthesized by a group of workers headed by M. Romaňuk of the Czechoslovak Academy of Sciences. Romaňuk and his associates estimate that from 10 to 100 grams of the material would clear all the insects from 2½ acres. Law's original mixture is of course even more potent, and so there is much interest in its other five components.

Another interesting development that preceded the isolation and identification of true juvenile hormone involved a team of investigators under W. S. Bowers of the U.S. Department of Agriculture's laboratory at Beltsville, Md. Bowers and his colleagues prepared an analogue of juvenile hormone that, as can be seen in the accompanying illustration [*top of opposite page*], differed by only two carbon atoms from the authentic Cecropia hormone (whose structure was then, of course, unknown). In terms of the dosage required it appears that the Beltsville compound is about 2 percent as active as Law's mixture and about .02 percent as active as the pure Cecropia hormone.

All the materials I have mentioned are selective in the sense of killing only insects. They leave unsolved, however, the problem of discriminating between the .1 percent of insects that qualify as pests and the 99.9 percent that are helpful or innocuous. Therefore any reckless use of the materials on a large scale could constitute an ecological disaster of the first rank.

The real need is for third-generation pesticides that are tailor-made to attack only certain predetermined pests. Can such pesticides be devised? Recent work that Sláma and I have carried out at Harvard suggests that this objective is by no means unattainable. The possibility arose rather fortuitously after Sláma arrived from Czechoslovakia, bringing with him some specimens of the European bug *Pyrrhocoris apterus*—a species that had been reared in his laboratory in Prague for 10 years.

To our considerable mystification the bugs invariably died without reaching sexual maturity when we attempted to rear them at Harvard. Instead of metamorphosing into normal adults they continued to grow as larvae or molted into adult-like forms retaining many larval characteristics. It was evident that the bugs had access to some unknown source of juvenile hormone.

Eventually we traced the source to the paper toweling that had been placed in the rearing jars. Then we discovered that almost any paper of American origin—including the paper on which *Scientific American* is printed—had the same effect. Paper of European or Japanese manufacture had no effect on the bugs. On further investigation we found that the juvenile hormone activity originated in the balsam fir, which is the principal source of pulp for paper in Canada and the northern U.S. The tree synthesizes what we named the "paper factor," and this substance accompanies the pulp all the way to the printed page.

Thanks again to Bowers and his associates at Beltsville, the active material of the paper factor has been isolated and characterized [*see lower illustration on opposite page*]. It proves to be the methyl ester of a certain unsaturated fatty-acid derivative. The factor's kinship with the other juvenile hormone analogues is evident from the illustrations.

Here, then, is an extractable juvenile hormone analogue with selective action against only one kind of insect. As it happens, the family Pyrrhocoridae includes some of the most destructive pests of the cotton plant. Why the balsam fir should have evolved a substance against only one family of insects is unexplained. The most intriguing possibility is that the paper factor is a biochemical memento of the juvenile hormone of a former natural enemy of the tree—a pyrrhocorid predator that, for obvious reasons, is either extinct or has learned to avoid the balsam fir.

In any event, the fact that the tree synthesizes the substance argues strongly that the juvenile hormone of other species of insects can be mimicked, and perhaps has been by trees or plants on which the insects preyed. Evidently during the 250 million years of insect evolution the detailed chemistry of juvenile hormone has evolved and diversified. The process would of necessity have gone hand in hand with a retuning of the hormonal receptor mechanisms in the cells and tissues of the insect, so that the use as pesticides of any analogues that are discovered seems certain to be effective.

The evergreen trees are an ancient lot. They were here before the insects; they are pollinated by the wind and thus, unlike many other plants, do not depend on the insects for anything. The paper factor is only one of thousands of terpenoid materials these trees synthesize for no apparent reason. What about the rest?

It seems altogether likely that many of these materials will also turn out to be analogues of the juvenile hormones of specific insect pests. Obviously this is the place to look for a whole battery of third-generation pesticides. Then man may be able to emulate the evergreen trees in their incredibly sophisticated self-defense against the insects.

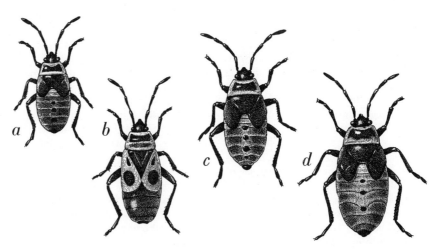

EFFECT OF PAPER FACTOR on *Pyrrhocoris apterus* is depicted. A larva of the fifth and normally final stage (*a*) turns into a winged adult (*b*). Contact with the paper factor causes the insect to turn into a sixth-stage larva (*c*) and sometimes into a giant seventh-stage larva (*d*). The abnormal larvae usually cannot shed their skin and die before reaching maturity.

19

MECHANICAL HARVESTING

CLARENCE F. KELLY
August 1967

Last year in the U.S. there was harvested more than 200 million tons of grains, 220 million tons of hay and silage, 21 million tons of vegetables, 20 million tons of fruit, 62.5 million tons of milk, 3.7 million tons of cotton, 106,-500 tons of wool and similarly large quantities of other agricultural products. If all this produce had had to be harvested by human hands, probably the entire U.S. population would not have sufficed to perform the job.

The fact that the production of these amounts of food and fiber engages only 5 percent of the U.S. labor force is primarily due to the mechanization of farming. Other technological developments—chemical fertilizers, pesticides, plant breeding and so on—make essential contributions, but mechanization is still the outstanding factor. Among the functions of mechanization none is more important than its role in the harvesting of crops. The picking and winnowing of a crop usually accounts for at least half of the total cost of production. It is also by far the most difficult part of the agricultural process to mechanize. Nevertheless, the mechanization of harvesting in the U.S. has made such strides that, in spite of the costliness of the machines and other technical aids, the cost of food to American families, in terms of its percentage (18 percent) of their income, is the lowest in the world.

We are still a long way from complete mechanization of this sector of agricultural activity. The harvesting of fruit, for example, is still done largely by hand. Research and ingenious new ideas, however, are rapidly adding one important crop after another to the list of those that can be picked by machine. Harvesting is a critical field for research. The economic condition of the nations and peoples of the world is determined in considerable part by the stage they have reached in harvesting efficiency.

The Saving of Labor

Man began his agricultural career by picking wild fruits and nuts and digging roots. When he progressed to planting grain, he harvested it by hand, stalk by stalk, and centuries may have passed before he advanced to using a crude knife to lop the head of grain from the plant. This tool evolved into a curved knife, or sickle (whose use depended on cultivating grain plants that would ripen together instead of at separate times), then a scythe, then a scythe with an attached cradle that would catch the stalks as they were cut and thus collect them in a sheaf for threshing.

This hand tool essentially represented the state of harvesting technology less than two centuries ago. In 1790 agriculture occupied 90 percent of the U.S. population. Half a century later the proportion had dropped to 70 percent, and by 1890 only 43 percent of the people were farmers. The principal events responsible for this trend were John Deere's production of the steel plow, Cyrus McCormick's invention of the reaper (which he began to manufacture in 1847) and John Appleby's development (in 1879) of the grain binder. After 1890 the substitution of machines for manpower on U.S. farms proceeded at a much more rapid rate. The farm population declined to 30 percent in 1915 and to 18 percent by 1940. The gasoline engine was, of course, the prime factor, and it became the motive power of a variety of mass-production harvesting machines: the self-propelled wheat combine, the hay baler, the cotton and corn pickers. Since World War II a proliferation of other harvesting devices, handling perishable as well as nonperishable crops, has been partly responsible for the further reduction of the farm labor force to the present 5 percent.

The harvesting of grains is now totally mechanized, and that of roots and tubers is almost wholly so. For efficiency's sake a harvesting machine should not be a mere accessory for human hands; it should perform the full task. This sometimes calls for a high degree of specialization in the machine. For certain crops there are now machines that are guided by sensing mechanisms and duplicate almost exactly the motions of the human hand. The design of a harvester must often be adapted not only to the crop but also to the vagaries of nature. For example, a sugar beet harvester that works well in the peat soils of the Sacramento River delta is helpless in the heavy, sticky soils of Ireland; a grain combine that is beautifully suited to harvesting wheat in the North Dakota prairie cannot operate in the wet rice fields of Arkansas unless its drive is changed to the crawler type; a berry harvester that picks raspberries efficiently in Oregon cannot be used in the San Joaquin Valley of California because of differences in the climate and in the berry varieties.

Let us look at the principal machines and see what their development entailed. By far the most important, of course, is the grain combine.

Grain

Analyzing the functions to be performed by the reaper (the original element of the combine), McCormick found that he had to solve seven problems: avoiding side draft (the tendency of the machine to pull sideways), holding the stalks for cutting, separating them from the stalks left standing at the end of the

COTTON PICKER harvests two rows of cotton at a time at a maxi- mum speed of 3.3 miles per hour. The machine uses rotating spin- dles to twist the cotton fibers out of the bolls. The basket at rear can hold up to 3,000 pounds of cotton before it has to be dumped.

cutter bar, pushing the stalks to be cut against the knives, cutting them, catching the fallen stalks and providing a power drive for the necessary operations. The apparatus McCormick developed to deal with these problems consisted of a main ground wheel that furnished the power and carried most of the machine's weight (to minimize side draft), a vibrating horizontal knife, a divider at the end of the cutter bar, fingers to hold the stalks, a reel to push the stalks against the knife and a platform to catch the cut stalks, with the heads of the grain falling toward the rear of the machine.

The reaper was only a first step toward fully mechanical harvesting. The grain stalks still had to be bundled in the field by hand, then be picked up and hauled to the stack or threshing machine. Mechanization advanced a substantial step when the reaper was equipped with a binder that bundled the grain mechanically. The final step to complete mechanization was the invention of the modern combine, which combines reaping and threshing in the same machine. This apparatus eliminates several operations: the grain no longer has to be bundled, stacked and transported to the thresher. The grain combine, to be sure, introduced new problems. Because the grain is threshed as it is collected, it must be fully ripe. The ripened grain in the field therefore runs the risk of being shaken from the stalks by wind before the harvester can cut it. Moreover, if the threshed grain is not uniformly ripe, pockets of moist grain in the storehouse may quickly spoil. These problems have been considerably reduced, however, by countermeasures in plant breeding and cultivation.

Today the combine has become so versatile a machine that it is used to harvest not only grains such as wheat but also grass seed, beans and even corn. The machine carries attachments and a variety of adjustments that enable the operator to adapt its operations to the crop or even the particular condition of a crop. By raising or lowering the cutter bar, by changing the speed of the various machine elements and by other adjustments he can allow for the height of the plants, the slope of the land, the length of the stalks, the moisture content of the ker-

nels, the size of the kernels (from the smallest grass seed to the largest beans), the presence of weeds, even the direction and force of the wind. The machine is designed to garner the full yield of the field without damaging the crop or collecting extraneous material.

Research is going forward on further refinements of the combine. The aims are to reduce the cost of the machine and its power requirements, to design automatic controls that will keep it operating at its most efficient capacity and thresh out every last kernel of grain and to incorporate a drier that will reduce the moisture content of the grain as it is reaped. Meanwhile plant breeders are working to adapt the plants more closely to the machine, seeking to develop varieties that will have a higher ratio of grain to straw, that will dry more evenly in the field, that will withstand more wind without shattering and that will produce more grain per acre.

The usefulness of the present combine can be measured by the fact that with this machine in California the harvesting of rice (reaping, threshing and hulling) requires less than one man-hour per acre, whereas in Japan, where the work is done largely by hand, the average labor expenditure according to a recent study is 258 man-hours per acre. It is a significant commentary on the world food problem that rice, the largest crop on our planet, is still harvested by hand, stalk by stalk, in most of the countries around the world.

Hay

Just as the combine has revolutionized the harvesting of grain, other machines have transformed the production of hay. In acreage and tonnage (more than 120 million tons a year) hay is the largest U.S. crop, and in dollar value (about $2 billion) it ranks second only to wheat. An important harvester of this crop, principally alfalfa, is of course the cow. Animal husbandry has always depended in large degree, however, on the mowing and storage of dried grass as feed, not only for seasons when pastures are dormant but also for areas where there are no pastures.

The traditional method of making hay,

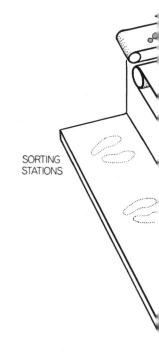

SORTING STATIONS

TRUCK

TOMATO HARVESTER has transformed tomato growing from total reliance on hand picking to largely mechanized harvesting within the past four years. The machine, known as the U.C.-Blackwelder for the University of California, which developed it, and the Blackwelder Manufacturing Company, which makes it, cuts the whole tomato plant and conveys it to the shaker-sifter, where the tomatoes are shaken loose. The vines are blown back onto the field; human workers stationed along the sorting belts remove tomatoes that are defective.

BLOWER

VINE DISCHARGE

VINES

SHAKER-SIFTER

TOMATOES

SORTING
STATIONS

DELIVERY
ELEVATOR

ROTATING
DIRT-REMOVER

DRIVER
STATION

SUPERVISOR
STATION

CUTTING
KNIVES

FEED ROLLERS

FIELD TECHNIQUE of harvesting tomatoes mechanically uses tractor-drawn bulk bins in connection with the harvesting machine. The three rollers on the front of the machine are directly behind the vine-cutters and are used to feed the cut plants into the machine. Harvesting is done only once per crop; it was therefore necessary to breed tomato plants that all ripened at virtually the same time.

LETTUCE HARVESTER was developed by the University of California at Davis. The machine gauges the heads of lettuce and cuts those that are firm enough and large enough for harvesting. Spokes of the wheellike device, which is a rotary elevator, grip the cut heads and lift them to the conveyor belt on the front of the machine. The belt delivers them to baskets or to a packing machine.

described by Roman writers as early as 2,000 years ago, was to cut the grass, let it dry in the sun for a few days and then store it for protection from rain and snow. Mechanization of the process began with the replacement of the hand scythe by mowing machines (using the reciprocating knife) in the second half of the 19th century. The mown grass was stacked by hand in small haycocks before it had lost its green color and then, after further drying, was piléd in large stacks outdoors or stored in a barn. Within the past century machines of various types have been developed to handle and pack the hay in convenient forms.

One simple method depends on the use of the so-called bull rake. The swath of grass felled by the mower is raked into windrows, and after it has dried in the field the bull rake collects it in piles for storage. The system is inefficient in use of storage space: the comparatively loose hay takes up 500 cubic feet of space per ton. Moreover, a strong human handler is still required to wrest the hay from the pile with a pitchfork and deliver it to the mangers. On most American farms the bull-rake collection method has been replaced by machines that bale the hay, and various systems for mechanical delivery of the hay in palatable form to the animals are being tested.

An automatic baler was developed about 1940. The machine, manned only by a driver, picks up the dried hay from the windrows in the field and rams it into bales weighing from 85 to 150 pounds each. In this packing the hay takes up only 200 cubic feet of space per ton—less than half the space of the bull-raked piles. A vehicle to pick up the bales and transport them to the barn follows the baler, and there are now devices for fully mechanical handling or packing of the bales in this task. One device is a bale thrower: it picks up the bale (made conveniently small in this case) and like a mortar tosses the package into the cart. A more orderly version of the device, developed within the past five years, picks up the bales (which can be larger in size) and packs them neatly in the wagon instead of throwing them in. At the storage destination the wagon tilts and deposits the load of bales in a tight stack. Some of these machines can deliver 14 tons of baled hay an hour.

The problem of delivering baled hay as loose hay to the feed manger by machine has not yet been solved satisfactorily, but there are some promising approaches. One is a machine that opens the bales, fluffs up the hay and delivers it by conveyor to the feed troughs, but the bales have to be lifted manually onto

the conveyor. Another device is a system that chops the hay into small pieces at the outset. A chopping machine cuts the hay into pieces one to three inches long as it picks up the hay from the windrow in the field; the machine blows the chopped hay into a following vehicle, which transports the hay to the barn and there blows it into bins. The chopped hay is economical in storage space (taking up only about 170 cubic feet per ton), but the system has some unacceptable drawbacks: it produces a great deal of dust; the nutritious leaves of the alfalfa or clover hay are partly lost, and animals do not like the chopped hay as well as they like long hay. The experimenters working on the method hope, however, to solve these problems. One approach under study is a machine that will strip the nutritious leaves from the grass plants in the field, leaving the woody stalks to grow more leaves.

Perhaps the most promising innovation in haymaking is the delivery of the feed in capsule form. Two such devices are already on the market: they produce small, dense cubes of packed hay—two and a half inches or an inch and a quarter in size. A huge machine goes down the windrow in the field picking up the dry hay and squeezing it under tremendous pressure into the little cubes. They have a density equivalent to 25 to 40 pounds per cubic foot, and when packed together they would occupy only 50 to 80 cubic feet of storage space per ton. The cubes flow fairly satisfactorily by gravity, can be shoveled and are quite acceptable to the cattle. The machine in its present form, however, is expensive, and it will work only on hay that has been dried to a low moisture level; consequently so far it seems the method will not be suitable for hay grown in humid regions.

Sugar Beets and Cotton

Sugar is another crop whose harvesting has been largely mechanized. The U.S. consumes about 10 million tons of sugar a year (nearly 100 pounds per capita), and more than 60 percent of this is now produced within the U.S. and its Caribbean island possessions. A little more than half of the production comes from sugarcane, the rest from sugar beets. The harvesting of cane still depends heavily on hand labor, as no satisfactory machine has yet been developed to cut the cane efficiently in the field. Sugar beets, however, are harvested entirely by machine, and the 1965 crop in the U.S. was 21 million tons.

The production of sugar from beets

was established in 1811 by the development of a refining method by the French investigator Benjamin Delessert and by a decree of Napoleon that subsidized the founding of a sugar beet industry. Sugar beets yield a good sugar content (12 percent or more), but their cultivation and harvesting presented formidable labor problems. For one thing, the beets grow too thickly and in clumps, because each seed has several nuclei from which plants grow. In cultivation by hand one worker would go down the row of young plants with a short-handled hoe thinning down the clumps to one every 10 or 12 inches, and another worker would follow, on hands and knees, to reduce each clump to a single plant.

By the 1920's American beet farmers were using horse-drawn machines that seeded and cultivated the plants and lifted the grown beets out of the soil at harvest time. Human workers still had to pick the beets, top them (cut off the tops with a knife) and load them into wagons. In the 1930's U.S. agricultural agencies and beet growers, seeking to make beet sugar competitive with sugar produced by foreign cane growers, undertook an intensive program to mechanize the entire process of beet harvesting. The most difficult problem proved to be topping: mechanical toppers, failing to distinguish between large and small beets, cut the tops at the wrong levels. Eventually a University of California at Davis agricultural engineer, John B. Powers, solved the problem by devising a gauging mechanism that adjusted the topper to make the cut correctly through the crown of each plant. After World War II two highly effective beet harvesters were developed. One of these machines tops the plant, then lifts the beets out of the ground by means of a pair of rotating disks, cleans the soil from them and conveys them to a trailer cart. The other type of machine employs a different method: a large, spiked wheel turns into the ground along the row of plants, impales each beet on a spike and brings it up to a mechanism that tops the beet, pulls it off the spike and drops it in a carrier.

These machines, and the development of devices (genetic and mechanical) that reduce each sugar beet seed to a single nucleus, have succeeded in fully mechanizing the harvesting of beets. In 1944 only 7 percent of the sugar beet crop in the U.S. was harvested by machine; within 13 years the mechanical takeover was complete and machines harvested the entire crop.

One of the last of the major U.S. crops to yield to mechanization was cotton. The long search for a workable mechanical cotton picker is a familiar story. Over the course of the past century more than 1,800 patents were issued in the U.S. to hopeful inventors who proposed a great variety of schemes for removing the cotton fiber from the boll: sucking or blowing it off pneumatically, threshing it off, pulling it off by static electricity, extracting it chemically, combing it off, brushing it off or picking it off by means of mechanical fingers. In 1885 a Scottish engineer named Angus Campbell found a road that was to lead to success. He devised a machine that swept rotating spindles through the cotton and twisted the loose fibers off the plant. The spindle idea, promptly recognized as sound in principle, attracted many inventors, but it proved to be difficult to translate into an effi-

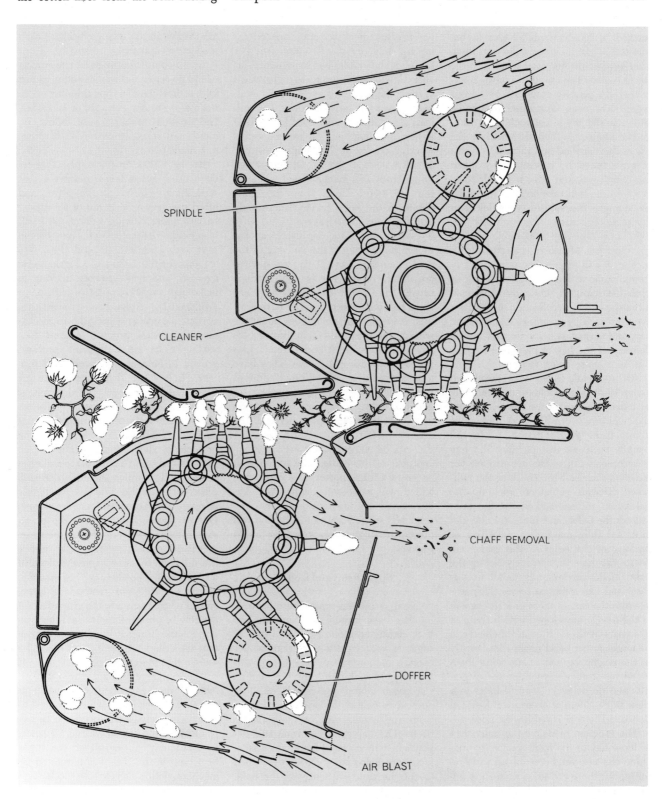

SPINDLE

CLEANER

CHAFF REMOVAL

DOFFER

AIR BLAST

KEY SECTION of the cotton picker includes spindles, which remove cotton fiber from the bolls, and doffers, which remove the fibers from the spindles. Blowers take the cotton from the doffers to the conveyor system. After a spindle has been doffed it goes through a cleaner that wipes off gum and plant juices with moistened pads. Spinning dries the spindle before it enters the next boll.

cient machine. The International Harvester Company spent 40 years and more than $5 million in the effort to develop a satisfactory spindle cotton picker. Success was finally achieved by the Rust brothers, John and Mack, who in 1927 produced a machine that proved its worth in field tests. The chief problem had been that, although barbed or serrated spindles were effective in pulling off the cotton, there still remained the difficulty of removing the fibers from the spindles. The Rusts solved this problem by using a smooth spindle and moistening it so that the fibers would stick and wind around it. The Rust picker, aided by the breeding of plants specifically designed to facilitate its use, has made cotton picking a machine operation and has radically reduced farm employment in the South (as the Rusts feared it would).

Fruits and Vegetables

The crops I have discussed so far—grain, hay, sugar beets, cotton—can withstand rough handling. Far more difficult is the problem of mechanizing the harvesting of the "perishable" crops—fruits and watery vegetables—that must be handled gently. These products, whose acceptability depends on their coming to the table ripe, fresh and undamaged, are incompatible with machine picking not only because of their structural delicacy but also because they characteristically ripen on the plant not all at one time but over the course of the plant's bearing period.

Let us consider first the progress that has been made in the machine harvesting of deciduous-tree fruits, such as prunes, apples, cherries, peaches and apricots. The basic approach being pursued is to develop machines that will shake the ripe fruit off the trees and catch it without damaging the tree or the fruit. Investigators from various points of view have been working on this problem for many years. They have accumulated a great deal of information about the properties of fruit trees and fruits—probably more than is known about any other class of crop plants. These studies have gone into details such as the amount of energy required to shake a given fruit off the branches under various conditions of fruit maturity and air temperature, the relation between the movements of the shaken branch and the kinetic energy imparted to the fruit itself, the power needed to shake tree limbs of various diameters effectively, the relation of the vibration frequency and amplitude to this power

EXPERIMENTAL CITRUS HARVESTER shakes the limbs of the fruit trees with blasts of air. The fruit drops to a catching frame attached to the harvester. The machine, which was developed by the Florida Manufacturing Corporation, clears a tree in two to three minutes.

requirement. The requirement is smallest when the frequency is comparatively low and the amplitude is comparatively large. It has been learned, for example, that only a third as much power is needed to shake most of the prunes off a tree with vibrations at the frequency of 400 cycles per minute and two inches in amplitude as with vibrations of one-inch amplitude at 1,100 cycles per minute.

Leaders in the research on mechanical fruit harvesting have been Robert B. Fridley of the University of California at Davis and P. A. Adrian of the U.S. Department of Agriculture. A workable system has now been developed. Its main elements are an efficient mechanical shaker that is attached to the tree trunk or a limb, a "catching frame" for receiving the falling fruit and a pruning system that minimizes the chances of the fruit hitting branches as it falls and does not appreciably reduce the tree's production. The shaker is of the "inertia" type: it shakes the tree without seriously shaking the tractor on which

the shaker is mounted. In effect it works something like unbalanced front wheels on an automobile, which can impart considerable shake to the steering mechanism. Power from a hydraulic motor is applied to an assembly of unbalanced weights that rotates in eccentric fashion and vibrates the tree by way of a housing attached to the tree. The falling fruit drops into a well-padded catcher placed below the lowest branches; the fall is so cushioned that the fruit does not bounce and is not damaged. The entire operation is housed in a self-propelled vehicle that automatically collects the fruit in bins.

Mechanical fruit picking is still only in its infancy. Within the past five years a more dramatic advance in machine harvesting has taken place in another field: the picking of tomatoes. In California, where 125,000 acres were devoted to tomato growing, the growers used to have to recruit 40,000 workers at the season's peak to harvest the crop by hand. Two investigators at the California Agricultural Experiment

CHERRY PICKER has two mechanical arms that grip the limbs of the cherry tree and shake them gently. The cherries fall onto the fabric collector at the base of the trunk and roll into the conveyor system. The cherry harvester is now in commercial production.

HAY-CUBER was designed to reduce the bulk of hay. It picks hay up from a windrow and compresses it into cubes of a size that cattle can eat readily. The machine was designed for alfalfa and other legumes rather than for hay crops consisting predominantly of grass; the legumes contain natural adhesives that, with water added in the machine, provide the bond that holds the cube together.

Station in Davis—an agricultural engineer and a plant biologist—undertook to develop a system for mechanizing tomato picking. They attacked the problem on two fronts. Gordie C. Hanna, the biologist, would breed a tomato plant designed for machine handling. The plant would bear tomatoes that were of uniform size and all ripened at the same time, that could easily be detached from the vine but would not drop off prematurely, that had a skin tough enough to withstand mechanical handling, that would store well and that would be pleasing to the consumer in flavor and other qualities. Coby Lorenzen, Jr., the agricultural engineer, meanwhile would work on the design of a machine that would harvest this tomato rapidly, efficiently and at reasonable cost.

After 10 years of study, experiments and development the two men achieved their objective in 1962. The plant was ready and so was the machine: a harvester that cut off the plant at ground level, lifted it, shook off the tomatoes and deposited them in a bin in which they were hauled to the processing plant [*see illustration on pages 170 and 171*]. The tomato "combine" won remarkably rapid acceptance. Within three years this machine was harvesting 24 percent of the California tomato crop; last year 800 of the machines were available and they picked almost 80 percent of the crop; this fall, with at least four major manufacturers now producing machines, a large percentage of the tomatoes grown in the U.S. for processing will be harvested by machine.

Machine Economics

How far can the mechanization of harvesting be carried? Fundamentally that will depend a great deal on the comparative costs of performing the task by hand and by machine. In most cases mechanical harvesting is less expensive, even allowing for the capital cost of the machine. For example, the Tennessee Agricultural Experiment Station recently studied the costs of harvesting snap beans by hand and by machine. Hand harvesting on farms of substantial size (200 acres) with an average yield (two tons of beans per acre) averaged $103 per acre; the cost of machine harvesting under the same conditions averaged only about $25 per acre, taking into account the cost of the machine for a five-year lifetime. The size of the farm, of course, makes a difference: harvesting the beans from a

small farm of only 25 acres yielding two tons per acre cost about the same for machine as for hand picking ($95); on the other hand, on a farm of 500 acres the cost by machine was only $11.32 per acre. In any case, aside from the money, there is a tremendous saving of labor; whereas the average bean picker picked 1.06 bushels per hour by hand, a single operator could pick 95.8 bushels per hour by machine. Moreover, there are a number of side benefits from harvesting by machine, including delivery of cleaner fruits and vegetables to the consumer.

It seems likely that, economics permitting, we shall see further extension and increasing sophistication of harvesting by machine in the U.S. Under study and test are various ideas for selective machines that will be able to discriminate between ripe and unripe fruits or vegetables (picking only the red tomatoes, for example, and leaving the green). Among these projects are machines for picking citrus fruits and strawberries. Harvesting machines for lettuce, asparagus and grapes are now entering the stage of development for manufacture by equipment companies.

Some Limitations

Yet it is obvious that, in a country with a highly developed state of agriculture, harvesting by machine must eventually reach a level of diminishing returns. There is a limit, after all, to the amount of food a population will consume, and consequently a limit to the economic incentive for employing machines to increase production. For example, a lettuce-harvesting machine has been designed that can pick four rows at a time, and I am told that 600 of these machines could harvest all the lettuce now grown in the world. Obviously it would not pay a large manufacturer to go to the expense of developing and tooling up for a machine with so limited a prospective sale. Usually in such a situation a publicly supported university does the research and a small manufacturing firm carries out the development and production.

It is in the developing countries of the world, where harvesting is still done mainly by hand, that the development and use of harvesting machines now has its most important future. For those countries, 90 percent of whose population is occupied in farming, harvesting machinery would provide a prime means of releasing people from the soil and thereby making possible the industrialization of the countries.

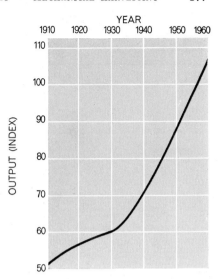

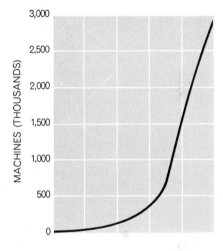

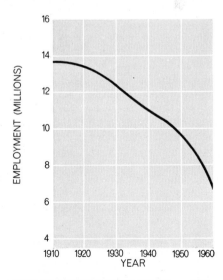

EFFECT OF MECHANIZATION on U.S. agriculture is indicated by a comparison of farm production, the extent of mechanization and the number of people employed in farm work. The charts cover the 50 years in which harvesting by machine developed to its present major role in agriculture.

20

WINE

MAYNARD A. AMERINE
August 1964

Wine is a chemical symphony composed of ethyl alcohol, several other alcohols, sugars, other carbohydrates, polyphenols, aldehydes, ketones, enzymes, pigments, at least half a dozen vitamins, 15 to 20 minerals, more than 22 organic acids and other grace notes that have not yet been identified. The number of possible permutations and combinations of these ingredients is enormous, and so, of course, are the varieties and qualities of wines. Considering the complexity of the subject, it is not surprising that perhaps more nonsense has been written about the making, uses and appreciation of wine than about any other product of man or nature.

Nevertheless, it can be said that in the 20th century wine making has become a reasonably well-understood art. The chemical processes involved are now sufficiently known so that the production of a sound wine is no longer an accident (although the production of a great wine may still be). For this we are indebted primarily to Louis Pasteur, who founded the modern technology of wine making along with several branches of chemistry, microbiology and medicine. Pasteur put the making of wine (and of beer as well) on a rational basis by explaining fermentation, which for thousands of years had been an unsolved mystery.

It seems likely that man's discovery of wine came later than that of beer (a fermentation product of grain) or of mead (a fermentation product of honey), because grapes grow only in certain climates and environments. By Neolithic times, however, the peoples of the Middle East were well acquainted with the fermented juice of the grape, and one of the oldest inscriptions in Egypt (on the tomb of Ptahhotep, who lived about 2500 B.C.) depicts the making of wine. The "blood of the grape" attracted ancient man not only as a beverage but also as a medicine and a symbolic offering to the gods.

The grape is its own wine maker. One simply pressed out the juice, let it stand, and its sugars turned into alcohol. Not until the 19th century did chemists begin to unravel the nature of this process. In 1810 Joseph Louis Gay-Lussac made the first crucial contribution toward solution of the mystery by discovering the general chemical formula of the breakdown of sugar into alcohol and carbon dioxide: $C_6H_{12}O_6 \rightarrow 2\ C_2H_5OH + 2\ CO_2$. Plainly this change did not take place spontaneously. What caused the sugar to break down? Gay-Lussac conjectured from his experiments that the process was stimulated somehow by oxygen. The German chemist Justus von Liebig put forward another hypothesis: that the fermentation arose from the "vibrations" of a decomposing "albuminoid" substance. Liebig's authority was so powerful that his view was not seriously challenged until the young Pasteur embarked on his studies of fermentation in the 1850's.

The Role of Yeast

"How account," Pasteur asked, "for the working of the vintage in the vat?" With his gift for designing experiments that went to the heart of the matter, Pasteur soon demonstrated that the working was produced by the microscopic organisms known as yeast. "Fermentation," he concluded, "is correlative with life." He showed that an infusion of yeast would convert even a simple sugar solution into alcohol, and he went on to identify some of the factors, such as acidity or alkalinity, that controlled the metabolic activities of the yeast organisms and thus determined the properties of a wine. Pasteur announced his main discoveries in two historic papers: *Mémoire sur la fermentation appelée lactique* (published in 1857) and *Études sur le vin* (1866).

How does the grape acquire its yeast? As every gardener knows, the skin of growing grapes is covered with a delicate natural bloom. It consists of a waxy film that collects cells of molds and wild yeasts, which are deposited on the grape by agencies such as the wind and insects. The skin of a single grape may bear as many as 10 million yeast cells. Of these, 100,000 or more are cells of the varieties called wine yeasts, of which the principal one is *Saccharomyces cerevisiae* var. *ellipsoideus*. It is the enzymes of the wine yeasts that are responsible for the fermentation of the grape's sugars to alcohol and for the creation of the numerous by-products that partially account for the flavor and other properties of the wine. The nature of the activity of the yeasts importantly affects the wine's quality, consequently it is one of the factors modern wineries are careful to control. In some old European vineyards the grapes and yeasts seem to have established over the centuries a natural harmony that brings out the grapes' best qualities in the wine. But most wineries, even in Europe, now improve on nature by adding pure cultures of desirable yeasts and using chemicals to sup-

CALIFORNIA VINEYARDS cover the hills surrounding the Napa Valley. Varieties of *Vitis vinifera*, the species of grape from which most European wines are made, adapt readily to the warm California environment.

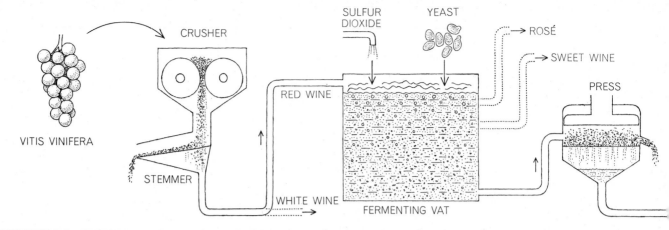

WESTERN U.S. METHOD of producing red wine duplicates the European process. The grapes are crushed between rollers (*left*), forming an intermediate product known as "must." The must is piped to a fermenting vat where yeasts speed the transformation

press the growth of undesirable yeasts present on the grape skins.

The Effect of Climate

The making of a wine starts long before the grapes reach the winery—indeed, long before the grapes are harvested from the vine. The grape is a complex product of soil, water, sun and temperature. Of these factors, the most significant single one is temperature. Grapes will grow only within the belts of the Northern and Southern hemispheres where the average annual temperature is between 50 and 68 degrees Fahrenheit [*see lower illustration on page 187*]. Even in these regions the European grape *Vitis vinifera* does not survive in areas marked by certain unfavorable conditions: summer temperatures not warm enough to ripen the fruit (as in most of Britain), high summer humidity that excessively exposes it to mold diseases or insect predators (as in the southeastern U.S.) or late

spring frosts (as in the northwestern U.S.).

The ideal climate for wine grapes is one that is warm but not too warm, cool but not too cool. On the one hand, a long growing season is required so that the grapes will produce a high content of sugar for conversion into alcohol. On the other hand, comparatively cool temperatures are desirable because they produce grapes with high acidity, an important contributor to the quality of wine, particularly the dry table wines. Both of these climatic conditions are well fulfilled in areas such as the Bordeaux district of France, northern Spain, central and northern Italy, Yugoslavia and northern California—and those areas produce fine red table wines. In areas with cooler or shorter growing seasons, such as Germany, Switzerland, Austria, the eastern U.S. and even the Burgundy district of France, the grapes in some years do not develop enough sugar, and sugar must be added when they are brought to the winery. This

addition cannot, however, replace flavor components that are missing when the grapes have not ripened fully. The variability of the summer climate in Europe is the main reason for the fluctuation in the quality of its wines from year to year and for the emphasis on vintage years.

Although a warm climate (such as that of southern Spain, Sicily, Cyprus and southern California) produces grapes with a high sugar content, they have the handicap of comparatively low acidity. These grapes are suitable for the sweet dessert wines, but they lack the subtle flavors and color of grapes grown in cooler areas. Moreover, they are sometimes overripe when they come to the fermenting vats, with sad effects on quality if one attempts to produce a table wine from them.

The Grape

No less important than the characteristics of the climate are the char-

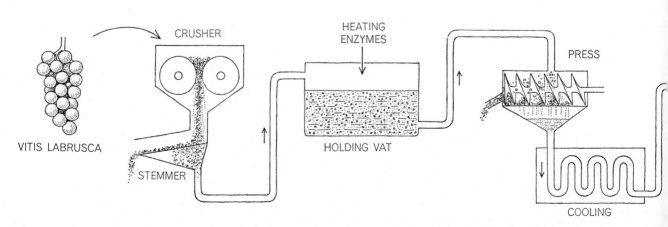

EASTERN U.S. METHOD of producing red wine begins with the crushing (*left*) of *Vitis labrusca* grape, a species low in sugar. Must is piped into a holding vat, where enzymes are added to break down mucilaginous substances in and around the pulp. The desired color

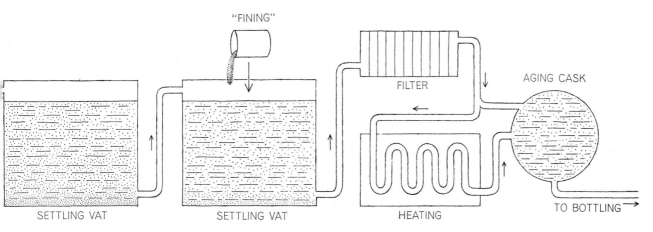

"FINING"

FILTER

AGING CASK

SETTLING VAT   SETTLING VAT   HEATING   TO BOTTLING

of sugars into alcohol, and then to a press where skin and seeds are separated out. The juice proceeds through two settling vats, wherein the "fining" process removes impurities. It is filtered, sometimes heated and cooled, and aged in casks prior to bottling.

acteristics of the grape. One of the benign aspects of the grape plant—which holds much promise for future wines—is its great variability. One species alone, *Vitis vinifera*, has some 5,000 known varieties, and even the less popular species are available in about 2,000 varieties. Grape breeders have also produced many hybrids between the species. The grape varieties differ in color (white, green, pink, red or purple), in the size of the grape clusters, in the texture of the grape (firm and pulpy or soft and liquid), in sugar content, in acidity, in earliness or lateness of ripening and in susceptibility to insects and diseases. With this variability in the material, plant geneticists look forward to breeding new varieties of grapes that will be tailored to specific climates, to the types of wine and to new heights of taste, aroma and bouquet. (As wine experts define the terms, aroma refers to the fragrance of the grape; bouquet, to the fragrance imparted by fermentation and aging.)

Vitis vinifera is by far the preponderant species of wine grape grown in vineyards throughout the world. The plant is believed to have originated near the shores of the Caspian Sea in what is now the southern U.S.S.R. From there early travelers and traders spread it around the Mediterranean, then to northern Europe and eventually explorers transported it to continents overseas. (More than 81 percent of the world's vineyard acreage and wine production are still concentrated, however, in Europe and North Africa, with France the leader.) In the U.S. the *vinifera* species has found a hospitable home in California, and some 100 varieties of this species are cultivated commercially there. *Vinifera* is vulnerable to the diseases and insects that thrive in a hot and humid summer climate; for this reason many vineyards in the eastern U.S., Canada, Brazil and certain areas in Europe cultivate other species, such as *Vitis labrusca* or *Vitis rotundifolia*.

Now let us examine the wine-making process. To follow it in detail we shall consider the typical procedure in a modern California winery.

The Wine-making Process

To begin, let us analyze the raw material. In a mature grape about 10 to 20 percent of the material by weight is accounted for by the skin, stem and seeds, and the remaining 80 to 90 percent is pulp and juice. The pulp and juice, when piped into the fermenting vat, is called "must." Chemically the grape must is mostly water, but between 18 and 25 percent by weight is sugar (the amount varying with the variety and ripeness of the grape). The sugar consists mainly of dextrose (that is, glucose that rotates polarized light to the right) and levulose (or fructose, which rotates polarized light to the left). The grapes from which table wines are made usually contain dextrose and levulose in about equal

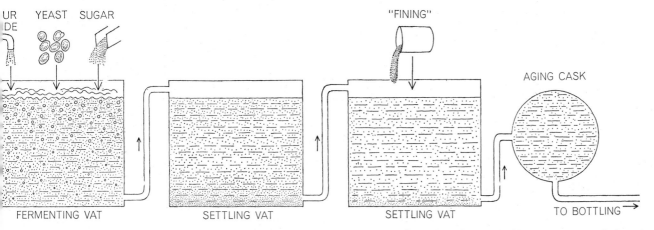

UR
DE   YEAST   SUGAR   "FINING"

AGING CASK

FERMENTING VAT   SETTLING VAT   SETTLING VAT   TO BOTTLING

is attained by heating. Must proceeds to a fermenting vat where sugar as well as yeast and sulfur dioxide are added. Removal of im- purities by fining takes place in settling vats, and the wine is then aged. Some Eastern wines are pasteurized before bottling.

amounts; for sweet wines vintners would prefer grapes with a higher proportion of levulose, because it is nearly twice as sweet as dextrose. In addition to these two principal sugars, grapes also contain small quantities of other carbohydrates, such as sucrose, pentoses and pentosans.

Acids make up between .3 and 1.5 percent of the grape must by weight. The two principal acids again are op-

tically opposite forms: dextrorotatory tartaric acid and levorotatory malic acid. There are also small amounts of other acids: citric, oxalic, glucuronic, gluconic and phosphoric. The pH, or active acidity, of mature *Vitis vinifera* grapes in California runs between 3.1 and 3.9.

Among the many other substances that have been identified in analyses of grape must are 20 amino acids (found

in the free state as well as in proteins), 13 anthocyanins (the pigments of many colored flowers), other pigments, tannins, odoriferous compounds and the various vitamins, enzymes, minerals and other ingredients already mentioned. Obviously many of these substances contribute to the making of wine by providing nutrient for the fermenting yeasts. The contributions of individual ingredients to the quality of wine, however, are imperfectly understood; presumably no one will ever be able to write a formula for a perfect wine, because personal taste is an indispensable part of the equation.

The fermentation process is enormously complicated [*see illustration on page 185*]. The breakdown of glucose alone involves no fewer than 22 enzymes, six or more coenzymes and magnesium and potassium ions. A number of other sequences, including the well-known Krebs cycle, participate in the process. From these many reactions emerges a mixed collection of other products in addition to alcohol, among them acetaldehyde, glycerol, succinic acid, esters and other aromatic compounds. The problem of the wine maker is to control the production and accumulation of this multitude of diverse products. In a modern winery this is done by various chemical and physical means.

Grapes have to be taken from the vine to the winery as quickly and carefully as possible in order to minimize their loss of water and sugar after picking and to prevent spoilage. At the winery they are immediately put in a crusher, which crushes the skins, freeing the pulp and juice (without breaking the seeds), and removes the stems. In the case of a white wine the juice is pressed out at this point and sent alone to the fermenting vat. For the making of red wine the entire contents of the crusher—juice, pulp, skins and seeds—go into the fermentation process. The red wine will take its color from the pigment in the skins and its strong flavor and astringency from tannins and other substances in the skin and seeds. (The rosé wines that have become more popular in recent years are made by starting the fermentation with the skin and pulp present, then, after about 24 hours, pressing out the juice and letting it complete the process alone.)

Wine in the Vat

In the fermenting vat (in California it is usually constructed either of red-

DELICATE BLOOM of grape skin consists of a waxy film that collects molds and yeasts. A single grape may accumulate 100,000 yeast cells with enzymes responsible for fermentation. Where the waxy film has been brushed off several grapes (*center*) a bright shine results.

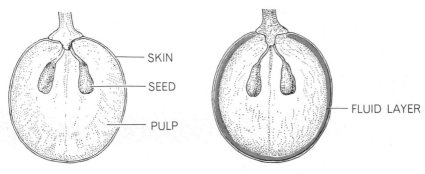

CABERNET FRANC, shown here in cross section, is an Old World grape of relatively low acidity that flourishes in California.

CONCORD GRAPE of the northeastern U.S. has a mucilaginous layer separating skin and pulp, hence its "slip skin" classification.

wood or of concrete) the first step is treatment of the must with liquefied sulfur dioxide or a sulfurous acid or salt. The main function of this chemical is to inhibit the growth of the wild yeasts on the grape skins. They are replaced by the addition of pure cultures of yeasts that will produce a better wine. Besides suppressing the deleterious yeasts the sulfur dioxide reduces oxidation (which may have a baneful effect, particularly on the quality of white wines) and also helps to acidify and clarify the wine. Sulfur dioxide is a dangerous tool—an excess of it will ruin the wine—but all in all its use has been a major 20th-century benefit to wine making, contributing in various ways to better regulation of the fermentation, a higher yield of alcohol from the sugar and a more flavorful product. When sulfur dioxide is used, the natural yeast flora from the grape are largely inhibited and an actively fermenting culture of yeast must be added.

Another recent innovation is careful control of temperature in the fermenting vat. Cooling systems are used to carry off the heat produced by fermentation so that the temperature in the vat is kept below 85 degrees F. (for red table wines) or below 60 degrees (for white wines). The slow fermentation at low temperatures produces more esters and other aromatic compounds, a higher yield of alcohol and a wine that is easier to clear and that is less susceptible to bacterial infection. In the opinion of most enologists it results in a better bouquet and aroma. The duration of the fermentation in a modern winery varies from a few days to a few weeks, depending on the temperature, the type of yeast used, the sugar content of the grapes and the kind of wine to be produced.

All wine is divided into two general classes, defined by the alcohol content. The table wines (also called "dinner," "dry" or "light" wines) contain not more than 14 percent of alcohol by volume. The "aperitif" and "dessert" wines (sherry, port, muscatel and the like) have a higher content, usually about 20 percent. They are given this high alcohol content by the addition of brandy distilled from wine. Added during the fermentation, the brandy stops the action of the yeast, and the wine is then left with some of its sugar unconverted to alcohol. In the making of muscatel, for example, the brandy is added and the fermentation halted when the juice still contains 10 to 15 percent of grape

RECEIVING TANKS at left transfer must from a crusher on the floor above to the holding vats at right, enabling the winery to process the harvest of two types of grape. This photograph and the one below were made at the Taylor Winery in Hammondsport, N.Y.

PRESSES receive crushed grapes from holding vats on the floor above through pipes (top). The black rubber bag visible inside the press in the foreground will be inflated with air, forcing residual skins and seeds to cling to the sides of the stainless steel cylinder.

"STORMY" STAGE of fermentation is under way in this vat at the United Vintners winery in Asti, Calif. Approximately 36 hours after yeast is added the temperature of the juice rises as high as 85 degrees and carbon dioxide bubbles violently to the surface.

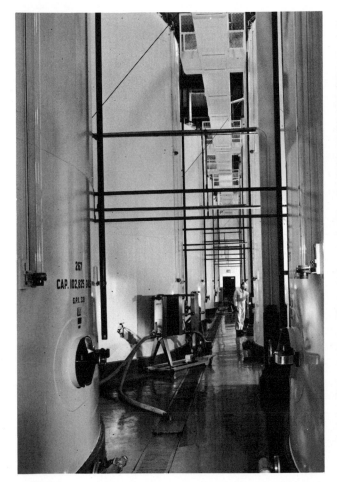

FERMENTING TANKS shown here can hold 100,000 gallons. They are made of concrete with a glass lining. The thin pipes between the tanks exchange heat to maintain a uniform temperature.

REDWOOD VATS house the aging wine and facilitate mellowing by admitting oxygen through redwood planks. These vats, photographed at the Taylor Winery, have a capacity of 63,000 gallons.

185

sugars by weight; the result is a very sweet wine. For port the fermentation is stopped a little later (at a sugar level of 9 to 14 percent) and for a dry sherry it may be allowed to proceed until the sugar content is 2.5 percent or less.

For the sake of simplicity let us proceed with the more typical case of a red table wine. When part of its sugar has been converted to alcohol and adequate color has been extracted from the skins, the partially fermented juice is separated from the pulp. At this time the skins are mainly free and floating on top; the liquid is drained off as "free run" and is considered to make the best wine. The rest of the juice is pressed out of the pulp by the familiar wine press (which most people confuse with the machine used to crush the grapes before they are put in the fermenting vat). The press used in many modern wineries still looks much as it has always looked—a hardwood container with a plunger—but nowadays a hydraulic ram replaces the old screw contrivance turned by hand. Recently developed cylindrical presses and roller presses are also in use.

The juice now proceeds to the completion of its fermentation and to the clearing and aging stages. Not to be guilty of omitting entirely from this account the important category of sparkling wines, I shall merely mention here that they are made from dry table wines by means of a secondary fermentation in a closed container, involving the addition of a calculated amount of sugar and 1 percent of a pure yeast culture. This fermentation produces the extra carbon dioxide—amounting to an internal pressure of four or five atmospheres in the bottle—that accounts for the fizz of champagne.

For clarification of the wine the fermented juice goes to settling vats. There the suspended yeast cells, cream of tartar and small particles of skin and pulp rapidly settle out of the liquid. Various chemical processes and a form of fermentation still continue, however. Wine, it has been said, is a living thing, and indeed in a sense it does go on growing and maturing—in the settling vats and later in its aging periods in cask and bottle. In the vats the yeast cells, as they break down, particularly in a wine juice of high acidity, stimulate the growth of *Lactobacillus* bacteria. Enzymes from these bacteria decarboxylate the wine's malic acid (that is, remove COOH groups) and convert it to lactic acid. This malo-lactic "fermentation," replacing a strong acid with

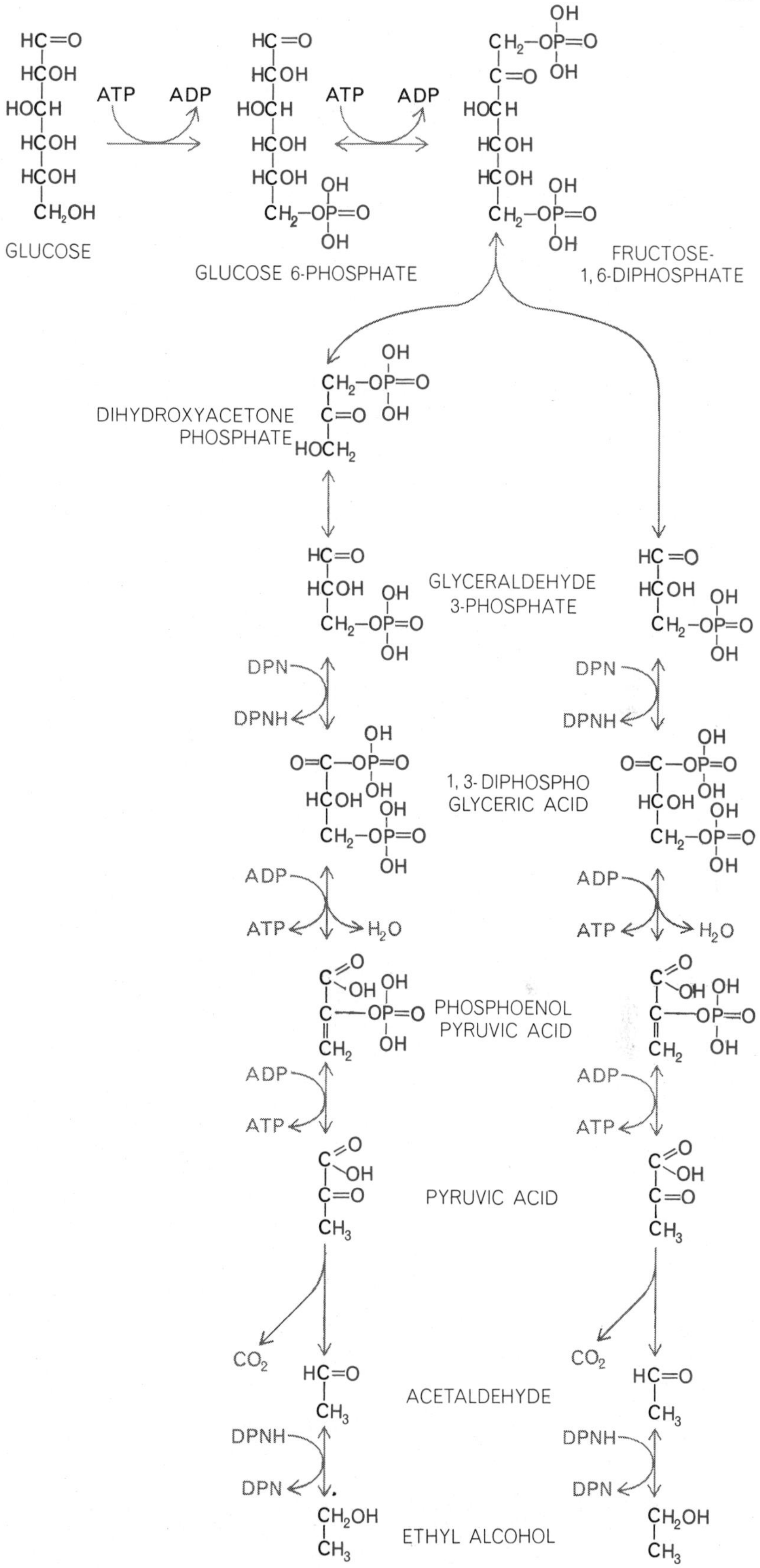

FERMENTATION entails the breakdown of the six-carbon sugar, glucose (*top left*) and the consequent production of alcohol. The splitting of the carbon backbone occurs when the intermediate product, fructose (*top right*), gives way to two molecules of glyceraldehyde phosphate. The major intermediate products are shown from top to bottom. The enzymes and coenzymes needed to power the process are represented by ATP and ADP, and DPN and DPNH. The reversible steps in the process are indicated by two-way arrows.

FILTERING UNIT shown at right in this photograph removes sediment from the wine in settling tanks (*left center*). Below the filter is a trough into which residue is dumped.

NEW BOTTLES containing domestic U.S. champagne are stacked on a "riddling" shelf of ash, counterpart of the French A-frame. Sediment accumulates in the neck and can be discarded by briefly uncorking the bottle. Both photographs were made at the Taylor Wineries.

a weak one, mellows the high-acid wine. Without it the high-quality wines of northern Europe could not be made.

As soon as possible the clearing wine is racked, or drawn off, from the settling lees to prevent excessive working and protect its flavor. The racking is repeated again and again, leaving behind lees at each step. During these off-pourings the wine also sheds the carbon dioxide with which it was charged in the fermentation process and absorbs oxygen from the air, which will help in its aging. To assist the clearing of the wine when racking alone does not suffice, wineries commonly inject "fining" substances (such as bentonite clay, gelatin, isinglass or egg white) that clump and precipitate the tiny particles in the wine; they may also apply pressure filtration, heat or chilling as aids to clearing.

Wine in Cask and Bottle

The aging of the wine begins in an oak cask. It is an extremely complex process of oxidation, reduction and esterification. The new wine gradually loses its yeasty flavor and harshness, declines in acidity and acquires a complex, delicate bouquet. As its pigments and tannin are oxidized, red wine turns a tawny color and white wine develops an amber hue. The amount of oxidation of its ingredients, by means of oxygen absorbed through the pores of the cask, is crucial to the eventual quality of the wine: the length of time it is left in the cask may make the difference between allowing a great wine to attain its potential and turning it into an ordinary one. If wine is bottled too soon, it may spoil or mature too slowly; if it is bottled too late, it will be vapid and off-color. The decision as to when to bottle is one of the most important in the wine maker's art. In present practice fine red table wines are kept in wooden cooperage for at least two years; white wines, from a few months to two years. Lesser-quality wines are stored in redwood, concrete or lined iron tanks.

After bottling, wine does not cease to "work." Aging in the bottle serves to eliminate the aerated odor the wine acquired at the time of bottling, reduce the wine's content of free sulfur dioxide and improve its bouquet. It is a mistake, however, to suppose the older the wine, the better, or that a bottle encrusted with the grime of many years is likely to contain a wine of rare distinction. The contents may, in fact, have become worthless long ago. Only a few very fine red wines benefit from prolonged aging. As a general rule, for a good red wine five to 10 years in the bottle is long enough, and a white wine will have reached its peak after two to five years. Wines of lesser quality require less time.

To summarize, the modern technology of wine making began with Pasteur's discovery that fermentation was produced by yeasts and that the process was far more complicated, with many more by-products, than Gay-Lussac's simple formula for the conversion of sugar to alcohol had suggested. The major modern developments have been the use of selected pure yeasts, the breeding and cultivation of supe-

rior varieties of grapes, the control of fermentation by certain chemicals and physical conditions (such as sulfur dioxide and cooling) and a gradual accumulation of more exact knowledge about the chemistry of the fermentation and aging processes. For all these advances, a truly great wine is still more or less a happy accident arising from time to time out of a particularly fortunate blend of the weather, the grape and the vintner's intuitive art. Much of the guesswork has been eliminated, however, from commercial wine making, and the quality of wines is a great deal more uniform than it used to be.

Even a brief account of wine making, which can touch only on the highlights,

cannot pass over the fascinating subject of the consumption of the product. The wine maker and the wine consumer are themselves partners in a peculiarly intimate symbiosis; indeed, historically they used to be one and the same person! Modern enology sheds interesting light on some of the folklore of wine drinking.

The matching of wines to food (red wine with red meat, white wine with fish) cannot be defended, objectively speaking, as much more than a superstition. It is true that red wine shares with meat a complexity of taste and texture, and that the high acidity of white wine may add spice to the blandness of fresh fish and, in earlier times

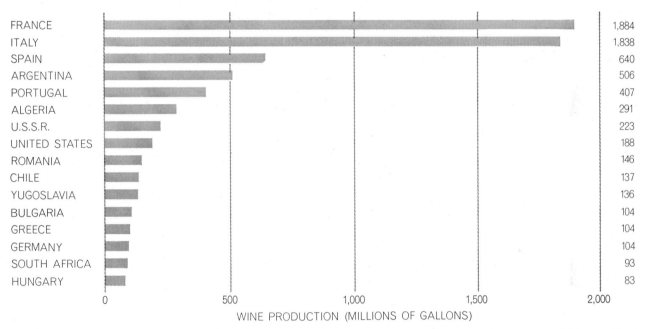

LEADING PRODUCERS of wine are listed according to 1962 output in millions of gallons. The figures for Algeria, the U.S.S.R. and Chile are estimates. No statistics are available for China. France and Italy together produced about half of the world supply.

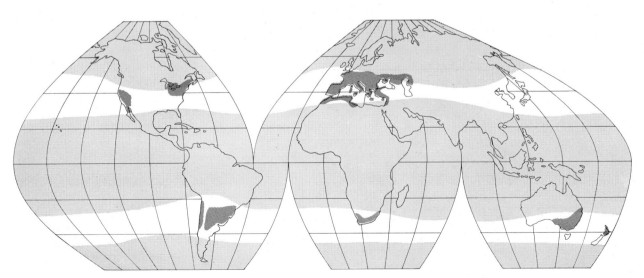

WINE-GROWING REGIONS of the world lie within belts where average annual temperature is between 50 and 68 degrees Fahrenheit. The hot summer of the southwestern U.S. and the humidity in the Southeast preclude the cultivation of *Vitis vinifera* grapes.

of nonrefrigeration, may have helped to mask the odor and taste of decaying fish. Most likely, however, the traditional ideas about food-wine pairing grew originally out of the simple geographical fact that a particular type of wine happened to be grown in a region that favored a particular food; that is, the coupling developed from agricultural rather than epicurean considerations.

The use of wine as medicine is another and much more interesting story. The medical historian Salvatore P. Lucia, of the University of California Medical School in San Francisco, asserts in his *A History of Wine as Therapy* that it is "the oldest of medicines." Salves made with wine were used in Sumer as early as the third millennium B.C., according to a clay tablet found in the ruins of Nippur. Virtually every culture has employed wine for medicinal purposes, either directly or as a solvent. It used to be listed in the U.S. *Pharmacopeia,* but it was dropped during prohibition (which all but killed the appreciation of wine in the U.S.) and has not been reinstated since. Many physicians, however, have resumed prescribing it for various ailments.

Wine is considered a specific for certain disorders because its alcohol is absorbed from the digestive tract into the bloodstream slowly (as opposed to the rapid absorption of pure ethyl alcohol) and because some of its ingredients may be metabolically helpful to the body. The physicians who believe in its therapeutic powers recommend it variously as an analgesic for minor pain, as a tranquilizer or sedative, as a vasodilator for hypertensive patients, as a diuretic, as a nutritional supplement for diabetics and as an aid to the absorption of fat by the intestines after an operation for ulcers or stomach cancer. The noted medical teacher William Dock, professor of medicine at the Downstate Medical Center of the State University of New York, has remarked: "It is useful to think what would happen if alcohol should be discovered all over again.... The sales for all other sedatives and tranquilizers would go down; there would be four-page spreads with color in all the medical journals...and the stock of the patent licensees would go right through the ceiling on Wall Street. The lucky discoverers would get every possible honor, as did the men who discovered insulin."

V

FOOD NEEDS AND POTENTIALS

V

FOOD NEEDS AND POTENTIALS

INTRODUCTION

During the past several decades the Western world has witnessed a virtually explosive increase in its capacity to produce food and in its food-production potentials. More recently the "green revolution" has been spreading to many of the less-developed nations, where new biological and chemical innovations have dramatically increased the yield potentials of rice, wheat, and maize. In spite of these advances the world as a whole remains ill-fed. Nutritional deficiencies are not confined to the poor countries. Malnutrition is found among children in Mississippi as well as in Mysore. Thus the efficient distribution and utilization of the food that is now being produced pose a challenge not only to food science and technology but to the social sciences as well.

The efforts of most agricultural and food scientists are directed toward expanding food-production potentials and increasing the efficiency with which food upplies are utilized. The ultimate success or failure of their efforts will depend upon the long-run relationship between population and food supply on the planet Earth. One way to gain an insight into the biological problem of world food supply is to take an accountant's point of view: that is, "balance the budget" in terms of the energy that flows through the living community. In "The Ecosphere"—an article that is essentially a broad outline for a calculus of survival—LaMont C. Cole demonstrates how a portion of the enormous energy from the sun is transformed on the earth's surface by energy accumulators (green plants) and is then dissipated, some by growth and respiration, some with the flow through a series of energy "consumers," as from plant-eating animals to carnivore to parasites to microorganisms that decompose the waste. The amount of food energy available to any organism thus depends on the organism's position in the system.

The problem of feeding the human population adequately is complicated not just by man's proclivity to reproduce but also by his eating habits—"Man eats what he likes and likes what he eats." In "The Human Population," Edward S. Deevey, Jr., stresses that human numbers are determined not only by food supply but by social forces and technology. Mankind has experienced three surges of population growth, each of which followed a revolutionary cultural change. The first was brought on by the invention of toolmaking, the second by the development of agriculture, and the third by the scientific-industrial revolution. The earth can still support substantial increases in human population, but not without drastic changes in both the food habits and the social behavior of man. Man's appetite for delicacies at the end of the food chain (steak and lobster, for example) and his seeming unconcern about overpopulating the environment are, in the long run, biologically incompatible.

Traditional peasant societies are characterized by a slowly changing equilibrium between food supplies and population. Sidney W. Mintz, in his article "Peasant Markets," describes the organization of present-day peasant markets—the central economic institution that links peasant producers to domestic and international markets. As long as agricultural production is relatively static, or urbanization proceeds slowly, traditional peasant markets perform the complex marketing functions fairly efficiently.

When technology becomes available to permit rapid expansion of production, however, or when the pace of urban-industrial development quickens, more efficient marketing and processing institutions are required to coordinate supply with demand. The essential characteristics of a modern food industry, capable of responding to the nutritional needs of both rich and poor societies, are outlined in Nevin S. Scrimshaw's article, "Food." Efficiency in food production is one essential element, another is the capacity to convert indigenous plant and animal products into nutritious and pleasing foods, and a third is the capacity of agricultural and food scientists to make use of new sources of food supply. In "Orthodox and Unorthodox Methods of Meeting World Food Needs," N. W. Pirie presents the case for intensive exploration of radically new means of meeting world food needs by using both natural and synthetic materials. We can expect that, to meet future demands, an ever-increasing part of food production will be based on raw materials not heretofore used for food.

A combination of rapid population growth and rising per capita demand for more food and food of higher quality, will place unprecedented demands on the world's agricultural producers and food industries in the next few decades. Realization of the optimal food-production potential will require the evolution of a world-wide agricultural research sytem that will encompass each crop variety and each animal species in each ecological region. This agricultural research system will have to be closely linked to a food-research system designed to explore the food-production potential of plants and animals now considered unpalatable and of synthetic food sources.

The rate at which it will be technically and economically feasible for man to exploit successive increments in the food-production potential of the ecosphere will depend on the rate at which he is able to reduce the cost of energy and on the rate at which human populations continue to increase. If energy-conversion systems continue to become more efficient and population growth rates can be sharply reduced, it may be possible to create a new, dynamic equilibrium in the relationship between man and his environment that will permit science to concentrate more of its attention on the realization of the full potential of mankind.

21

THE ECOSPHERE

LAMONT C. COLE
April 1958

Probably I should apologize for using a coined word like "ecosphere," but it seems nicely to describe just what I want to discuss. It is intended to combine two concepts: the "biosphere" and the "ecosystem."

The great 19th-century French naturalist Jean Lamarck first conceived the idea of the biosphere as the collective totality of living creatures on the earth, and the concept has been taken up and developed in recent years by the Russian geochemist V. I. Vernadsky. The word "ecosystem" means a self-sustaining community of organisms—plants as well as animals—taken together with its inorganic environment.

Now all these are interdependent. Animal life could not exist without plants nor plants without animals, which supply them with carbon dioxide. Even the composition of the inorganic environment depends upon the cyclic activity of life. Photosynthesis by the earth's plants would remove all of the carbon dioxide from the atmosphere within a year or so if it were not returned by fires and by the respiration of animals and other consumers of plants. Similarly nitrogen-fixing organisms would exhaust all of the nitrogen in the air in less than a million years. And so on. The conclusion is that a self-sustaining community must contain not just plants, animals and nitrogen-fixers but also decomposers which can free the chemicals bound in proto-

THE AMAZON, one of whose mouths is shown in this aerial photograph, plays an important role in the earth's circulation of water. Together with the Congo it carries more than 10 per cent of the 9,000 cubic miles of water that flow into the sea every year.

plasm. It is very fortunate from our standpoint that some microorganisms have solved the biochemical trick of decomposing chitin, lignins and other inert organic compounds that tie up carbon.

A community must consist of producers or accumulators of energy (green plants), primary consumers (fungi, microorganisms and herbivores), higher-order consumers (carnivorous predators, parasites and scavengers), and decomposers that regenerate the raw materials.

Communities vary, of course, all over the world, and each ecosystem is a composite of the community and the features of the inorganic environment that govern the availability of energy and essential chemicals and the conditions that the community members must tolerate. But the system that I wish to consider here is not a local one but the largest possible ecosystem: namely, the sum total of life on earth together with the global environment and the earth's total resources. This is what I call the ecosphere. My purpose is to reach some conclusions on such questions as how much life the earth can support.

Organisms living on the face of the earth as it floats around in space can receive energy from several sources. Energy from outside comes to us as sunlight and starlight, is reflected to us as moonlight, and is brought to earth by cosmic radiation and meteors. Internally the earth is heated by radioactivity, and it is also gaining heat energy from the tidal friction that is gradually slowing our rotation. On top of this man is tapping enormous amounts of stored energy by burning fossil fuels. But all these secondary sources of energy are infinitesimal compared to our daily sunshine, which accounts for 99.9998 per cent of our total energy income.

This supply of solar energy amounts to 13×10^{23} gram-calories per year, or, if you prefer, it represents a continuous power supply at the rate of 2.5 billion billion horsepower. About one third of the incoming energy is lost at once by being reflected back to space, chiefly by clouds. The rest is absorbed by the atmosphere and the earth itself, to remain here temporarily until it is re-radiated to space as heat. During its residence on earth this energy serves to melt ice, to warm the land and oceans, to evaporate water, to generate winds and waves and currents. In addition to these activities, a ridiculously small proportion—about four hundredths of 1 per cent—of the solar energy goes to feed the metabolism of the biosphere.

Practically all of this energy enters the biosphere by means of photosynthesis. The plants use one sixth of the energy they take up from sunlight for their own metabolism, making the other five sixths available for animals and other consumers. About 5 per cent of this net energy is dissipated by forest and grass fires and by man's burning of plant products as fuel.

When an animal or other consumer eats plant protoplasm, it uses some of the substance for energy to fuel its metabolism and some as raw materials for growth. Some it discharges in broken-down form as metabolic waste products: for example, animals excrete urea, and yeast releases ethyl alcohol. And a large part of the plant material it ingests is simply indigestible and passes through the body unused. Herbivores, whether they are insects, rabbits, geese or cattle, succeed in extracting only about 50 per cent of the calories stored in the plant protoplasm. (The lost calories are, however, extractable by other consumers: flies may feed on the excretions or man himself may burn cattle dung for fuel.)

Of the plant calories consumed by an animal that eats the plant, only 20 to 30 per cent is actually built into protoplasm. Thus, since half of its consumption is lost as waste, the net efficiency of a herbivore in converting plant protoplasm into meat is about 10 to 15 per cent. The secondary consumers—i.e., meat-eaters feeding on the herbivores—do a little better. Because animal protoplasm has a smaller proportion of indigestible matter than plants have, a carnivore can use 70 per cent of the meat for its internal chemistry. But again only 30 per cent at most goes into building tissue. So the maximum efficiency of carnivores in converting one kind of meat into another is 20 per cent.

Some of the consequences of these relationships are of general interest and are fairly well known. For example, 1,000 calories stored up by the algae in Cayuga Lake can be converted into protoplasm amounting to 150 calories by small aquatic animals. In turn, smelt eating these animals produce 30 calories of protoplasm from the 150. If a man then eats the smelt, he can synthesize six calories worth of fat or muscle from the 30; if he waits for the smelt to be eaten by a trout and then eats the trout, the yield shrinks to 1.2 calories. If we were really dependent on the lake for food, we would do well to exterminate the trout and eat the smelt ourselves, or, better yet, to exterminate the smelt and live on planktonburgers. The same principles, of course, apply on land. If man is really determined to support the largest possible populations of his kind, he will have to shorten the food chains leading to himself and, so far as practicable, turn to a vegetarian diet.

The rapid shrinkage of stored energy as it passes from one organism to another serves to make the study of natural communities a trifle more simple for the ecologist than it would otherwise be. It explains why food chains in nature rarely contain more than four or five links. Thus in our Cayuga Lake chain the trout was the third animal link and man the fourth. Chains of the same sort occur in the ocean, with, for example, a tuna or cod as the third link and perhaps a shark or a seal replacing man as the fourth link. Now if we look for the fifth link in the chain we find that it takes something like a killer whale or a polar bear to be able to subsist on seals. As to a sixth link—it would take quite a predator to make its living by devouring killer whales or polar bears.

We could, of course, trace food chains in other directions. Each species has its parasites that extort their cut of the stored energy, and these in turn support other parasites down to the point where there is not enough energy available to support another organism. Also, we should not forget the unused energy contained in the feces and urine of each animal. The organic matter in feces is often the basic resource of a food chain in which the next link may be a dung beetle or the larva of a fly.

I estimate that the maximum amount of protoplasm of all types that can be produced on earth each year amounts to 410 billion tons, of which 290 billion represent plant growth and the other 120 billion all of the consumer organisms. We see, then, that the availability of energy sets a limit to the amount of life on earth—that is, to the size of the biosphere. This energy also keeps the nonliving part of the ecosphere animated, largely through the agency of moving water, which is the single most important chemical substance in the physiology of the ecosphere.

Each year the oceans evaporate a quantity of water equivalent to an average depth of one meter. The total evaporation from land and bodies of fresh water is one sixth of the evaporation from the sea, and at least one fifth of this evaporation is from the transpiration of plants growing on land. The grand total of water evaporated annually is roughly 100,000 cubic miles, and this must be roughly the annual precipitation. The precipitation on land exceeds the evaporation by slightly over 9,000 cubic

ENERGY CYCLES of the ecosphere are powered by the sun. Land plants bind solar energy into organic compounds utilized successively (*gray arrows*) by herbivore, carnivore and scavenger; residual compounds are decomposed by bacteria. Energy fixed by micro-scopic sea plants through a similar "food chain" (*color arrows*). In the water cycle (*broken gray arrows*) water evaporated from the sea is precipitated on land and used by living organisms, eventually returning to the sea bearing minerals and organic matter.

miles, which therefore represents the annual runoff of water from land to sea. It is astonishing to me to note that more than one tenth of this total runoff is carried to the sea by just two rivers—the Amazon and the Congo.

Precipitation supplies nonmarine organisms with the water which they require in large quantities. Protoplasm averages at least 75 per cent water, and plants require something like 450 grams of water to produce one gram of dry organic matter. The water moving from land to sea also erodes the land surface and dissolves soluble mineral matter. It brings to the plants the chemical nutrients that they require and it tends to level the land surface and deposit the minerals in the sea. At present the continents are being worn down at an average world-wide rate of one centimeter per century. The leveling process, however, apparently has never gone on to completion on the earth. Geological uplift of the land always intervenes and brings marine sediments above sea level, where the cycle can begin again.

The rivers of the world are now washing into the seas some four billion tons of dissolved inorganic matter a year, about 400 million tons of dissolved organic matter and about five times as much undissolved matter. The undissolved matter represents destruction of the land where organisms live, but the dissolved material is of greater interest, because it includes such important chemicals as 3.5 million tons of phosphorus, 100 million tons of potassium and 10 million tons of fixed nitrogen. In order to say what these losses may mean to the biosphere we must review a few facts about the chemical composition of the earth and of organisms.

E very organism seems to require at least 20 chemical elements and probably several others in trace amounts. Some of the organisms' requirements are rather surprising. *Penicillium* is said to need traces of tungsten, and the common duckweed demands manganese and the rare earth gallium. There is a European pansy which needs high concentrations of zinc in the soil, and several plants in different parts of the world are so hungry for copper that they help prospectors to find the mineral. Many organisms have fantastic abilities to concentrate the necessary elements from dilute media. The sea-squirts have vanadium in their blood, and the liver of the edible scallop contains on a dry-weight basis one tenth of 1 per cent of cadmium, although the amount of this element in sea water is so small that it cannot be de-

tected by chemical tests.

But the exotic chemical tastes of organisms are comparatively unimportant. Their main needs can be summed up in just five words—oxygen, carbon, hydrogen, nitrogen and phosphorus, which account for more than 95 per cent of the mass of all protoplasm. Oxygen is the most abundant chemical element on earth, so we probably do not need to be concerned about any absolute deficiency of oxygen. But nitrogen is a different matter. Whereas protein, the main stuff of life, is 18 per cent nitrogen, the relative abundance of this element on the earth is only one 10,000th of the earth's mass. It is apparent that our land forms of life could not long tolerate a net annual loss of 10 million tons of fixed nitrogen to the sea. Fortunately this nitrogen loss from land is reversible, so that we can speak of a "nitrogen cycle." Organisms in the sea convert the fixed nitrogen into ammonia, a gas which can return to land via the atmosphere.

Carbon also is not in too abundant supply, for it amounts to less than three parts in 10,000 of the total mass of the earth's matter. But once again the biosphere profits from the fact that carbon can escape from the oceans as a gas—carbon dioxide. This gas goes through a complex circulation in the atmosphere, being released from the oceans in tropical regions and absorbed by the ocean waters in polar regions. Because some carbon is deposited in ocean sediments as carbonates, there is a net loss of carbon from the ecosphere. But there seems to be no danger that a shortage of this element will restrict life. The atmosphere contains 2,400 billion tons of carbon dioxide, and at least 30 times that much is dissolved in the oceans, waiting to be released if the atmosphere should become depleted. Volcanoes discharge carbon dioxide, and man is burning fossil fuels at such a rate that he has been accused of increasing the average carbon dioxide content of the atmosphere by some 10 per cent in the last 50 years. In addition, lots of limestone, which is more than 4 per cent carbon dioxide, has been pushed up from ancient seas by uplifts of the earth.

The story of phosphorus appears somewhat more alarming. This element accounts for a bit more than one tenth of 1 per cent of the mass of terrestrial matter, is enriched to about twice this level in plant protoplasm and is greatly enriched in animals, accounting for more than 1 per cent of the weight of the human body. As a constituent of nucleic

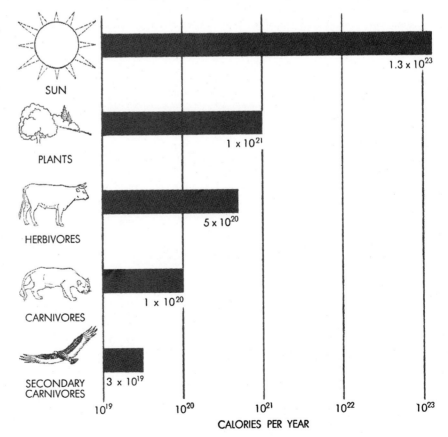

SUN

PLANTS

HERBIVORES

CARNIVORES

SECONDARY CARNIVORES

1.3×10^{23}

1×10^{21}

5×10^{20}

1×10^{20}

3×10^{19}

10^{19} 10^{20} 10^{21} 10^{22} 10^{23}

CALORIES PER YEAR

UTILIZATION OF SOLAR ENERGY decreases with each step along the food chain. These bars (on a logarithmic scale) show that plants use only .08 per cent of energy reaching the atmosphere; plant-eaters use only part of this fraction and flesh-eaters even less.

196

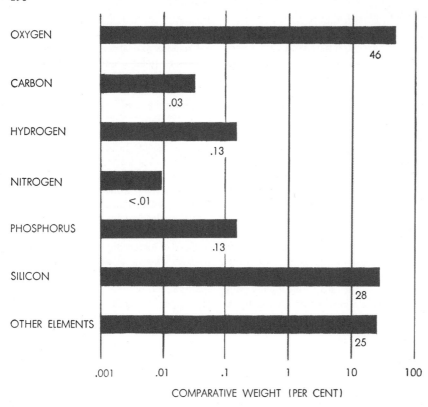

OXYGEN 46

CARBON .03

HYDROGEN .13

NITROGEN <.01

PHOSPHORUS .13

SILICON 28

OTHER ELEMENTS 25

.001 .01 .1 1 10 100

COMPARATIVE WEIGHT (PER CENT)

ESTIMATED RELATIVE ABUNDANCE of elements in the earth and its atmosphere (*above*) and in living matter (*below*) is compared in these charts; the scale is logarithmic. Silicon, with many stable compounds, is abundant on earth but rare in living organisms. Nitrogen, rare on earth, is important to life, making up as much as 18 per cent of proteins.

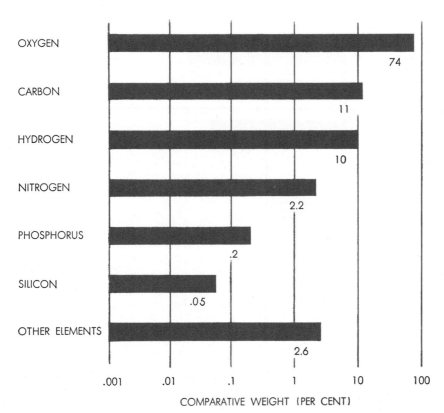

OXYGEN 74

CARBON 11

HYDROGEN 10

NITROGEN 2.2

PHOSPHORUS .2

SILICON .05

OTHER ELEMENTS 2.6

.001 .01 .1 1 10 100

COMPARATIVE WEIGHT (PER CENT)

acids it is indispensable for all types of life known to us. But many agricultural lands already suffer a deficiency of phosphorus, and a corn crop of 60 bushels per acre removes 10 per cent of the phosphorus in the upper six inches of fertile soil. Each year 3.5 million tons of phosphorus are washed from the land and precipitated in the seas. And unfortunately phosphorus does not escape from the sea as a gas. Its only important recovery from the sea is in the guano produced by sea birds, but less than 3 per cent of the phosphorus annually lost from the land is returned in this way.

I must agree with agriculturalists who say that phosphorus is the critical limiting resource for the functioning of the ecosphere. The supply is at least shrinking (if dwindling is too strong a word) and there seems to be no practical way of improving the situation short of waiting for the next geological cycle of uplift to bring phosphate rock above sea level. Perhaps we should also worry about other essential elements, such as calcium, potassium, magnesium and iron, which behave much like phosphorus in the metabolism of the ecosphere, but the evidence clearly indicates that if present trends continue phosphorus will be the first to run out.

This brings me to the close of a very superficial summary of some of the physiological processes of the ecosphere. There are drastic oversimplifications in this treatment; the importance of some processes may be overestimated, and others (*e.g.*, dumping sewage in rivers and oceans) may not have received enough attention. The figures for the total quantity of energy received by the earth, for total annual precipitation and for the total supply of some chemical elements may overlook the very irregular distribution of these resources in time and space. Much solar energy falls on deserts and fields of snow and ice where it cannot be used by plants, and much precipitation arrives at unfavorable seasons or in such torrents that it does more harm than good to organisms. Yet I believe that there may be some merit, both intellectual and practical, in attempting to scan the entire picture.

Our survey suggests that man may be justified in feeling some real concern about the problem of erosion. It should also make us aware of the important role played by organisms that we might otherwise ignore or even regard as pests. The dung beetles, the various scavengers and the termites and other decomposers all play important bit parts in this great production. At least six diverse groups

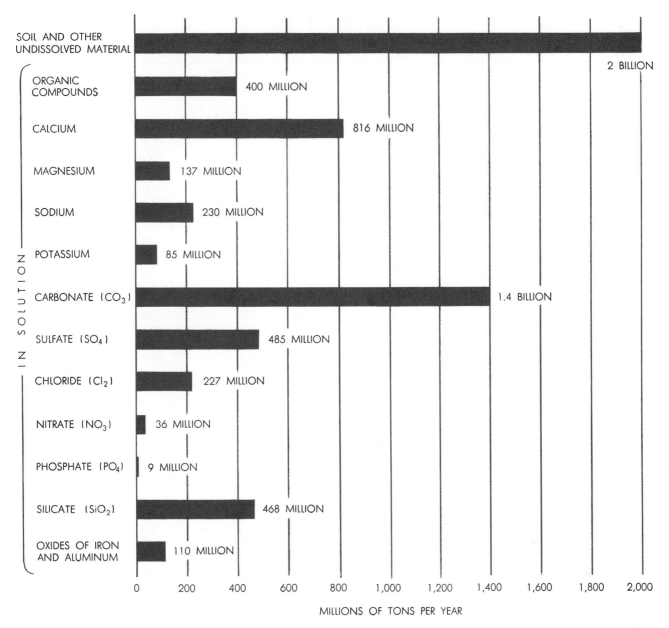

ANNUAL LOSS of minerals and organic matter washed into the sea amounts to billions of tons. Much nitrogen and carbon even-tually return to the land via the atmosphere; the loss of phos-phate is more serious since almost all of it remains in the oceans.

of bacteria are absolutely essential for the proper physiological functioning of the nitrogen cycle alone. Man in his care-lessness would probably neither notice nor care if by some unlikely chance his radioactive fallout or one of his chemi-cal sprays or fumes should exterminate all of the microorganisms that are cap-able of decomposing chitin. Yet, as we have seen, such a tragedy would even-tually mean an end to life on earth.

Finally, it is interesting to ask how large a role man plays in the physiology of the ecosphere. The Statistical Office of the United Nations estimates the present human population of the earth at 2.7 billion persons. Each of these is supposed to consume at least 2,200 metabolizable kilocalories per day. This

makes a total food requirement of 22×10^{14} kilocalories per year. I have estimated that all of the plant growth in the world amounts to an annual net of 5×10^{17} kilocalories, of which not more than 50 per cent is metabolizable by any primary consumer. Thus if man were to feed exclusively on plants he would require almost exactly 1 per cent of the total productivity of the earth.

To me this is a very impressive figure. There are more than one million species of animals, and when just one of these million species can corner 1 per cent of the total food resources, this form is truly in a position of overwhelming dominance. The figure becomes even more impressive when we reflect that 70 per cent of the total plant production

takes place in the oceans, and that our figure for productivity includes inedible materials such as straw and lumber.

If human beings were to eat meat ex-clusively, the present world population would require 4 per cent of all of the flesh of primary consumers of all types that the earth could support—and this means that much of our meat would be insects and tiny crustaceans. I suspect that the human population is already so large that no conceivable technical ad-vances could make it possible for all mankind to live on a meat diet. Speaking as one who would like to live on a meat diet, I can't see very much to be opti-mistic about for the future. This opinion, however, cannot be expected to alter the physiology of the ecosphere.

22

THE HUMAN POPULATION

EDWARD S. DEEVEY, JR.
September 1960

Almost until the present turn in human affairs an expanding population has been equated with progress. "Increase and multiply" is the Scriptural injunction. The number of surviving offspring is the measure of fitness in natural selection. If number is the criterion, the human species is making great progress. The population, now passing 2.7 billion, is doubling itself every 50 years or so. To some horrified observers, however, the population increase has become a "population explosion." The present rate of increase, they point out, is itself increasing. At 1 per cent per year it is double that of the past few centuries. By A.D. 2000, even according to the "medium" estimate of the careful demographers of the United Nations, the rate of increase will have accelerated to 3 per cent per year, and the total population will have reached 6.267 billion. If Thomas Malthus's assumption of a uniform rate of doubling is naive, because it so quickly leads to impossible numbers, how long can an accelerating annual increase, say from 1 to 3 per cent in 40 years, be maintained? The demographers confronted with this question lower their eyes: "It would be absurd," they say, "to carry detailed calculations forward into a more remote future. It is most debatable whether the trends in mortality and fertility can continue much longer. Other factors may eventually bring population growth to a halt."

So they may, and must. It comes to this: Explosions are not made by force alone, but by force that exceeds restraint. Before accepting the implications of the population explosion, it is well to set the present in the context of the record of earlier human populations. As will be seen, the population curve has moved upward stepwise in response to the three major revolutions that have marked the evolution of culture [*see bottom illustration on page 202*]. The tool-using and toolmaking revolution that started the growth of the human stem from the primate line gave the food-gatherer and hunter access to the widest range of environments. Nowhere was the population large, but over the earth as a whole it reached the not insignificant total of five million, an average of .04 person per square kilometer (.1 person per square mile) of land. With the agricultural revolution the population moved up two orders of magnitude to a new plateau, multiplying 100 times in the short span of 8,000 years, to an average of one person per square kilometer. The increase over the last 300 years, a multiplication by five, plainly reflects the first repercussions of the scientific-industrial revolution. There are now 16.4 persons per square kilometer of the earth's land area. It is thus the release of restraint that the curve portrays at three epochal points in cultural history.

But the evolution of the population size also indicates the approach to equilibrium in the two interrevolutionary periods of the past. At what level will the present surge of numbers reach equilibrium? That is again a question of restraint, whether it is to be imposed by the limitations of man's new command over his environment or by his command over his own nature.

The human generative force is neither new nor metabiological, nor is it especially strong in man as compared to other animals. Under conditions of maximal increase in a suitable environment empty of competitors, with births at maximum and deaths negligible, rats can multiply their numbers 25 times in an average generation-time of 31 weeks.

For the water flea *Daphnia,* beloved by ecologists for the speedy answers it gives, the figures are 221 times in a generation of 6.8 days. Mankind's best efforts seem puny by contrast: multiplication by about 1.4 times in a generation of 28 years. Yet neither in human nor in experimental populations do such rates continue unchecked. Sooner or later the births slow down and the deaths increase, until—in experiments, at any rate—the growth tapers off, and the population effectively saturates its space. Ecologists define this state (of zero rate of change) as equilibrium, without denying the possibility of oscillations that average out to zero, and without forgetting the continuous input of energy (food, for instance) that is needed to maintain the system.

Two kinds of check, then, operate to limit the size of a population, or of any living thing that grows. Obviously the environment (amount of space, food or other needed resources) sets the upper limit; sometimes this is manipulatable, even by the population itself, as when it exploits a new kind of food in the same old space, and reaches a new, higher limit. More subtly, populations can be said to limit their own rates of increase. As the numbers rise, female fruit-flies, for example, lay fewer eggs when jostled by their sisters; some microorganisms battle each other with antibiotics; flour beetles accidentally eat their own defenseless eggs and pupae; infectious diseases spread faster, or become more virulent, as their hosts become more numerous. For human populations pestilence and warfare, Malthus's "natural restraints," belong among these devices for self-limitation. So, too, does his "moral restraint," or voluntary birth control. Nowadays a good deal of attention is being given, not only to voluntary methods,

ROMAN TOMBSTONE from the first century A.D. records the death of Cominia Tyche, aged 27 years, 11 months, 28 days. Tombstones are a source of information on life expectancy in the ancient world. Stone is in the Metropolitan Museum of Art in New York.

YEARS AGO	CULTURAL STAGE	AREA POPULATED	ASSUMED DENSITY PER SQUARE KILOMETER	TOTAL POPULATION (MILLIONS)
1,000,000	LOWER PALEOLITHIC		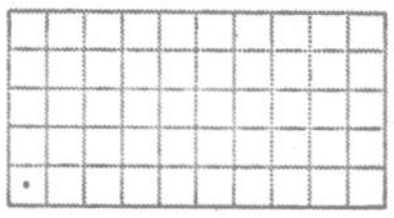 .00425	.125
300,000	MIDDLE PALEOLITHIC		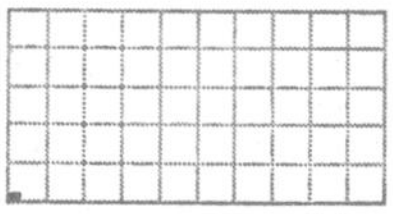 .012	1
25,000	UPPER PALEOLITHIC		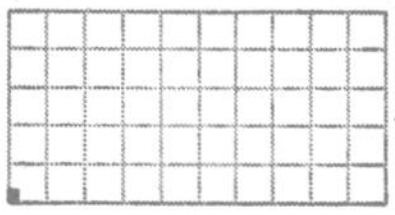 .04	3.34
10,000	MESOLITHIC		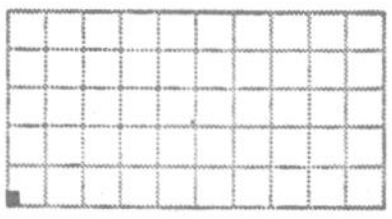 .04	5.32
6,000	VILLAGE FARMING AND EARLY URBAN		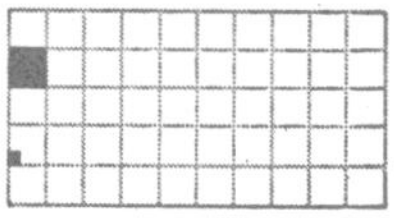 1.0 / .04	86.5
2,000	VILLAGE FARMING AND URBAN		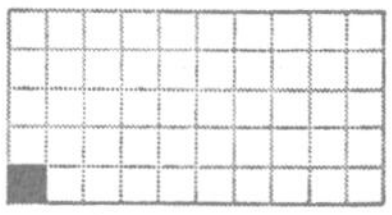 1.0	133
310	FARMING AND INDUSTRIAL		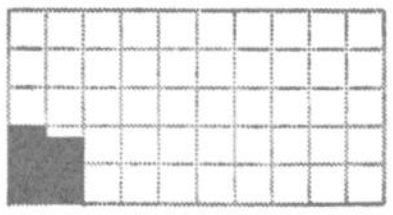 3.7	545
210	FARMING AND INDUSTRIAL		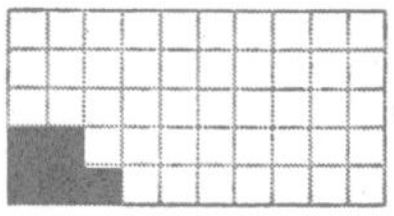 4.9	728
160	FARMING AND INDUSTRIAL		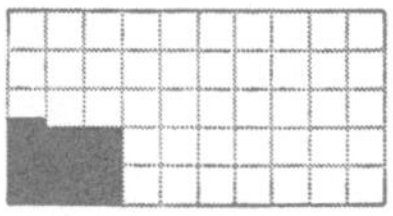 6.2	906
60	FARMING AND INDUSTRIAL		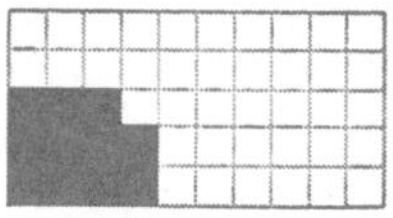 11.0	1,610
10	FARMING AND INDUSTRIAL		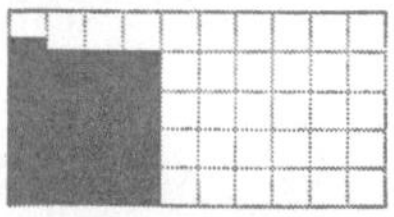 16.4	2,400
A.D. 2000	FARMING AND INDUSTRIAL		46.0	6,270

but also to a fascinating new possibility: mental stress.

Population control by means of personality derangement is probably a vertebrate patent; at least it seems a luxury beyond the reach of a water flea. The general idea, as current among students of small mammals, is that of hormonal imbalance (or stress, as defined by Hans Selye of the University of Montreal); psychic tension, resulting from overcrowding, disturbs the pituitary-adrenal system and diverts or suppresses the hormones governing sexuality and parental care. Most of the evidence comes from somewhat artificial experiments with caged rodents. It is possible, though the case is far from proved, that the lemming's famous mechanism for restoring equilibrium is the product of stress; in experimental populations of rats and mice, at least, anxiety has been observed to increase the death rate through fighting or merely from shock.

From this viewpoint there emerges an interesting distinction between crowding and overcrowding among vertebrates; overcrowding is what is perceived as such by members of the population. Since the human rate of increase is holding its own and even accelerating, however, it is plain that the mass of men, although increasingly afflicted with mental discomfort, do not yet see themselves as overcrowded. What will happen in the future brings other questions. For the present it may be noted that some kind of check has always operated, up to now, to prevent populations from ex-

POPULATION GROWTH, from inception of the hominid line one million years ago through the different stages of cultural evolution to A.D. 2000, is shown in the chart on the opposite page. In Lower Paleolithic stage, population was restricted to Africa (*colored area on world map in third column*), with a density of only .00425 person per square kilometer (*fourth column*) and a total population of only 125,000 (*column at right*). By the Mesolithic stage, 10,000 years ago, hunting and food gathering techniques had spread the population over most of the earth and brought the total to 5,320,-000. In the village farming and early urban stage, population increased to a total of 86,500,000 and a density of one person per square kilometer in the Old World and .04 per square kilometer in the New World. Today the population density exceeds 16 persons per square kilometer, and pioneering of the antarctic continent has begun.

ceeding the space that contains them. Of course space may be non-Euclidean, and man may be exempt from this law.

The commonly accepted picture of the growth of the population out of the long past takes the form of the top graph on the next page. Two things are wrong with this picture. In the first place the basis of estimates, back of about A.D. 1650, is rarely stated. One suspects that writers have been copying each other's guesses. The second defect is that the scales of the graph have been chosen so as to make the first defect seem unimportant. The missile has left the pad and is heading out of sight—so it is said; who cares whether there were a million or a hundred million people around when Babylon was founded? The difference is nearly lost in the thickness of the draftsman's line.

I cannot think it unimportant that (as I calculate) there were 36 billion Paleolithic hunters and gatherers, including the first tool-using hominids. One begins to see why stone tools are among the commonest Pleistocene fossils. Another 30 billion may have walked the earth before the invention of agriculture. A cumulative total of about 110 billion individuals seem to have passed their days, and left their bones, if not their marks, on this crowded planet. Neither for our understanding of culture nor in terms of man's impact upon the land is it a negligible consideration that the patch of ground allotted to every person now alive may have been the lifetime habitat of 40 predecessors.

These calculations exaggerate the truth in a different way: by condensing into single sums the enormous length of prehistoric time. To arrive at the total of 36 billion Paleolithic hunters and gatherers I have assumed mean standing populations of half a million for the Lower Paleolithic, and two million for the Middle and Upper Paleolithic to 25,000 years ago. For Paleolithic times there are no archeological records worth considering in such calculations. I have used some figures for modern hunting tribes, quoted by Robert J. Braidwood and Charles A. Reed, though they are not guilty of my extrapolations. The assumed densities per square kilometer range from a tenth to a third of those estimated for eastern North America before Columbus came, when an observer would hardly have described the woods as full of Indians. (Of course I have excluded any New World population from my estimates prior to the Mesolithic climax of the food-gathering and hunting phase of cultural evolution.) It is only

because average generations of 25 years succeeded each other 39,000 times that the total looms so large.

For my estimates as of the opening of the agricultural revolution, I have also depended upon Braidwood and Reed. In their work in Mesopotamia they have counted the number of rooms in buried houses, allowing for the areas of town sites and of cultivated land, and have compared the populations so computed with modern counterparts. For early village-farmers, like those at Jarmo, and for the urban citizens of Sumer, about 2500 B.C., their estimates (9.7 and 15.4 persons per square kilometer) are probably fairly close. They are intended to apply to large tracts of inhabited country, not to pavement-bound clusters of artisans and priests. Nevertheless, in extending these estimates to continent-wide areas, I have divided the lower figure by 10, making it one per square kilometer. So much of Asia is unirrigated and nonurban even today that the figure may still be too high. But the Maya, at about the same level of culture (3,000 or 4,000 years later), provide a useful standard of comparison. The present population of their classic homeland averages .6 per square kilometer, but the land can support a population about a hundred times as large, and probably did at the time of the classic climax. The rest of the New World, outside Middle America, was (and is) more thinly settled, but a world-wide average of one per square kilometer seems reasonable for agricultural, pre-industrial society.

For modern populations, from A.D. 1650 on, I have taken the estimates of economic historians, given in such books as the treatise *World Population and Production*, by Wladimir S. and Emma S. Woytinsky. All these estimates are included in the bottom graph on the next page. Logarithmic scales are used in order to compress so many people and millennia onto a single page. Foreshortening time in this way is convenient, if not particularly logical, and back of 50,000 years ago the time-scale is pretty arbitrary anyway. No attempt is made to show the oscillations that probably occurred, in glacial and interglacial ages, for example.

The stepwise evolution of population size, entirely concealed in graphs with arithmetic scales, is the most noticeable feature of this diagram. For most of the million-year period the number of hominids, including man, was about what would be expected of any large Pleistocene mammal—scarcer than horses, say, but commoner than elephants. Intellectual superiority was simply a successful adaptation, like longer legs; essential to stay in the running, of course, but making man at best the first among equals. Then the food-gatherers and hunters became plowmen and herdsmen, and the population was boosted by about 16 times, between 10,000 and 6,000 years ago. The scientific-industrial revolution, beginning some 300 years ago, has spread its effects much faster, but it has not yet taken the number as far above the earlier base line.

The long-term population equilibrium implied by such base lines suggests

something else. Some kind of restraint kept the number fairly stable. "Food supply" offers a quick answer, but not, I think, the correct one. At any rate, a forest is full of game for an expert mouse-hunter, and a Paleolithic man who stuck to business should have found enough food on two square kilometers, instead of 20 or 200. Social forces were probably more powerful than mere starvation in causing men to huddle in small bands. Besides, the number was presumably adjusted to conditions in the poorest years, and not to average environments.

The main point is that there were ad-

justments. They can only have come about because the average female bore two children who survived to reproduce. If the average life span is 25 years, the "number of children ever born" is about four (because about 50 per cent die before breeding), whereas a population that is really trying can average close to eight. Looking back on former times, then, from our modern point of view, we might say that about two births out of four were surplus, though they were needed to counterbalance the juvenile death toll. But what about the other four, which evidently did not occur? Unless the life expectancy was very much less

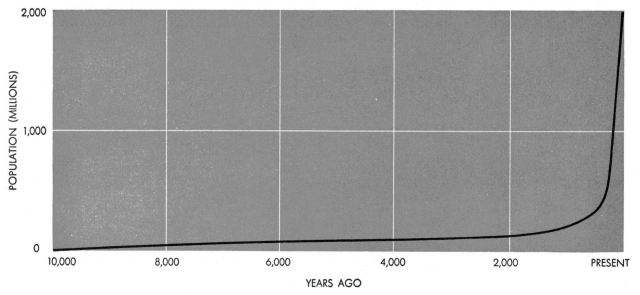

ARITHMETIC POPULATION CURVE plots the growth of human population from 10,000 years ago to the present. Such a curve suggests that the population figure remained close to the base line for an indefinite period from the remote past to about 500 years ago, and that it has surged abruptly during the last 500 years as a result of the scientific-industrial revolution.

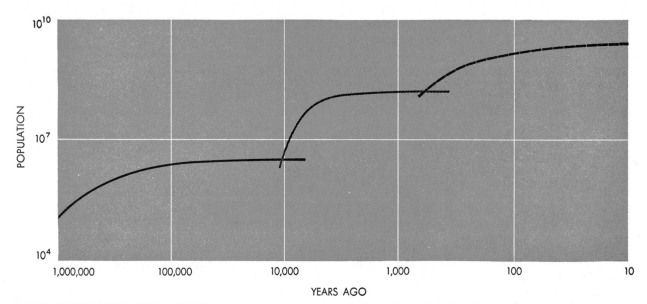

LOGARITHMIC POPULATION CURVE makes it possible to plot, in a small space, the growth of population over a longer period of time and over a wider range (from 10^4, or 10,000, to 10^{10}, or 10 billion, persons). Curve, based on assumptions concerning relationship of technology and population as shown in chart on page 200, reveals three population surges reflecting tool-making or cultural revolution (*solid line*), agricultural revolution (*gray line*) and scientific-industrial revolution (*broken line*).

than I have assumed (and will presently justify), some degree of voluntary birth control has always prevailed.

Our 40 predecessors on earth make an impressive total, but somehow it sounds different to say that nearly 3 per cent of the people who have ever lived are still around. When we realize that they are living twice as long as their parents did, we are less inclined to discount the revolution in which we are living. One of its effects has just begun to be felt: The mean age of the population is increasing all over the world. Among the more forgivable results of Western culture, when introduced into simpler societies, is a steep drop in the death rate. Public-health authorities are fond of citing Ceylon in this connection. In a period of a year during 1946 and 1947 a campaign against malaria reduced the death rate there from 20 to 14 per 1,000. Eventually the birth rate falls too, but not so fast, nor has it yet fallen so far as a bare replacement value. The natural outcome of this imbalance is that acceleration of annual increase which so bemuses demographers. In the long run it must prove to be temporary, unless the birth rate accelerates, for the deaths that are being systematically prevented are premature ones. That is, the infants who now survive diphtheria and measles are certain to die of something else later on, and while the mean lifespan is approaching the maximum, for the first time in history, there is no reason to think that the maximum itself has been stretched. Meanwhile the expectation of life at birth is rising daily in most countries, so that it has already surpassed 70 years in some, including the U. S., and probably averages between 40 and 50.

It is hard to be certain of any such world-wide figure. The countries where mortality is heaviest are those with the least accurate records. In principle, however, mean age at death is easier to find out than the number of children born, the frequency or mean age at marriage, or any other component of a birth rate. The dead bones, the court and parish records and the tombstones that archeology deals with have something to say about death, of populations as well as of people. Their testimony confirms the impression that threescore years and ten, if taken as an average and not as a maximum lifetime, is something decidedly new. Of course the possibilities of bias in such evidence are almost endless. For instance, military cemeteries tend to be full of young adult males. The hardest bias to allow for is the deficiency of in-

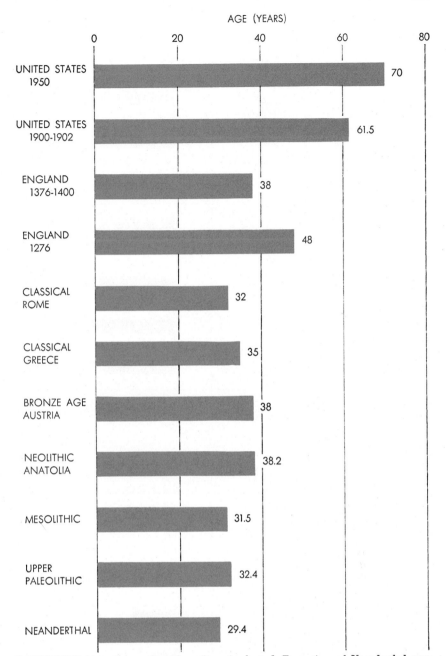

LONGEVITY in ancient and modern times is charted. From time of Neanderthal man to 14th century A.D., life span appears to have hovered around 35 years. An exception is 13th-century England. Increase in longevity partly responsible for current population increase has come in modern era. In U.S. longevity increased about 10 years in last half-century.

fants and children; juvenile bones are less durable than those of adults, and are often treated less respectfully. Probably we shall never know the true expectation of life at birth for any ancient people. Bypassing this difficulty, we can look at the mean age at death among the fraction surviving to adolescence.

The "nasty, brutish and short" lives of Neanderthal people have been rather elaborately guessed at 29.4 years. The record, beyond them, is not one of steady improvement. For example, Neolithic farmers in Anatolia and Bronze Age Austrians averaged 38 years, and even the

Mesolithic savages managed more than 30. But in the golden ages of Greece and Rome the life span was 35 years or less. During the Middle Ages the chances of long life were probably no better. The important thing about these averages is not the differences among them, but their similarity. Remembering the crudeness of the estimates, and the fact that juvenile mortality is omitted, it is fair to guess that human life-expectancy at birth has never been far from 25 years— 25 plus or minus five, say—from Neanderthal times up to the present century. It follows, as I have said, that about half

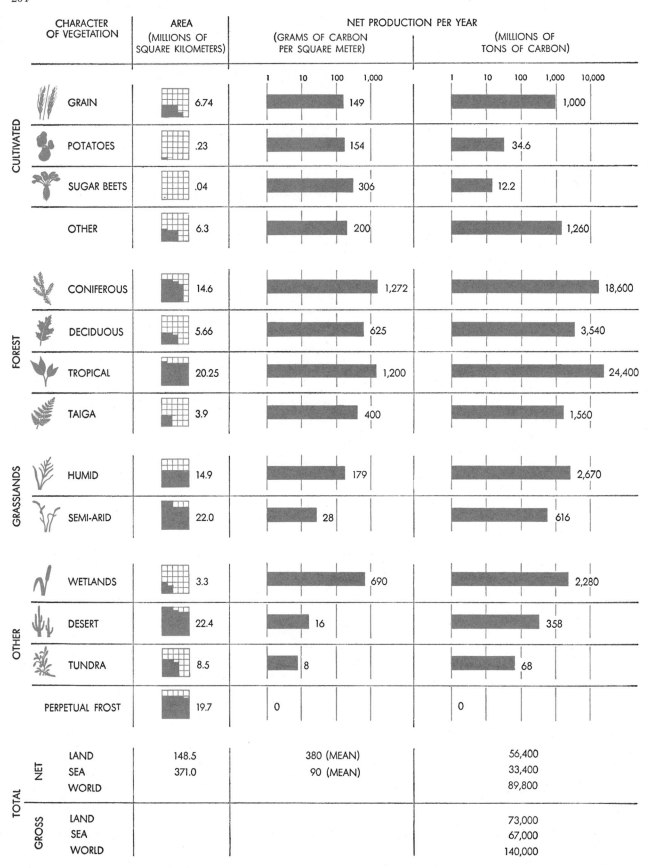

CHARACTER OF VEGETATION		AREA (MILLIONS OF SQUARE KILOMETERS)	NET PRODUCTION PER YEAR	
			(GRAMS OF CARBON PER SQUARE METER)	(MILLIONS OF TONS OF CARBON)
CULTIVATED	GRAIN	6.74	149	1,000
	POTATOES	.23	154	34.6
	SUGAR BEETS	.04	306	12.2
	OTHER	6.3	200	1,260
FOREST	CONIFEROUS	14.6	1,272	18,600
	DECIDUOUS	5.66	625	3,540
	TROPICAL	20.25	1,200	24,400
	TAIGA	3.9	400	1,560
GRASSLANDS	HUMID	14.9	179	2,670
	SEMI-ARID	22.0	28	616
OTHER	WETLANDS	3.3	690	2,280
	DESERT	22.4	16	358
	TUNDRA	8.5	8	68
	PERPETUAL FROST	19.7	0	0
TOTAL NET	LAND	148.5	380 (MEAN)	56,400
	SEA	371.0	90 (MEAN)	33,400
	WORLD			89,800
TOTAL GROSS	LAND			73,000
	SEA			67,000
	WORLD			140,000

PRODUCTION OF ORGANIC MATTER per year by the land vegetation of the world—and thus its ultimate food-producing capacity—is charted in terms of the amount of carbon incorporated in organic compounds. Cultivated vegetation (*top left*) is less efficient than forest and wetlands vegetation, as indicated by the uptake of carbon per square meter (*third column*), and it yields a smaller over-all output than forest, humid grasslands and wetlands vegetation (*fourth column*). The scales at top of third and fourth columns are logarithmic. Land vegetation leads sea vegetation in efficiency and in net and gross tonnage (*bottom*). The difference between the net production and gross production is accounted for by the consumption of carbon in plant respiration.

the children ever born have lived to become sexually mature. It is not hard to see why an average family size of four or more, or twice the minimum replacement rate, has come to seem part of a God-given scheme of things.

The 25-fold upsurge in the number of men between 10,000 and 2,000 years ago was sparked by a genuine increase in the means of subsistence. A shift from animal to plant food, even without agricultural labor and ingenuity, would practically guarantee a 10-fold increase, for a given area can usually produce about 10 times as much plant as animal substance. The scientific-industrial revolution has increased the efficiency of growing these foods, but hardly, as yet, beyond the point needed to support another 10 times as many people, fewer of whom are farmers. At the present rate of multiplication, without acceleration, another 10-fold rise is due within 230 years. Disregarding the fact that developed societies spend 30 to 60 times as much energy for other purposes as they need for food, one is made a little nervous by the thought of so many hungry mouths. Can the increase of efficiency keep pace? Can some of the apparently ample energy be converted to food as needed, perhaps at the cost of reducing the size of Sunday newspapers? Or is man now pressing so hard on his food supply that another 10-fold increase of numbers is impossible?

The answers to these questions are not easy to find, and students with different viewpoints disagree about them. Richard L. Meier of the University of Michigan estimates that a total of 50 billion people (a 20-fold increase, that is) can be supported on earth, and the geochemist Harrison Brown of the California Institute of Technology will allow (reluctantly) twice or four times as many. Some economists are even more optimistic; Arnold C. Harberger of the University of Chicago presents the interesting notion that a larger crop of people will contain more geniuses, whose intellects will find a solution to the problem of feeding *still* more people. And the British economist Colin Clark points out that competition for resources will sharpen everyone's wits, as it always has, even if the level of innate intelligence is not raised.

An ecologist's answer is bound to be cast in terms of solar energy, chlorophyll and the amount of land on which the two can interact to produce organic carbon. Sources of energy other than the sun are either too expensive, or nonrenewable or both. Land areas will continue for a very long time to be the places where food is grown, for the sea is not so productive as the land, on the average. One reason, sometimes forgotten, is that the plants of the sea are microscopic algae, which, being smaller than land plants, respire away a larger fraction of the carbon they fix. The culture of the fresh-water alga *Chlorella* has undeniable promise as a source of human food. But the high efficiencies quoted for its photosynthesis, as compared with agricultural plants, are not sustained outdoors under field conditions. Even if Chlorella (or another exceptionally efficient producer, such as the water hyacinth) is the food plant of the future, flat areas exposed to sunlight will be needed. The 148.5 million square kilometers of land will have to be used with thoughtful care if the human population is to increase 20-fold. With a population of 400 per square kilometer (50 billion total) it would seem that men's bodies, if not their artifacts, will stand in the way of vital sunshine.

Plants capture the solar energy impinging on a given area with an efficiency of about .1 per cent. (Higher values often quoted are based on some fraction of the total radiation, such as visible light.) Herbivores capture about a 10th of the plants' energy, and carnivores convert about 10 per cent of the energy captured by herbivores (or other carnivores). This means, of course, that carnivores, feeding on plants at second hand, can scarcely do so with better than 1 per cent efficiency ($1/10 \times 1/10$ equals $1/100$). Eugene I. Rabinowitch of the University of Illinois has calculated that the current crop of men represents an ultimate conversion of about 1 per cent of the energy trapped by land vegetation. Recently, however, I have re-examined the base figure—the efficiency of the land-plant production—and believe it should be raised by a factor of three or four. The old value came from estimates made in 1919 and in 1937. A good deal has been learned since those days. The biggest surprise is the high productivity of forests, especially the forests of the Temperate Zone.

If my new figures are correct, the population could theoretically increase by 30 or 40 times. But man would have to displace all other herbivores and utilize all the vegetation with the 10 per cent efficiency established by the ecological rule of tithes. No land that now supports greenery could be spared for nonagricultural purposes; the populace would have to reside in the polar regions, or on artificial "green isles in the sea, love"—scummed over, of course, by 10 inches of Chlorella culture.

The picture is doubtless overdrawn. There is plenty of room for improvement in present farming practice. More land could be brought under cultivation if a better distribution of water could be arranged. More efficient basic crops can be grown and used less wastefully. Other sources of energy, notably atomic energy, can be fed back into food production to supplement the sun's rays. None of these measures is more than palliative, however; none promises so much as a 10-fold increase in efficiency; worse, none is likely to be achieved at a pace equivalent to the present rate of doubling of the world's population. A 10-fold, even a 20-fold, increase can be tolerated, perhaps, but the standard of living seems certain to be lower than today's. What happens then, when men perceive themselves to be overcrowded?

The idea of population equilibrium will take some getting used to. A population that is kept stable by emigration, like that of the Western Islands of Scotland, is widely regarded as sick—a shining example of a self-fulfilling diagnosis. Since the fall of the death rate is temporary, it is those two or more extra births per female that demand attention. The experiments with crowded rodents point to one way they might be corrected, through the effect of anxiety in suppressing ovulation and spermatogenesis and inducing fetal resorption. Some of the most dramatic results are delayed until after birth: litters are carelessly nursed, deserted or even eaten. Since fetuses, too, have endocrine glands, the specter of maternal transmission of anxiety now looms: W. R. Thompson of Wesleyan University has shown that the offspring of frustrated mother mice are more "emotional" throughout their own lives, and my student Kim Keeley has confirmed this.

Considered abstractly, these devices for self-regulation compel admiration for their elegance. But there is a neater device that men can use: rational, voluntary control over numbers. In mentioning the dire effects of psychic stress I am not implying that the population explosion will be contained by cannibalism or fetal resorption, or any power so naked. I simply suggest that vertebrates have that power, whether they want it or not, as part of the benefit—and the price —of being vertebrates. And if the human method of adjusting numbers to resources fails to work in the next 1,000 years as it has in the last million, subhuman methods are ready to take over.

23

FOOD

NEVIN S. SCRIMSHAW
September 1963

Nearly half the world's population is underfed or otherwise malnourished. The lives of the people in the underdeveloped areas are dominated by the scramble for food to stay alive. Such people are perpetually tired, weak and vulnerable to disease—prisoners of a vicious circle that keeps their productivity far below par and so defeats their efforts to feed their families adequately. Because their undernourishment begins soon after birth, it produces permanently depressing and irremediable effects on the population as a whole. Malnutrition and disease kill a high proportion of the children by the age of four; the death rates for these young children are 20 to 60 times higher than in the U.S. and western Europe. Among those who survive, few escape physical or mental retardation or both.

Obviously the first necessity, if the underdeveloped countries are to develop, is more and better food. Much has been said about the need for industrialization of these countries as the quickest and most effective way to raise their incomes and level of living. But they cannot industrialize successfully without a substantial improvement in their nourishment and human efficiency. This must depend primarily on improvement of their agriculture and utilization of food. In these countries from 60 to 80 per cent of the people are engaged in farming, but their productivity is so low that it falls far short of feeding the population. That stands as a roadblock against their advance. Unless they improve their food-producing efficiency, any diversion of their working force to industry will only make their food problem more desperate.

Moreover, during the coming decades their food requirements will rise astronomically, both because of their rapid population growth and because of the demand for a better scale of living that comes with industrialization. The Food and Agriculture Organization of the United Nations has estimated that to provide a decent level of nutrition for the world's peoples the production of food will have to be doubled by 1980 and tripled by 2000.

Can the developing nations make the grade? Is our planet capable of feeding the hungry half of the world and supporting its vast, growing population? This is a complex question that involves many issues other than the volume of food production. Just as important are the conservation of food, the kinds of foods produced and the ways in which food is used. Food supply is not merely a matter of the number of bushels of grain the farmer harvests or the number of chickens he raises. Other vital elements in the equation are the selection, handling, processing, storage, transportation and marketing of the food crops. Each factor allows opportunities for improvement of efficiency that can greatly enhance the food supply.

Let us consider what science and technology have to contribute to the food problem.

The simplest way to increase food production, one might suppose, is to bring more land under cultivation and put more people to work on it. The U.S.S.R. and some of the underdeveloped countries have resorted to this straightforward approach, without notable success. It contains several fallacies. For one thing, it usually means moving into marginal land where the soil and climatic conditions give a poor return. Cultivation may quickly deplete this soil, ruining it for pasture or forest growth. It is often possible, of course, to turn such lands into useful farms by agricultural know-how; for instance, a so-phisticated knowledge of how to use the available water through an irrigation system may reclaim semiarid grasslands for crop-growing. But the cultivation of marginal lands is in any case unsuccessful unless it is carried out by farmers with a centuries-old tradition of experience or by modern experts with a detailed knowledge of the local conditions and the varieties of crops that are suitable for those conditions. Such knowledge is conspicuously absent in the underdeveloped countries.

Furthermore, we know that the highly developed countries have not increased the number of acres under cultivation but on the contrary have abandoned their marginal lands and steadily reduced the proportion of the population engaged in farming. Efficient farming calls for concentration on the most efficient lands, and it also results in greater production with fewer people. The U.S., for example, produces a huge surplus of food with only about 10 per cent of its people working on the farms.

The problem of the underdeveloped countries, then, is to increase the productivity of their farms and farmers. This would allow them to industrialize and to feed their people more adequately. It is not easy to accomplish, however. The peasant farmers are conservative and resistant to change in their methods of cultivation. The entire population

CHICAGO STOCKYARDS are a focal point in the distribution of protein in the U.S., handling more than three million head of cattle, hogs and sheep a year. Only a few of its hundreds of pens appear in the aerial photograph on the opposite page. The brown cattle are Herefords; the black, Angus; the white, Holsteins. Green areas are unused pens. Near center men are herding cattle out of a pen through long runway.

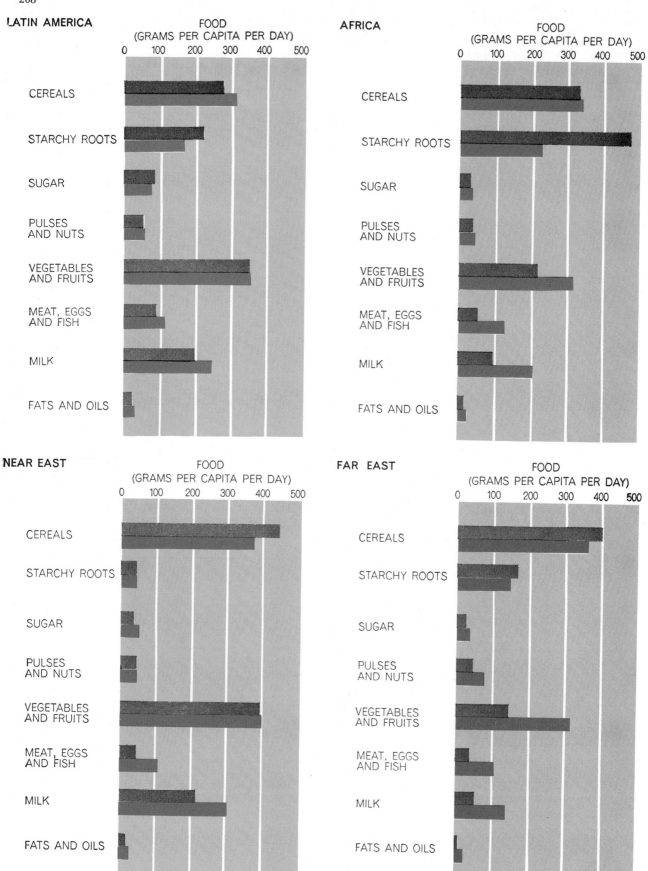

LATIN AMERICA

FOOD
(GRAMS PER CAPITA PER DAY)

0   100   200   300   400   500

CEREALS

STARCHY ROOTS

SUGAR

PULSES
AND NUTS

VEGETABLES
AND FRUITS

MEAT, EGGS
AND FISH

MILK

FATS AND OILS

AFRICA

FOOD
(GRAMS PER CAPITA PER DAY)

0   100   200   300   400   500

CEREALS

STARCHY ROOTS

SUGAR

PULSES
AND NUTS

VEGETABLES
AND FRUITS

MEAT, EGGS
AND FISH

MILK

FATS AND OILS

NEAR EAST

FOOD
(GRAMS PER CAPITA PER DAY)

0   100   200   300   400   500

CEREALS

STARCHY ROOTS

SUGAR

PULSES
AND NUTS

VEGETABLES
AND FRUITS

MEAT, EGGS
AND FISH

MILK

FATS AND OILS

FAR EAST

FOOD
(GRAMS PER CAPITA PER DAY)

0   100   200   300   400   500

CEREALS

STARCHY ROOTS

SUGAR

PULSES
AND NUTS

VEGETABLES
AND FRUITS

MEAT, EGGS
AND FISH

MILK

FATS AND OILS

FOOD SUPPLIES available (*gray bars*) and needed (*colored bars*) vary widely in four underdeveloped regions, according to studies by the Food and Agriculture Organization (FAO). All the regions suffer from both shortages of food and badly unbalanced diets. The lack of proteins is particularly acute and plays an important role in malnutrition. Pulses include leguminous crops such as peas and beans. The relatively well-fed countries of Paraguay, Uruguay and Argentina are omitted from the Latin America chart.

needs to be indoctrinated in the possibilities offered by scientific agriculture, including the officials who must provide the necessary funds, planning, legislation, training and research programs. The underdeveloped countries are greatly in need of studies and experiments to help them to adapt modern agricultural methods to their own conditions.

During the past two decades some of these countries have increased their food production, but their populations have in the meantime grown faster; therefore they are farther behind than before. Furthermore, the food increase has been gained at the expense of using up marginal lands. In productivity per acre or per man they have not gained at all.

Meanwhile the efficiency of farming in the developed countries has progressed phenomenally. In the U.S. the productivity per farm worker has tripled since 1940 [see illustration on page 214]. With a 7 per cent reduction in the total acreage under cultivation, U.S. production of cereal grains has jumped 50 per cent; the increase in the corn output, thanks to hybrid corn, has been even greater.

The "secret" of these improvements can be summed up in a few words: chemicals, mechanization, breeding and feeding.

Fertilizers are an old story to farmers, even in backward countries, but the practitioners of modern farming have raised the use of chemical fertilizers to a high art. To these they have added a pharmacopoeia of chemicals for special purposes: poisons to kill insects, fungi and other pests; plant-growth regulators to control weeds, force early sprouting, stimulate ripening and prevent premature dropping of fruit; soil-conditioners to improve the physical characteristics of the soil. Most of these techniques and materials could easily be introduced on the farms of the underdeveloped areas. They require capital investment, but they would pay for themselves many times over in higher yields.

The mechanization of farming has become so familiar in Western countries that we have forgotten the many changes it has brought about. It has released for human food a great deal of land formerly devoted to growing feed for draft animals. Feeding fuel to a machine is cheaper than feeding a horse, and the machine needs less care and maintenance. The machine not only plows and cultivates but also digs ditches and postholes, loads and handles heavy materials, harvests, threshes, chops forage, cleans vegetables and does many other things the intelligent horse could never do. It does all these things swiftly and virtually at a moment's notice, so that the farmer no longer has to worry about whether or not he can get a job done before threatening weather ruins his planting or his harvest.

The machine has also facilitated the building and development of irrigation systems. It makes easy work of the construction of dams, the digging of water channels and the pumping of water. In the U.S. irrigation has made it possible to increase the crop yield of Western lands by 50 to 100 per cent. In the arid zones of India and the Middle East, which for centuries have been entirely dependent on irrigation for their farming, extension of their systems with machinery would be a great boon. In some areas where enough water could be furnished by irrigation, two or three crops a year could be produced and the crops could be diversified.

Finally, a combination of selective breeding and efficient feeding has generated astonishing bounties in both plant and animal production. For most of the major plant crops, thanks to modern genetics, we have seen the development of new varieties that give a higher yield and are more resistant to disease. The same is true of the animals that supply our meat, milk and eggs. "Hybrid vigor" has become a magic phrase in the U.S. farm belt. Furthermore, the farmer today can buy selected seeds he knows will do certain specific things with high reliability: produce plants that mature faster or are adapted to a wide range of conditions or grow to a uniform height and all ripen at the same time so that they can be harvested by machine.

We now have wilt-resisting peas and cabbages, mosaic-resisting snap beans, virus-resisting potatoes, mildew-resisting cucumbers and lima beans, anthracnose-resisting watermelons and leaf-spot-resisting strawberries. We have new cereal grains rich in high-quality protein, special squash rich in vitamin A, cottonseed from which the toxic pigment called gossypol has been bred out. We have cows that give richer milk, hogs that grow exceptionally fast on less feed, hogs with more lean meat and less fat, poultry with a high ratio of lean meat.

To improvement of the animal breeds the advanced countries have coupled scientific husbandry: finely calculated diets and rations, synthetic hormones, pesticides and sanitary stalls, drugs and vaccines to control disease and many other measures that have heightened the efficiency of production. The results are most strikingly shown in poultry raising. There are now breeds of hens that lay more than 200 eggs a year and broilers that grow to a three-pound market size within 10 weeks. Diseases, waste motion and costs have been sharply reduced. Raised in individual cages arrayed in batteries of hundreds or thousands, the chickens minimize the expenditure of energy by themselves and their caretakers and facilitate record-keeping, so that the less productive birds can easily be eliminated.

In general it would not be difficult to apply most of the agricultural improvements to the countries that need them so urgently. The main biological problem would be to select the right plants and animals for transfer to those countries. For instance, Temperate Zone varieties of corn and soybeans do not grow well in hotter areas; prize pigs from mild climates are often unable to nurse their young in the Tropics; plants and animals that are successful in one region may quickly succumb to diseases in another. But analysis of the ecological conditions and testing can resolve these problems. It is known, for example, that certain plants can readily be transplanted from areas in the U.S. to areas in Japan because the climatic conditions are much the same. The identification and classification of such ecological analogues on a world-wide scale would greatly facilitate the transfer of agricultural techniques to the underdeveloped regions.

Aside from more efficient methods, however, those countries need a sounder over-all policy, which is to say, in most cases, more diversification of crops. Many of the underdeveloped nations are enslaved by a single cash crop, such as rubber, hemp, cotton, coffee, tea, sugar or olive oil, with deadly effects on their basic food supply. It is true that the export of the single crop provides cash with which to buy food, but it places the country at the mercy of crop failure and price fluctuations in the world market. There have been periods when it has meant mass starvation for a whole region.

Without giving up its profitable crop, each country should be able to expand its own food production and achieve a better-balanced agricultural economy. In some cases it could improve its food supply immediately without radical changes. For example, the cotton-raising countries usually export the cottonseed-oil meal along with the fiber;

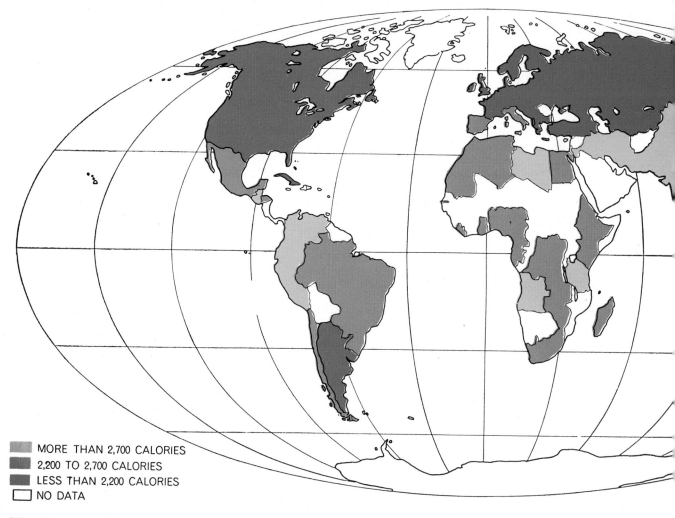

MORE THAN 2,700 CALORIES

2,200 TO 2,700 CALORIES

LESS THAN 2,200 CALORIES

NO DATA

NUMBER OF CALORIES per person per day as estimated by the FAO in 1962 is plotted on a map of the world. Several heavily populated countries, such as India and China, fall into the lowest classification, which means that a large part of the world's

instead they could keep the meal and use it for animal feed and even as human food.

Thus in their campaign against hunger the developing countries need first of all to increase their food production. The second way in which they could make great strides is by better food conservation or preservation. In this field the advanced countries have achieved improvements fully as spectacular as in production.

For food-raising *Homo sapiens* the perishability of foodstuffs has always been a major problem. Gradually he learned that his food supply would go further if he kept it edible longer by smoking, drying or salting it, or by keeping it cool in caves, wells, snow or ice from ponds. With limited effectiveness, these devices have served man for many centuries. But general food conservation on a large scale did not begin until the 19th century, with the arrival of the insulated refrigerator.

Within the past two decades we have seen freezing become a major means of preserving food in the U.S. Still newer is the recent development of freeze-drying—a system of vacuum dehydration of frozen food that makes it possible to store many foods without refrigeration and still retain their fresh flavor and characteristic properties. This method is ideal for keeping food in tropical areas, but it is still comparatively expensive. Vacuum-drying without freezing, however, is less costly and can preserve certain foods with little change in their flavor or texture. Also cheaper than freeze-drying is the new process of foam-mat drying, which is particularly good for fruit juices and purées.

Sterilization of food by ionizing radiation, which once seemed very promising, now looks impractical, because it damages the flavor and nutritional value. But irradiation with smaller doses, in the pasteurization range, may help to prolong the storage life of foods, although they will have to be refrigerated. Bacon

preserved by this process has recently been approved for sale in the U.S. Another new technique is dipping the food in an antibiotic bath; this works well for fresh fish and meat. Of course there are also the chemical preservatives and other additives that have been used in food for some time, such as propionates to inhibit molds in bread, antioxidants to slow down the process by which fats become rancid, emulsifying agents, bleaching agents and so on.

In many other ways, some obvious and some subtle, modern food industries have contrived to reduce the attrition of food between its harvesting in the field and its delivery to the consumer. These include scientific storage at the right temperature and humidity with protection from rodents and insects, protective packaging (for which polyethylene and other synthetic wrappings have been particularly useful) and rapid transportation in refrigerated ships, cars and airplanes. Today there is virtually no food that cannot be delivered fresh and

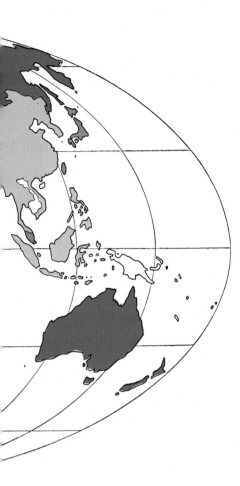

population is undernourished. While production is rising, so is the number of people.

with only minor losses to consumers everywhere.

Better food production and better food conservation are the prime requirements of the ill-fed countries. There is a third modern development that could also help them tremendously—artificial enrichment of their food with vitamins and other substances.

Everyone knows the story of the dramatic conquest of goiter in the U.S. and elsewhere by the simple device of adding iodine to the salt. Iodized salt in the 1920's practically eliminated goiter in the U.S. Middle West and Switzerland, where iodine is missing from the normal inland diet. In recent years Guatemala by the iodization of salt has abruptly reduced the incidence of goiter to less than 10 per cent in areas where it was formerly 30 to 60 per cent, and Colombia has achieved similar results. Salt iodization is now an officially sponsored practice in a number of underdeveloped countries.

Other deficiency diseases, such as pellagra and beriberi, can be eliminated by the simple addition of vitamins to wheat flour and polished rice. Here it is usually a case of restoring valuable food elements that are lost in the processing of the whole grain into the "refined" food. Since 1941 the enrichment of wheat flour with thiamine, riboflavin and niacin has been a general practice in the U.S. and Canada, and it is required by law in Puerto Rico and the Central American countries. Such legislation should be adopted by all countries depending on refined wheat flour as a basic food. The same goes for polished rice. In a test on a large scale in the Philippines from 1948 to 1950 it was shown that enrichment of polished rice with vitamins was very effective in combating beriberi in this rice-eating population. Corn meal also needs to be enriched; a diet consisting mainly of corn may produce pellagra, the disease resulting from a deficiency of the vitamin niacin combined with a diet low in the amino acid tryptophan.

Enrichment of a nation's wheat, corn and rice with the vitamins thiamine, riboflavin and niacin, plus calcium and iron, costs only a few cents per person per year. It would produce significant improvements in the health of most ill-fed populations, and it is strongly recommended by international health organizations.

A great part of the hungry half of the world suffers primarily from a deficiency of protein. In most vegetables and other plant foods the protein content is low in quantity and poor in quality—meaning that it is only partly metabolized by the body. High-quality protein is hard to come by. In many of the underdeveloped countries it would require the relatively wasteful allocation of land to pasture for animals, whereas it is often more efficient at present to devote the land to the direct growing of food for human beings.

Fortunately, however, low-protein foods can be enriched economically by adding a source of the missing amino acids that are essential to the synthesis of proteins by the body. The nutritive value of corn meal, for example, can be greatly improved by adding to it a supplement of 3 per cent fish flour, 3 per cent egg powder, 3 per cent food yeast, 5 per cent skim milk, 8 per cent soybean flour or 8 per cent cottonseed flour. Any of these supplements will supply material for protein synthesis and also improve the efficiency of utilization of the protein in the corn meal.

A most promising development is the progress that is being made in the artificial synthesis of amino acids themselves. Synthetic methionine is already being fed to animals in the U.S. on a considerable scale. The addition of lysine to wheat flour or bread can raise the proportion of the wheat's usable protein from about a half to two-thirds, and the amino acid threonine could make grain protein almost as fully usable as the proteins of meat and milk. The main problem so far is the cost of the synthetic amino acids. As more of them are synthesized and the price is brought down, these products of laboratory chemistry will make it possible to turn grain into meat for the meatless regions of the world.

Already nutritionists, using only natural sources, have concocted mixtures that can make a purely vegetarian diet richer in protein. The basic ingredients are a cereal grain, such as corn, rice or wheat, and an oilseed meal. This meal, or flour, is made from the cakes that are left when the oil is pressed out of the seed. It is consequently less expensive than comparable animal protein because it is a dividend remaining after sale of the oil. It generally contains about 50 per cent protein. Good sources of oilseed meal are cottonseed, soybean seeds, sesame seeds, sunflower seeds and peanuts.

When a properly processed oilseed meal is mixed with a grain in the ratio of one part meal to two parts grain, the combination contains about 25 per cent protein of meatlike quality. With the addition of a small amount of yeast and vitamin A it makes a highly nutritious food. In tropical and subtropical areas it could serve as a complete basic food lacking only vitamin C (which is supplied in abundance by tropical fruits and vegetables) and sufficient calories. The latter are obtained readily from sugar, starchy vegetables and such fruits as bananas and plantains.

Low-cost mixtures of this kind have been developed by the Institute of Nutrition of Central America and Panama. Under the generic name of Incaparina, they are already being manufactured and sold as basic foods in Guatemala, El Salvador, Mexico and Colombia and will soon be available in other Latin-American countries. Incaparina has been found to be almost as good a protein source for young children as milk, and it has proved to be effective in preventing or curing protein malnutrition in children. Almost every region of the world either has already or can grow the raw materials for this food. The basic formula is about

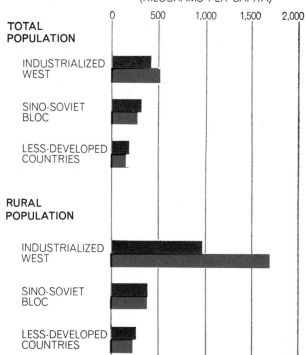
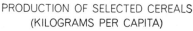

YIELD OF SELECTED CEREALS
(QUINTALS PER HECTARE)

PRODUCTION OF SELECTED CEREALS
(KILOGRAMS PER CAPITA)

WESTERN EUROPE

JAPAN

U.S.

U.S.S.R.

EASTERN EUROPE

COMMUNIST ASIA
INCLUDING CHINA

TOTAL POPULATION

INDUSTRIALIZED WEST

SINO-SOVIET BLOC

LESS-DEVELOPED COUNTRIES

RURAL POPULATION

INDUSTRIALIZED WEST

SINO-SOVIET BLOC

LESS-DEVELOPED COUNTRIES

EFFICIENCY OF AGRICULTURAL PRODUCTION has grown in industrialized areas but has remained static or even declined elsewhere. Gray bars represent years 1935–1939, colored bars the year 1959. Chart at left shows yields of selected cereals. A quintal is 100,000 grams, or about 220 pounds. A hectare is 2.47 acres. Charts at right show production of selected cereals per capita of total population (*top*) and of rural population (*bottom*). The latter includes people living in both rural areas and small villages.

55 per cent grain (corn, sorghum, rice, wheat or whatever other cereal is available locally), 38 per cent oilseed meal, 3 per cent torula yeast, 3 per cent leaf meal (as a source of vitamin A) and 1 per cent calcium carbonate.

Many other schemes for getting more protein from plants have been studied. One on which a great deal of work has been done is the growing in liquid culture of the single-celled alga *Chlorella*. Efforts have also been made to concentrate or extract protein from grass, vegetables, cereals and other plant materials. But so far all these investigations have been disappointing in one way or another: the food produced is either too expensive, unpalatable or low in nutritive value.

Then there is the sea, whose tremendous population of fish and other edibles continues to excite the imagination of those concerned with the world's food problem. The main obstacle here is the cost of storing the catch of fish; such storage requires mechanical refrigeration, not generally available in the underdeveloped countries. The grinding of fish to make a protein-rich flour looks like a promising answer to this problem, but it calls for technical skill and costs more than providing protein in the form of surplus dried skim milk or oil-

seed meal. Moreover, large quantities of fish flour are not attractive in a basic daily diet. All in all, it must be said that sea food offers possibilities one should not neglect but that it cannot be regarded as a panacea for prompt solution of the world's food problems.

Finally, there is the dream of manufacturing completely synthetic foods at a cost low enough to end all food worries. After all, the essential nutrients man requires are basically chemicals whose formulas are well known. Most of them can be synthesized in the laboratory, either by direct chemical manipulation or with the help of microorganisms. We already have synthetic vitamins, synthetic amino acids, hydrogenated fats, artificial flavoring and coloring agents and so on. From a concentrate of soybean protein the skill of the food chemist can prepare a meat-like product that with proper flavoring, coloring and molding can pass for pressed ham or chicken.

The cost of such creations is still exorbitant. But the progress of chemistry is steadily reducing the cost, and almost certainly we shall eventually have synthetic foods that will compete in cost, palatability and nutritive value with the products of the farm. Although that day is too far away to promise relief of the present food crisis in the underdeveloped

regions, it may help to forestall the crises threatened for the future by the growth of the world's population.

Along with modern food technology go modern dangers. As man takes a more active hand in shaping and extending his food supply he introduces new hazards in what he eats—mainly potentially dangerous new food additives. Thus the safety of our food has become a paramount issue in the industrial age.

Indeed, it has always been something of an issue. We tend to overlook the fact that there are toxic substances in most of the plants we use for food, even the common ones. Fortunately they are usually eliminated or reduced to harmless proportions by cooking or other processing.

Many legumes (notably soybeans) contain an inhibitor that interferes with the action of the protein-digesting enzyme trypsin. Some also have substances that clump the red blood cells. Cabbages and several other common vegetables contain materials that deny iodine to the thyroid gland and so tend to produce goiter. Certain vegetables and cereals have high concentrations of oxalates and phytates, which bind iron and calcium and prevent the use of these minerals by the body. There are also common plants that harbor some of the deadliest poisons known to man. The cassava root contains cyanides; lima beans, the common vetch and the broad bean have a glucoside that gives rise to cyanides; the broad bean also contains a compound that causes hemolytic anemia; the chick-pea contains an unknown substance that produces the disease lathyrism (spastic paralysis of the legs). Consequently man has always had to be careful, and still needs to be, in his choice and processing of natural foods.

The new dangers arise from the increasing and necessary use of chemicals at all stages in the production and handling of food, from the planting of the seed to the packaging of the final product. The hazard begins with the pesticides and other poisons used to protect and promote the growth of the plant. (One may hope that another contaminant—radioactive fallout from nuclear tests—will now effectively be eliminated.) After harvesting, grain and legumes become subject to poisoning by molds unless they are properly stored. Then there is the potential toxicity of residues of the hormones and antibiotics that have become a standard part of the feeding of meat animals. Next come the chemicals added to foods during the processing for flavor, color and preservative purposes. Along the way the food may pick up traces of toxic detergents that have been used to clean the tanks

or containers in which it is processed. Finally, the wrappings in which the food is packaged may inadvertently add some toxic contamination.

The whole sequence is imperfectly known; no one can be quite sure just where all the dangers lurk. Gradually the advanced countries have awakened to the need for vigilance in all stages of the handling of food. With respect to the chemical treatment of foods, U.S. legislation and the policy of the Food and Drug Administration are now based on the principle that "there are no harmless substances; there are only harmless ways of using them."

If the technically developed countries are concerned about the safety of their food supplies, obviously the less developed ones must be even more so as they attempt a rapid modernization of their food-producing and food-processing methods. The control of food contamination has become a world-wide problem, and the World Health Organization and the Food and Agriculture Organization of the United Nations have initiated conferences, committees and periodic reports on control regulations in the various countries. International research and standards will be helpful, but each country must take the responsibility for guarding the safety of its own food supply.

What can be done to help the hungry half of the world pull itself up from its undernourished state and speed up

the developments that would enable it to feed itself decently?

Even pessimists must note, first of all, that the prospects of the impoverished peoples are brightened by a most remarkable turn in human history. Whereas in the past men have been concerned only with feeding their own families and have fought long and bitter wars for food, we see today a new and remarkable world-wide concern for feeding the hungry wherever they are. Whether this arises out of advanced humanitarianism, the fears of the well-fed or the contest between the West and Communism is less important than the fact that the wealthy countries are taking an interest in the peoples of the poor countries.

During the past nine years the U.S. has sent more than $12 billion worth of its surplus food to these countries. The Food and Agriculture Organization, at the suggestion of Canada and the U.S., has launched an international effort for the same purpose with a $100 million fund as a starter, and it is now conducting a five-year Freedom from Hunger campaign.

This emergency help is not to be underestimated, and one hopes that it will be continued and even enlarged, preferably under international auspices.

A second way in which the developed countries are helping substantially is by example and by technical advice and assistance to the developing areas. The example, again, is important. The U.S. Department of Agriculture has estimated

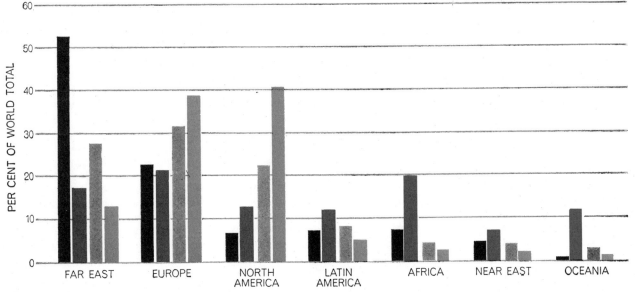

PER CENT OF WORLD TOTAL

FAR EAST   EUROPE   NORTH AMERICA   LATIN AMERICA   AFRICA   NEAR EAST   OCEANIA

■ POPULATION
■ AGRICULTURAL LAND
■ AGRICULTURAL PRODUCTION
■ INCOME

UNEQUAL DISTRIBUTION of world's income, agricultural land and agricultural production in relation to population shows up plainly on chart. Far East, with more than half the world's population, has less than 15 per cent of the income. Figures come from a study by the FAO in 1956. The category of Oceania embraces only Australia and New Zealand.

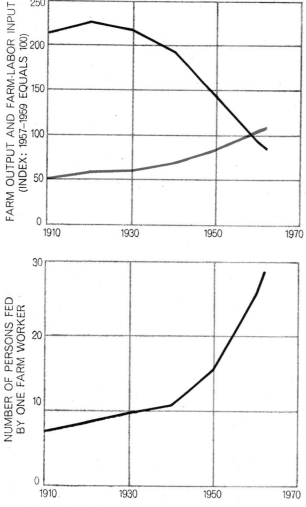

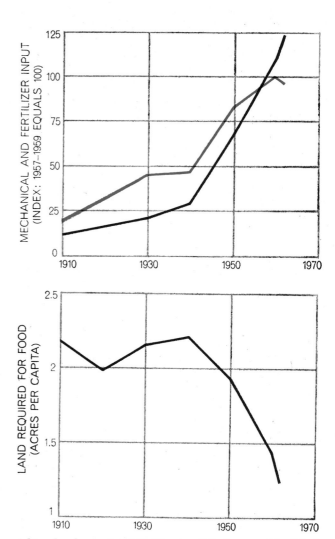

VAST IMPROVEMENT of agricultural production in U.S. demonstrates value of modern techniques. At top left, index of farm-labor input (*black curve*) declined from 212 in 1910 to 85 in 1962, while the farm output index (*colored curve*) rose from 51 to 108. The change was due in great part to increased use of machinery (*top right, colored curve*) and fertilizers and liming materials (*black curve*). The number of persons supplied by one farm worker in 1820 was 4.12. By 1910 it had risen to 7.07 (*bottom left*) and in 1962 it had jumped to 28.57. The number of acres needed to feed one person declined from 2.17 in 1910 to 1.23 in 1962 (*bottom right*).

that at the present rate of progress in agricultural productivity the developed countries will be able to produce almost twice as much food as they need by the year 2000. Such an advance cannot fail to infect and stimulate the backward countries.

Yet when all is said and done, these countries must themselves generate the means for their emancipation from hunger. To do so they will have to change long-established habits and attitudes. Neither well-meant exhortations nor government decrees are likely to persuade them—certainly not in a hurry. Concrete steps may, however, speed reforms by quickly convincing the people of their value.

It is easy to list effective projects that the governments of these countries might undertake. Make available to the farmers the seeds and stocks of improved plant varieties and animal breeds. Build chemical-fertilizer plants. Supply agricultural chemicals for pest control and other special purposes. Provide new implements and machinery suitable for the local types of farming. Extend credit to the farmers for their new seeds and equipment. Pay them subsidies to start urgently needed new crops. And above all, establish training programs that will show them how to handle their new materials and equipment and to farm more efficiently.

Education must receive the first priority for the advancement of these countries. The development of each one of the Western countries has been founded on the literacy and knowledge of its population. This applies to their progress in agriculture as well as to their achievements in industrial technology and professional services. To raise itself the underdeveloped country requires a population that understands modern agriculture and nutrition, is equipped with teachers and experts in all the fields of food technology and is led by political and administrative officials who appreciate the possibilities of science and technology.

This will be a long and difficult program for some of the poorly educated and ill-fed nations. But investment in education is a far more practical and effective program for them than investment in big buildings, dams, roads and factories that are put up mainly as visible symbols of progress. Just as the strength of the so-called developed countries lies in their educational systems and their culture, so the great hope and promise of the future for the underdeveloped countries resides in the fact that they too will come to share in the full wealth of mankind's knowledge and contribute to it themselves.

24

PEASANT MARKETS

SIDNEY W. MINTZ
August 1960

On market days in Haiti the towns and the country market-places gather thousands of peasants for hours of busy and noisy activity. The people come for gossip, courtship and the playing-out of personal rivalries, to visit a clinic or to register a birth; but above all they come for business—to sell the tiny surpluses of their little farms and to buy necessities. They press together in the ragged lanes among the stalls and the heaps of produce spread on the ground, inspecting and handling the displays of textiles, hardware, spices, soap and cooking oils, buying, selling and chaffering. Children push by hawking trays of sweets; farmers pull produce-laden animals through the crowds, calling loudly for the right of way. Trucks back up and turn around, their drivers honking horns, apparently oblivious of the people and the great piles of goods. There are vigorous arguments, sometimes ending in blows and arrests. In the very intensity of color, sound and smell the outsider is overwhelmed with an impression of confusion and disorder.

But for all its apparent anarchy the market place is characterized by an elaborate underlying order. Wherever they exist, peasant markets reveal a great deal about the societies they serve. They are a central economic institution in many countries where large numbers of small-scale farmers work their own land. To follow the movement of marketers and stock through the system is an ideal way to begin to study the economy and to trace the distribution of economic and political power in the society.

In the simpler economies, wherein producers merely exchange local commodities, the market place may do little more than facilitate barter. In societies that use money but have fixed or traditional prices, the market place reflects

that isolation from the world market; its transactions neither affect nor respond to economic events in the world at large. Where trade crosses national boundaries, links diverse regions and supports specialist traders, the market place takes on a new significance, joining local activities to the world outside.

As the underdeveloped areas of the world—for example, Jamaica, Haiti, Ghana, Nigeria, India, Burma, Indonesia —move more fully into the orbit of world trade, their market systems have been passing from the earliest of these stages to the next. The transition disrupts traditional relationships and creates new alignments and rivalries in the society, and these are nowhere more dramatically revealed than in the market place. It is here that the peasant trades his surplus for the necessities he cannot produce from his own holding, and it is the market place that determines, directly or indirectly, the prices at which the exporter will purchase the peasant's produce for delivery to the world market. In certain countries the connection that the market establishes between the peasant producer and the world market is the keystone of national development. Those who hold political power may use the peasant market-system to try to educate, persuade, coerce and manipulate the peasantry, particularly with the aim of maintaining or increasing export production. The market places are primarily loci of trade, but they are also the arena where the diverse interests of the peasantry, traders and officials are pitted and exposed.

The study of the tangle of interests that animates the peasant market thus brings into the open numerous connections between regions, classes and interest groups. Traditional anthropologi-

cal studies, which focus on small, local groups, cannot yield comparable insights into such large, differentiated societies as those of India and Nigeria or even Haiti. Courts and legislatures provide good settings for observation of the competing elements in a society. But the market place reveals far more because it allows these elements so much greater freedom to express themselves.

Though man has probably been a trader since the beginnings of society, his trading activities have not invariably produced market places. When the Spanish conquerors came to the New World, for example, they were stunned by the size and grandeur of such Aztec market places as Tlatelolco, where 50,-000 traders assembled on market day. A wealthy merchant group, the *pochteca*, controlled trade and wielded considerable power, and also served as efficient spies for the military. But in the great contemporary Andean empire of the Incas the conquistadors found neither market places nor merchants. Instead of trade they found royal monopolies in gold, silver, coca and fine textiles. Thus while market places are not found everywhere, their very absence tells us something about a society. The presence of markets does not necessarily imply a particular course of social development. Yet there are striking similarities among the peasant markets of the world, especially those of the new nations of Africa, Asia and tropical America.

Haiti is an older nation, with a history of political independence. But at its present stage of economic evolution this Caribbean republic is representative of the new nations that are emerging in the colonial regions of the world. Before the revolution of 1791-1804, Haitian slaves grew their food on plantation wastelands, selling surpluses in supervised

BUSTLING ACTIVITY OF MARKET DAY animates a clearing in rural Haiti. Tradeswomen are grouped by commodity they sell. In this photograph woman in left foreground inspects wares of grain seller. At left in middle distance is lean-to of tuber sellers.

market places. In the 1790's the French observer Moreau de Saint Méry described such a market place, where 15,000 slaves traded on market day. The revolution destroyed the plantations that had made the island of Saint-Domingue one of the richest colonies in history, and substantially eliminated the French planters. Gradually Haiti became a peasant country where small-scale landholders cultivated their subsistence crops for local sale and a few items for export. The cash they received paid for the soap, cloth, oil, metal tools and flour they needed and could not produce. The national government sustained itself almost entirely by taxes on imports and exports; the local government, by levies on dealings in the market place.

Today, 150 years later, nearly 90 per cent of the people live in the countryside, and 80 per cent of them work their own land. Haitian peasants still cultivate much of their own food and produce a small surplus destined for export or for consumption in the domestic economy through sale in the peasant market. By aiming at these three different production goals they try to minimize risk and to secure a reasonably stable subsistence. They further hedge their investment of time and capital by diversifying the cultivation of their land, and this accounts for the curiously cluttered look of their little plots. Like other Caribbean farmers, the Haitian peasant makes thorough use of his land: he grows root crops underground, vines and creepers on the surface, grains above ground and trees and climbing vines in the air. Though technologically backward, the method provides a constant trickle of varied produce for the household where storage is difficult or impractical, a supply of craft and medicinal materials as required and a small quantity of items for sale at various times. It is upon this foundation that the Haitian market-system rests.

The peasant's wife most often handles the market transactions of the family, selling what the land has produced for sale and using the cash received to buy household necessities. Many peasant women become professional traders in this way. This further distributes the family's economic risks, since the men do the farming and the women do the trading partly as separate ventures.

Most of the trade in Haiti goes on in the nearly 300 officially controlled market places. In each region one or more central market-places services other, smaller centers. The larger centers are established in the towns; the satellite country market-places spring up over-

HAITIAN FARMS CLING TO TERRACED HILLSIDES near rural village of Kenskoff. Characteristically cluttered, the Haitian farm grows small but widely diversified crops.

ROAD TO MARKET PLACE is crowded with peasants. Whole families often walk all night, carrying their crop surpluses, in order to arrive at the market in time for early trading.

night as little towns that last only through the day to gather in and to absorb the peasant buying-power. Market days are staggered, enabling itinerant buyers and sellers to move from one market to another. Most of the important market places are on well-traveled truck routes. Thus the markets form a network, and the produce bought in one is put up for resale—after bulking, processing and transport—in another. For example, pork purchased in one market is cut up, salted and shipped to the next, while rice, millet and maize are husked or ground between purchase and resale to increase their value. The whole system of market places constantly adjusts and readjusts to seasonal changes, to the success or failure of harvests, to the growth and contraction of production areas, to the expansion of roads and trucking.

Trade begins beyond the fringes of the market, where licensed tradesmen from the towns, called *spéculateurs,* maintain outposts at which they buy commodities for export. Peasant women on their way to market stop to sell their coffee, beeswax and sisal to the *spéculateurs,* and then proceed to market with the cash they have received. Because competition is heavy and supplies uncertain, *spéculateurs* do not always wait for the peasants to come to them. They often send illegal buyers called "zombies" or "submarines" to make purchases directly at the farms.

But it is in the tumult of the market place that most trading activity goes on. Only after many days of observing and classifying the actors and their activities does the underlying order become apparent. Sellers of the necessities that peasants come to buy are present each day in a given market place. Perishable foods come and go seasonally, but grains are nearly always available. Prices for different products fluctuate differently, perishables showing the greatest eccentricity, cloth and hardware changing very little from week to week, though perceptibly from season to season. Watching the market place each market day, one sees women dealing in the same goods always clustered together; grain sellers, corn-meal retailers and sellers of spice and sundries arrange themselves in rows. For the seller this permits a quicker check on the day's trade, on one's favored customers, and of course on prices. When sellers of the same stocks are together, the speed with which price is established, and with which it changes during the trading, is increased. Buyers of particular goods come regularly where the sellers are clustered.

PEASANT MEN WORK AS ARTISANS in the market, rarely as tradesmen. At left is a cobbler who rebuilds discarded shoes and sells them. At right a blacksmith hammers sheet metal into hoes. Artisans usually tend their farms on other than market days.

Behind the facade of apparently uniform and competitive prices, however, there exists a relationship called *pratique*, in which the retailer gives her favored customers certain concessions in price or quantity or in the terms of credit in return for assurances of the customer's patronage when the market is glutted and prices are low. The retailer also makes *pratique* with her suppliers, thus assuring herself of a stock when certain commodities are scarce. Since *pratique* is a clandestine relationship, it can only be understood by carefully noting the details of many transactions.

Even the casual observer soon notices that the important heavy trade in perishables and the small-scale retailing of imports are carried on entirely by women. Men rarely trade; both sexes believe women are commercially shrewd-er than men. There are, to be sure, male traders, but with the exception of the peasant who has come to market to sell livestock or craft articles they are almost always townsmen. The peasant woman makes her entrance into the market as a trader on the most modest terms, first as her household's representative in the market, then perhaps with a small stake borrowed from relatives or other traders. In a country where a handful of grain makes a meal and a bit of land a farmer, a few pennies constitute operating capital for the middleman, and what one can carry in one's hands is enough stock to begin trading. If the woman is resourceful she may parlay her small stake in a series of small trading transactions to a sum sufficient to secure her status as a *revendeuse* (literally reseller).

Thousands of these women move from market place to market place, each deal-ing in small amounts, but together buying and selling vast quantities of stock. They live by connecting centers of supply and demand; their potential profit rests in the price differentials between regions and in their ability to contribute to the value of products by carrying, processing, storing, bulking and breaking bulk. They often render services at incredibly low cost. Thus salt retailers in one market place interpose themselves between truckers and consumers, breaking bulk and retailing salt for earnings that sometimes fall below five cents a day. If these services were not provided, consumers would have to buy in uneconomically large quantities, or truckers would have to sell in uneconomically small ones. The fact that consumers buy from them even though they sit only a few feet from the trucks that bring the salt is proof that the service they sell is

TRUCKING AND TINKERING are other male occupations. Trucks carry resellers and their stock between markets. They serve a vital function because regular bulk shipments are unknown. Tinker (*right*) does brisk business because new pots are expensive.

WOMEN DOMINATE TRADE in perishables in Haiti. In photograph at left, purchaser holds a measuring can while bean seller fills it. They may fill and empty the can again before agreeing that it is properly filled. At right is section of the tuber market.

SUCCESSFUL TRADESWOMAN operates between markets. Her annual volume of business may amount to thousands of dollars.

SALT VENDOR'S CHILD arranges stock of coarse, unrefined salt for sale. Salt is among the few commodities shipped in bulk.

worth buying. In their intermediary activities, the *revendeuses* scour remote countrysides. They buy basic commodities at their sources, where they are cheap, because the economic integration of the back country with the national economy is incomplete. Thus, by servicing buyers and sellers both, they help unite the peasant plot and the local market place with national currents of exchange, stabilizing general price levels and contributing to economic growth. The path to success is uncertain, but some few reach the top. The volume of a *revendeuse's* business may approach that of the famous "market mammies" of Nigeria, whose transactions amount to thousands of dollars a year. Though all apparently aspire to become city retailers, women with rural family-attachments incline to remain in the countryside and usually identify themselves with the peasantry.

For transportation from market place to market place the *revendeuses* depend upon the truckers. Demand is not sufficiently firm and centralized to give the truckers bulk cargoes to haul. It is not surprising, therefore, to discover that they are essentially passenger carriers, whose business it is to transport the *revendeuses* and their modest stocks. Trucking is a risky enterprise in a country where roads are few, maintenance facilities are poor and high taxes are levied against fuel and passengers. The trucker is a relatively new figure in the economy. His economic interests are at present firmly identified with those of the *revendeuses* and opposed to the *rentiers,* merchants and officials of the towns and cities. With the *revendeuses* he is against any forces aimed at the restriction and centralization of trade in Haiti. On the other hand, if the growth and evolution of the economy should make it possible for the trucker to profit by bulk transport, this general accord might well vanish.

In such an eventuality the truckers might find themselves allied with the townsmen. The *spéculateurs,* coffee processors, wholesalers and merchants, separately and in combination, all aim to encompass as much of the peasants' economic activity as possible. For although each peasant may be poor, the wealth that changes hands when thousands of peasants shop in the market place is considerable. Successful market places outside the towns constantly tempt the town merchants, particularly cloth- and shoe-sellers, who carry large stocks from their town shops into the country on market day. The townsmen

MARKETING OF FISH is characteristic of way trade is conducted in Haiti. Resellers wade out to fishing boats to buy fish (*photograph at top*); return to shore to sell them to waiting consumers (*photograph at bottom*). Fishermen will not deal directly with consumers.

would of course prefer that all peasant trading took place in towns, where markets could be centralized, and the small reseller subjected to more control. To that end they have inspired repeated attempts at restrictive legislation. In this they are joined by the *rentiers*, the value of whose property would appreciate with increase in town trade. The importers and exporters among the city merchants would prefer to see export production rise, even at the cost of subsistence crops.

Officials of the national government are often similarly disposed. Unlike the local governments, which are largely supported by taxes on market-place transactions, the national government derives its revenues chiefly from taxes on imports and exports. Hence the state officials want to see peasant agriculture producing more exportable goods. One could say that their aim is to maximize the peasantry's taxable income.

In the market place one sees the whole structure of official power: police, military, judicial, executive. All market places in Haiti are under some supervision by state officials, who carry on two major and familiar functions: maintaining order and collecting taxes and license fees. At the top this structure is tied to the ministries in the capital; at the bottom it embraces notaries, justices of the peace, soldiery and local political leaders. It is within the market place, in the regulation of concrete economic transactions, that the penetration of political control is seen at its most complete as well as its most trivial. State officials supervise the workers who clean and maintain the market place; they catch and imprison thieves; they stop fights. They are supported by the lowest ranks of political officials, the *chefs de section*, who come from the rural areas to the market place to oversee peasants from their neighborhoods. The peasantry's name for any official, no matter how lowly, is always *l'état*.

Traders are taxed or licensed for taking livestock to market, for butchering, for selling animals, for selling meat, for selling foods of any kind, for selling alcoholic beverages and tobacco, for dealing as intermediaries in all other agricultural products, for tethering beasts of burden and for the stands and sheds they use to display meat and other products for sale. This revenue goes largely to governments of the *arrondissements*, though part is drained off by the national government. Tax revenues are used for the operation of local governments, and to pay for tax collectors' salaries and the administration of the tax system.

Just as the political and commercial elements of the towns seek to centralize and control the markets, so the rural tradesmen seek to maintain the status quo. Their interests dictate a diffuse and open market in which ingenuity and intelligence enable them to compensate for lack of capital, and where they may hope to make the transition from perishable-produce dealers to hard-goods wholesalers or credit merchants.

The contention of these various groups, however, is ultimately intelligible only in terms of the behavior and power of the peasant, whose best interests do not lie decisively in either camp. In determining how he may maximize his cash income by transactions in the market place, he weighs the demand and prices of the domestic market against the opportunities offered by the export market. In striking a balance between the two alternatives he may incline toward the production of export goods to supplement his cash, but he is wary of export-market fluctuations which can deeply affect him and which he cannot control. His choices are not entirely free, for the various factions of the nation, especially those that favor increased production for export, exert considerable pressure upon him. As the source of his cash income, the market places are the peasantry's first line of defense against greater dependence upon the world market and a greater involvement with the officialdom of the state.

Apparently the alignments of interest that may be discerned in the peasant markets today have characterized Haitian society for many years. During the 19th century Haiti's seacoast towns sought to maintain economic hegemony over the inland towns. In the struggles of town merchants against the peasantry, and of the seacoast against the interior, climaxed by the capital's economic dominion over the nation, there are startling parallels with conflicts waged during the growth of capitalism in the nations of Western Europe. In both cases groups with vested economic interests sought to restrict the spread of competitive trading activity.

Thus study of an internal market-system may provide a lively vision of relationships among key economic and political groups in a society. Eventually it may be possible to compare internal market-systems in different societies as total systems, thereby revealing similarities and differences among the societies themselves that might otherwise be difficult to discover.

25

ORTHODOX AND UNORTHODOX METHODS OF MEETING WORLD FOOD NEEDS

N. W. PIRIE
February 1967

The world has been familiar with famine throughout recorded history. Until the present century some people have been hungry all the time and all the people have been hungry some of the time. Now a few industrialized countries have managed, by a mixture of luck, skill and cunning, to break loose from the traditional pattern and establish systems in which most of the population can expect to go through life without knowing hunger. Instead their food problems are overnutrition (about which much is now being written) and malnutrition. Malnutrition appears when the food eaten is supplying enough energy, or even too much, but is deficient in some components of a satisfactory diet. Its presence continually and on a large scale is a technical triumph of which primitive man was incapable because he lacked the skill to process the food he gathered in a manner that would remove some of the essential components but leave it palatable and pleasing in appearance. Furthermore, until the development of agriculture few foods contained the excess carbohydrate that characterizes much of the world's food today. The right policy in technically skilled countries, however, is not to try to "go back to nature" and eat crude foods. Processing does good as well as harm. What we now need is

widespread knowledge of the principles of nutrition and enough good sense to use our technical skill prudently.

It is salutary to remember how recently this pattern was established. There was some hunger in Britain 50 years ago and much hunger 50 years before that. Still earlier many settlements in now well-fed regions of Australia and the U.S. had to be abandoned because of starvation. It is said that scurvy killed about 10,000 "forty-niners," and California was the scene of some of the classic descriptions of the disease. One has to learn how to live and farm in each new region; it cannot be assumed that methods that are successful in one country will work elsewhere. It is therefore probable that methods will be found for making the currently ill-fed regions productive and self-sufficient. The search for them should be started immediately and should be conducted without too much regard for traditional methods and preconceptions.

The problem can be simply stated: How can human affairs be managed so that the whole world can enjoy the degree of freedom from hunger that the industrialized countries now have?

It is well known that in many parts of the world not only is there a food shortage but also the population is in-

creasing rapidly. Some of the reasons for this situation are fairly easy to establish. When the conditions of life change slowly, compensating changes can keep pace with them. In Europe during the 16th century half of the children probably never reached the age of five. There are no general statistics for this period, but in the 17th century 22 out of the 32 British royal children (from James I to Anne) died before they were 21, and it is unlikely that the poor fared better than royalty. The establishment of our present standards of infant mortality had little to do with medical knowledge. Until this century the farther away one could keep from doctors, except for the treatment of physical injury, the better. It was increasing technical skill in bringing in clean water and getting rid of sewage that made communities healthy, and this skill was applied by people who had never heard of germs or, like Florence Nightingale, disbelieved what they were told. But the change came slowly enough for families to adjust the birthrate to suit the new conditions. Moreover, there was incompletely filled land to be used. What René Dubos calls the "population avalanche" is on us because it is now possible to undertake public health measures on a larger scale and finish them quicker than heretofore.

LATIN AMERICA

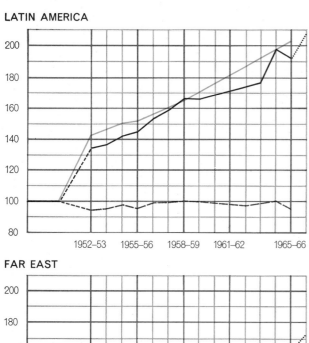

NEAR EAST

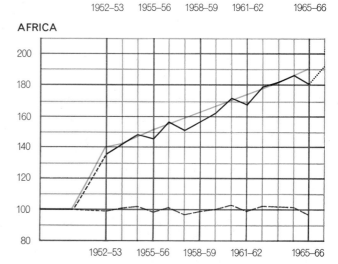

FAR EAST

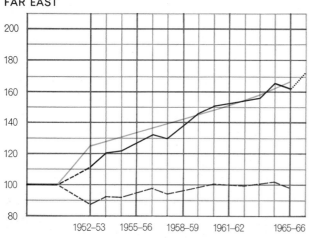

AFRICA

POPULATION AND FOOD PRODUCTION are compared for four developing regions in these charts prepared by the Food and Agriculture Organization of the United Nations. The colored curve is population; the solid black curve, food production; the broken gray curve, food production per caput. The figures for population are in millions; those for food production are given according to an index of 100 for the prewar average. The food production figures for 1965–1966 (July 15, 1965, to July 15, 1966) show the effects of adverse weather in many parts of the world. In that period world food production per caput fell 2 percent. The dots at end of food production curve show increase required to regain per caput level of 1964–1965. Mainland China is not represented in Far East figures.

Once the principles are understood the hygiene of an area can be improved quickly by a few people, and the population as a whole gets the advantage of improved health without having to take any very active steps to achieve it. Even where methods for improving conventional agriculture are known their application is of necessity slower, because it depends on a change in the outlook of most of the people in a farming community rather than in the outlook of the few who control water and sewage. Furthermore, fecundity is potentially unlimited but food production is not. Clearly, therefore, the "avalanche" will have to be stopped. It is important to remember, however, that it cannot be stopped in any noncoercive way without the cooperation of the people; that means more education, which means more hygiene and so, at least for a time, a still greater increase in the population. The

first result of an effective campaign for contraception will be an increase in population rather than a diminution. More effort should be put into the encouragement of contraception. And more research is needed on improved methods, leading to the ideal: that people should have to do something positive to reverse a normal state of infertility, so that no conception would be inadvertent. This, however, is a complement to, and not a substitute for, work on the production of more food. Strained as existing supplies are, it seems inevitable that they will be strained still further during the next half-century. After that the entire world may have established the population equilibrium that now exists in some industrialized countries.

The normal humane reaction in the presence of misery is pity, and this reaction is followed, where appropriate,

by charity. Hence the immense effort that is now being put into shipping the food surpluses, accumulated in some parts of the world, to areas of need. This is commendable and spiritually satisfying to the donor, but for two reasons it has little effect on the real problem. The amount of surplus food is not large enough to make much of a dent in the world's present need. The surplus could be increased, but the logistic problem of shipping still greater quantities of food would be formidable. The more serious objection to charity, except during temporary periods of crisis, is that it discourages the recipient. A century ago the philanthropist Edward Denison remarked: "Every shilling I give away does fourpence worth of good by keeping the recipients' miserable bodies alive and eightpence worth of harm by helping to destroy their miserable souls." Nearly 1,000 years ago Maimonides categorized the forms of charity and concluded that the most commendable form was to act in such a way that charity would become unnecessary.

Trade is the obvious alternative to charity. Unfortunately the developing countries are in a poor bargaining position. Since 1957 the prices paid for their primary products declined so much that the industrialized countries made a saving of $7,000 million and an extra profit of $3,000 million because of the increased cost of manufactured goods. The developing countries thus lost $10,000 million—about the same as the total "aid" they received from commercial, private and international sources. (The figures are from the *Financial Times* of London for July 19, 1965.) With the market rigged against them in this way it is not likely that they will soon be able to buy their food as countries such as Britain do. At present the industrialized countries are exporting about 30 million tons of grain a year, largely against credit. It is unlikely that this state of affairs can last; half of the world cannot permanently feed the other half.

The idea that the developing parts of the world should be fed by either charity or trade depends on the assumption that they are in some way unsuited for adequate food production. This idea is baseless. Once the methods have been devised, food can be produced in most places where there is sunlight and water; for political stability food must be produced where the mouths are. Any country dependent on imports for its main foodstuffs is to some extent controlled by others.

The problem can be more narrowly stated: How to produce enough food in the more populous parts of the world?

Food production can and will be increased in many orthodox ways. There is still some uncultivated but cultivable land, irrigation and drainage can be improved and extended, fertilizers can be used on a much greater scale and the general level of farming technique can be improved. If all the farmers in a region were as skilled as the best 10 percent of them, there would probably be enough food for everyone today. These improvements could be achieved by vigorous government action and without further research.

In the Temperate Zone plant breeders have greatly increased cereal yields during the past 20 years, and these improved varieties could be used more widely. There have been no comparable developments with food crops in the Tropical Zone, but there is no reason

to think that progress there could not be equally spectacular. This research should not be limited to cereals. In many parts of the wet Tropics yams (*Dioscoria* and *Colocasia*) are staple foods but the varieties used contain little protein. There is, however, some evidence that the protein content of yams varies; a New Guinea variety called Wundunggul contains 2.5 percent nitrogen. If this nitrogen is all in protein, the yam contains 15 percent protein and is worthy of serious study.

It is generally agreed that pests and

diseases rob us of as much as a third of our crops. When the improvements outlined above have been made, the proportional loss as well as the absolute one could become greater, since well-nourished crops, growing uniformly in large fields, are particularly susceptible. The cost of treatment may be only a tenth of the value of the crop saved; the methods are well publicized by firms making pesticides. There is no need to labor this aspect of the problem here. More attention should, however, be given to losses during storage; the need for

SYNTHETIC FOOD is represented by Incaparina, made of maize, sorghum and cottonseed by the Quaker Oats Company. The product has been skillfully promoted by Quaker Oats in Central America. The Spanish words at the top of this 500-gram package mean: "For 25 glasses or portions." Those below "Incaparina" mean: "It is very nourishing and costs little."

satisfactory storage techniques for use in primitive conditions is especially acute. So much mystical nonsense has been written by believers in the merits of "natural" foods that most scientists show understandable impatience at the idea that pesticide residues may be harmful to the ultimate consumer of the protected crop. Furthermore, a food shortage may well do more harm to a community than sensibly applied pesticides can do. There is, nevertheless, great scope here for research on improved techniques.

There are such good prospects that productivity can be increased by each, or even all, of these methods if they are assiduously developed that it seems to many experts that there is no immediate need for any more radical approach to the problem of world feeding. This is the attitude of the United Nations Food and Agriculture Organization (F.A.O.). One cannot praise too highly its work in compiling statistics and persistently calling attention to the need for agricultural improvements. On the other hand, while recognizing that the F.A.O. is not a research organization, one can deplore its equally persistent tendency to denigrate every unorthodox approach to the problem. History may partly excuse this attitude. Ever since the time of Malthus prophets have been making our flesh creep with warnings of impending famine. Conditions have remained much the same—or have improved. These prophecies remain unfulfilled because 400 years of explora-

tion enabled new land to be cultivated, 200 years of biological research laid the foundation for scientific agriculture, and 50 years of rational chemistry made it possible to produce fertilizers by fixing the nitrogen of the air. The cautious prophet should therefore not say that hunger is inevitable but that it is probable unless the relevant research is done on an adequate scale. The time to do it is now, before the need has become more acute.

The main product of agriculture is carbohydrate. The foods that make up the world's diet—the cereals, potatoes, yams, cassava and so on—are from 1 to 12 percent protein on the basis of dry weight. An adult man needs 14 percent protein in his food; children and pregnant or lactating women need from 16 to 20 percent. However great an increase there may be in the consumption of conventional bulk foods, there will be a protein deficit. Moreover, it will be exacerbated if food is made palatable by the addition of fats and sugar, which give energy but contain no protein. Too much stress cannot be laid on the fact that the percentage of protein in a diet is the vital thing; increased consumption of low-protein food makes the consumer fat but as malnourished as before.

Recognition of the importance of protein sources, and their deficiency in most of the world's diet, has come slowly. It is nonetheless gathering momentum. Fifteen years ago little attention was paid to protein sources by international agencies and gatherings such as the International Nutrition Congress. Now protein

is one of the main themes. Audiences at these gatherings are a step ahead of the management. At the International Congress of Food Science and Technology last year, for example, the session "Novel Protein Sources" proved more popular than those who had allocated the rooms had foreseen; that session was more uncomfortably overcrowded than any other. The remainder of this article will be exclusively concerned with protein. All the components of a diet are needed, but the need for protein will be the most difficult to meet.

Animal products—meat, milk, cheese, eggs, fish—are widely esteemed and are used as protein concentrates to improve diets that are otherwise mainly carbohydrate. About a third of the world's cattle population is in Africa and India; most of these animals are relatively unproductive and are maintained largely for reasons of prestige and religion. It is easy to sidestep the main problem and argue that the protein shortage in these countries could be ameliorated, even if it could not be abolished, if herds were culled and the remainder made fully productive. The more thoughtful Africans and Indians realize this, and the situation will doubtless change. But every community tends to devote an amount of effort to nonproductive activity that seems to outsiders unreasonable. In the Middle Ages cathedrals were built by people who lived in hovels, and we now spend more on space research than on research in agriculture and medicine. Change is inevitable, and contemporary forms of religious observance and prestige are certain to be modified; the transition will not be hastened by nagging from outside.

According to most forecasters, the need to grow crops on land now used to maintain animals will lead to a decline in meat consumption in industrial countries, and the essential disappearance of meat is sometimes predicted. Although the decline is probable, the disappearance is not. There is much land that is suitable for grazing but not for tillage. Furthermore, there will always be a great deal of plant residue that (perhaps after supplementation with urea) can be more conveniently used as animal feed than in any other way. It is by no means certain, however, that we will always use the ideal herbivore. There is good reason to think that several species of now wild herbivore, running together, give a greater return of human food in many areas of tropical bush or savanna than domesticated species [see "Wildlife Hus-

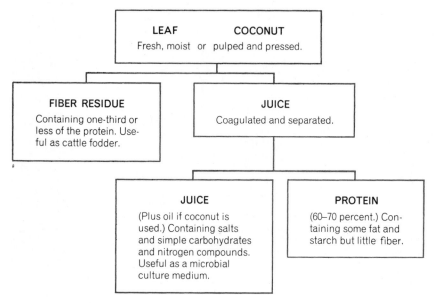

LEAVES AND COCONUT are a source of protein if fiber and juice are removed from them. This chart shows the steps of the process and also indicates the uses of by-products.

bandry in Africa," by F. Fraser Darling; SCIENTIFIC AMERICAN, November, 1960]. In addition, wild herbivores generally yield more protein per pound "on the hoof" than domestic species. These are matters that are being investigated by the International Biological Program. Even better results may be achieved after a few years of skilled breeding. Ruminants such as the antelope and the water buffalo are not the only species worthy of attention, and land is not the only site available for grazing. The capybara, a large rodent, is well adapted to South America and is palatable. Water weeds, and plants growing in swamps and on lake margins, contribute hardly anything to human nutrition. They could be collected and fed to land animals, but it would seem to be more efficient to domesticate the *Sirenia* (the freshwater manatee and the marine dugong) and use them as sources of meat. These herbivores are wholly aquatic and so, unlike semiaquatic species such as tapirs and hippopotamuses, do not compete for food with more familiar animals.

It is usual in articles such as this one to stress the importance of fish. This is admirable, but stress should not be allowed to drift into obsession. The more cautious forecasters estimate that the fish catch could be increased only two- or threefold without depleting stocks. Moreover, much of the world's population lives far from large bodies of water, and since fishing has an accident rate twice as high as coal mining it is likely to remain a relatively unattractive occupation. In the past decade the wet weight of fish caught annually increased from 29 million tons to 52 million, but the proportion used as human food decreased from 83 percent to 63. The remainder was used as fodder, mainly in the already well-fed countries. When still more fish are caught, the temptation to use a still larger proportion as fodder will be greater because much of the extra catch will consist of unfamiliar species. By grinding and solvent extraction these unfamiliar fishes can be turned into an edible product containing 80 percent protein. This process has suffered from every form of misfortune: the use of unsuitable solvents, commercial overstatement, excessive hygienic caution and political intrigue. It is nevertheless sound in principle and will do much to increase the amount of food protein made on an industrial scale and distributed through international channels.

One also hears of "fish farming." Activity that could properly be given this

MANATEE, an aquatic mammal, is an example of an unorthodox source of meat. It can also control aquatic weeds, which it eats. An adult manatee is between nine and 15 feet long.

CAPYBARA, a large rodent that lives in South America, has also been suggested as a source of meat. Like the manatee, it feeds on aquatic weeds. An adult is about four feet long.

ELAND, a large African antelope, is an accepted meat animal. Its importance as a food source is that it is adapted to grazing on marginal lands that are not suited to agriculture.

title is possible in lagoons but unlikely in the oceans because fish move too freely, and fertilizer, intended to encourage the growth of their food, spreads too easily into the useless unlit depths. With mollusks and crustaceans, farming becomes much more promising and deserves more scientific attention than it gets. Food for such marine invertebrates does not seem to be the limiting factor; many of them can use the world's largest biological resource, the million million tons of organic matter in suspension in the sea. Sedentary mollusks are limited by predators and attachment sites. The predators could be controlled and the sites, with modern materials, could be increased.

Animals that live on something that we could not use as food—forage growing in rough country, straws and other residues, phytoplankton and other forms of marine organic matter—cannot properly be said to have a "conversion efficiency." We either use an animal converter or these materials are wasted. Efficiency has a real meaning when we consider animals that live either on crops that people could eat or on crops grown on land that could have grown food. The inefficiency of animal conversion, expressed as pounds of protein the animal must eat to make a pound of protein in the animal product that people eat, is much greater than is generally realized. This unawareness probably arises from the tendency among animal feeders to present their results as the ratio of the dry food eaten to the wet weight (including all inedible parts) of the carcass produced. Furthermore, the figures generally relate only to one phase of the animal's life, without allowance for unproductive periods. It is unlikely that the true efficiency of protein conversion is often greater than one pound of food protein for every seven pounds of fodder protein; it is generally less. Although animal products are highly esteemed in most countries, their production is an extravagance when it depends on land or fodder that could have been used to feed people. The extravagance may be tolerated in well-fed countries but not in those that are short of food.

As I have indicated, the world's main food crops need to be supplemented with protein. Peas and beans, which are 25 to 40 percent protein, are traditionally used. Green vegetables and immature flowers are gaining recognition. They can yield 400 pounds of edible protein per acre in a three-to-four-month growing period, but because they contain fiber and other indigestible components a person cannot get more than two or three grams of protein from them in a day. This amount, however, is much more than is normally consumed, and such plants offer a rewarding field for research. The varieties cultivated in industrialized countries are often ill-adapted to other climatic conditions. The work of vegetable improvement that was done in Europe in the 18th and 19th centuries should now be replicated in the wet Tropics. Biochemical control would

be needed to ensure that what is being produced is not only nontoxic but also nutritionally valuable. This would be an excellent project for the International Biological Program; the raw materials have worldwide distribution and the need is also worldwide.

The residue that is left when oil is expressed from soya, groundnut, cottonseed and sunflower is now for the most part used as animal feed or fertilizer or is simply discarded. It contains about 20 million tons of protein, that is, twice the world's present estimated deficit. Because its potential value is not yet widely realized, most of this material is at present so contaminated, or damaged by overheating during the expressing of the oil, that it is useless as a source of human food. But methods are being devised, notably in the Indian state of Mysore and in Guatemala, for processing the oilseeds more carefully in order to produce an acceptable food containing 40 to 50 percent protein. The avoidance of damage during processing is not the only problem that arises with oilseeds; each species contains, or may contain, harmful components, for example gossypol in cottonseed, enzyme inhibitors in soya and aflatoxin in peanuts. Gossypol can be extracted or low-gossypol strains of cotton can be used (it is said, however, that these are particularly attractive to insect pests); enzyme inhibitors can be destroyed, and the infestation that produces aflatoxin can be prevented by proper harvesting and storage. The alternative of extracting a purified protein concentrate from the residues is often advocated. This approach seems mistaken; it increases the cost of the protein fivefold. In addition, the process is in the main simply the removal of starch or some other digestible carbohydrate, and carbohydrate has to be added to the protein concentrate again during cooking.

The residue left after expressing oil from an oilseed can be used because it contains little fiber. Coconuts and the leaves from many species of plants are also potential protein sources, but they contain so much fiber that it is essential to separate the protein if more than two or three grams is being eaten each day. The process of separation, although simple, is still in its infancy, and many improvements remain to be made. Units for effecting it are working in Mysore, at the Rothamsted Experimental Station and elsewhere. In wet tropical regions the conventional seed-bearing plants often do not ripen, but coconuts thrive and leaves grow exuberantly. It is in these

TANKS ARE FILLED WITH LITTLE FISH from a truck (*left*) at a fish meal plant in the Peruvian port of Callao. The fish are anchovetas, a variety of anchovy. By grinding and solvent extraction they are converted into a product that contains 80 percent protein.

BAGS OF FISH MEAL are piled in the yard of the same plant. Fish meal is currently used primarily as a supplement to feed for domestic animals. In 1965 Peru harvested 7.46 million metric tons of fish, a catch that makes it the world's leading fishing nation.

regions that protein separation has its greatest potentiality.

The protein sources discussed in the last two paragraphs would be opened up by handling conventional agricultural products in unusual ways. Attention is also being given to completely novel forms of production based on photosynthesis by unicellular algae and other microorganisms. The early work was uncritical and, considering the small increases in the rate of fixation of carbon dioxide given by these methods compared with conventional agriculture, the necessary expenditure on equipment was out of proportion. It is an illusion to think that algae have any special photosynthetic capacity. Their merit is that it is much easier to spread an algal suspension, rather than a set of slowly expanding seedlings, uniformly over a sunlit surface so as to make optimal use of the light. Recently more realistic methods, using open tanks and the roofs of greenhouses in which other plants can be grown during the winter, have been tried in Japan and Czechoslovakia. The product resembles leaf protein in many ways but contains more indigestible matter because the algal cell walls are not removed; it may prove possible either to separate the protein from the cell walls or to digest the walls with enzymes.

All the processes discussed so far depend on what might be called current photosynthesis. Microorganisms that do not themselves photosynthesize can produce foodstuffs from the products of photosynthesis in the immediate past (straw, sawdust or the by-product liquor from leaf-protein production) or in the remote past (petroleum, coal or methane). The former substrates would have to be collected over a wide area, whereas the latter are concentrated in a few places and so lend themselves to convenient large-scale industrial processing. At first sight this seems advantageous, and in fact it would be so were we merely concerned with increasing the amount of food in the world. That will be the problem later; now the important thing is, as I have said, to make food where the mouths are, and elaborate and sophisticated techniques are not well adapted to this end. The most valuable aspect of the research now being done in many countries on microbial growth on fossil substrates such as petroleum is that it will familiarize people with the idea of microbial food and so will hasten its acceptance when it is produced from local materials.

Finally, there is synthetic food. Plants make fats and carbohydrates so economically that it is unlikely synthesis could be cheaper. Many of the abundant plant proteins do not have an amino acid composition ideally suited to human needs. These proteins are sometimes complementary, so that the deficiencies of each can be made good by judicious mixing. When this is not possible, the deficient amino acids can be synthesized or made by fermentation and then added to the food. Production of amino acids for this purpose will probably be possible only in industrialized countries; their use may therefore seem to violate the principle that food must be where the mouths are. The quantities needed, however, are small. It is obviously better to upgrade an abundant local protein by adding .5 percent of methionine to it rather than import a whole protein to make up this one deficiency.

The food that is now needed or that will soon be needed in the underprivileged parts of the world might be supplied by charity, by the extension of existing methods of agriculture or by novel processes. I have argued that the first cannot be satisfactory and that it would be dangerous to assume that the second will suffice. Without wishing in any way to minimize the importance of what is being done in these two directions, it seems necessary to take novelties seriously. By definition a novelty is novel. That is to say, it may have an unfamiliar appearance, texture or flavor. In commenting on any of the proposals made—the use of strange animals, oilseed residues or leaf and microbial protein—it is irrelevant to say that they are unfamiliar. If the world is to be properly fed, products such as these will probably have to be used. Our problem is to make them acceptable.

Socrates, when one of his companions said he had learned little by foreign travel, replied: "That is not surprising. You were accompanied by yourself." Similarly, food technologists, accustomed to the dietary prejudices of Europe and the U.S., are apt to project their prejudices onto other communities. They have two opposite obsessions: to fabricate a "chewy" texture in their product and to produce a bland stable powder with an indefinite shelf life. Neither quality is universal in the familiar foods of most of the world. The former may have merits, although these probably do not outweigh the extra difficulties involved. It is odd that, at a time when people in industrialized countries are beginning to revolt against uniform and prepacked foods, we should be bent on foisting the latter quality off on others. Instead, novelties should be introduced into regions where they will most smoothly conform with local culinary habits. Novel forms of fish and mollusks are probably well adapted to Southeast Asia, where fermented fish is popular. Leaf, oilseed or microbial protein would fit smoothly into a culture accustomed to porridge, gruel and curry. It is important to remember the irrational diversity of our tastes. Even in Europe and the U.S. a flavor and appearance unacceptable in an egg is acceptable in cheese, a smell unacceptable in chicken is acceptable in pheasant or partridge, and a flavor unacceptable in wine is acceptable in grapefruit juice. These things are a matter of habit, and habits, although they will not change in a day or even a month, can readily be changed by suitable example and persuasion. The essential first step is to find out what is meant by the word "suitable."

Enough experience is now accumulating for us to define the parameters of success. The four most important are:

First, research on the novelty should be done privately and completely, so that when popularization starts there can be no rationally based doubts about the merits of the product.

Second, the novelty should manifestly be eaten by the innovators themselves. It is folly to ask people to practice what we only preach—we must practice it ourselves.

Third, example is the main factor leading to a change in habits; it is therefore essential to get the support of influential local people—from film stars to political leaders. Care should be taken that the first users are not underprivileged groups (prisoners, refugees and so on), because the stigma will not easily be removed.

Last, an adequate and regular supply of the product should be assured before there is any publicity, because it is hard to reawaken interest that has waned because the product is not obtainable.

All these proposals, except that of the simplest form of agricultural extension, call for research. It is worth considering who should do it and what form opposition is likely to take. Opposition to innovation is an interesting and underinvestigated part of psychopathology. It takes three main forms: total, quasi-logical and "instant."

Total opposition is the denial of the problem. Even today there are those who, in the course of condemning some specific proposal, sometimes deny that a protein shortage exists or is impending. This is a point that should be settled at the very beginning: "Do we have a problem or not?" Fortunately for research, and for humanity, the international agencies are in agreement that we do.

Quasi-logical opposition comes from economists. They may accept the problem but argue that some proposed solutions will be too expensive. There are two relevant questions: "Compared with what?" and "How do you know when it has not been tried?" When there are several equally feasible methods for getting the extra food that is needed in a region, a comparison of their probable costs is obviously worthwhile. But when all costs are, for various reasons, unknown, the exercise becomes futile because assumptions play a larger part in it than rigid economic argument, and scientists are better qualified than economists to make the assumptions; they know more of the facts, are aware of more possibilities and are less subject to romantic illusion.

"Instant opposition" arises because innovators are apt to irritate right-minded people, and enthusiasm invites skepticism. The innovator must therefore expect to run into trouble. When someone made the old comment that genius was an infinite capacity for taking trouble,

MUSSELS ARE GROWN in this floating "park" in Vigo Bay on the northwestern coast of Spain. Suspended from each of the large anchored raftlike structures are ropes on which the mussels are seeded and grow to maturity (see photograph on opposite page).

Samuel Butler replied: "It isn't. It is an infinite capacity for getting into trouble and for staying in trouble for as long as the genius lasts." In an attenuated form the principle applies even when genius is not involved. There are many different ways of getting into trouble, and it is as naïve and illogical to assume that an idea must be correct because it is meeting opposition as it is to take "instant opposition" seriously.

The governments of countries with a food shortage know that, for a decade at least, more people will be better fed if money is spent on importing food rather than on setting up a research project on means to make more food from local products for local consumption. The more farsighted statesmen realize that ultimately the research will have to be done, but it is hard to resist political pressure, and resistance is hampered by the high cost of primitive agriculture. In a market in New Guinea local sweet potatoes cost three times as much per calorie as imported wheat, and fresh fish cost twice as much per gram of protein as canned fish. So poor countries are hardly likely to mount research projects.

At the other extreme are the giants of private enterprise. They already do very well selling soft drinks and patent foods in underdeveloped countries, and their skill in creating a market, regardless of the real merits of their product, is unrivaled. Thus baby foods, which few experts regard as superior to mother's milk, were used in Uganda by 42 percent of the families in 1959, whereas only 14 percent had used them in 1950. Undoubtedly there are efficient firms that operate with strict integrity, and a few of them have ventured into the production of low-cost protein-rich foods. After the necessary research and preliminary publicity had been done with money from international sources, the Quaker Oats Company has done a masterly job in making, distributing and popularizing Incaparina (maize, sorghum and cottonseed) in Central America. And skilled advertising increased the sales of Pronutro (soya, peanuts and fish) tenfold in two years. Other attempts have failed because the possible profit, when one is selling to poor people without misleading them with meretricious advertising, is too small to cover the costs of the preliminary educational campaign. That, as I have suggested, can be managed only with the cooperation of governments and the local leaders of opinion.

Large-scale private enterprise will probably not find this activity lucrative, and from some points of view the methods of production that would be used may not be desirable. Already more than a third of the world's city population (12 percent of the total population) live in shantytowns on the fringes of cities, and rural depopulation is accelerating. This, together with transport difficulties, makes it at least arguable that research attention should be focused on simple techniques adapted for use in a large village or small town, rather than on fully industrialized techniques. The latter have their place, but they should not become our exclusive concern.

If neither the governments of needy countries nor private enterprise is likely to undertake the necessary research and development, it remains for the governments of industrialized countries, the international agencies and the foundations. So far these groups have been reluctant to admit that any radical changes in research policy will be needed, but times are changing. The novelties are now at least mentioned by the F.A.O. even if only to be gently damned with a few misstatements. On the Barnum principle, "I don't care what people say about me so long as they talk about me," this is a step forward. The governments of wealthy countries supply most of the support for the international agencies, they support other forms of aid, and much knowledge that is of use in poorer countries is an international by-product of their more parochial research. They may feel that they are already doing their share. Our best hope must therefore lie with the foundations. Several institutes of food technology are needed to undertake fundamental and applied research on the production of food from local products for local consumption. At least one of the institutes should be in the wet Tropics and all should give particular attention to protein sources. Using locally available material, each institute should study all the types of raw material discussed here. This will ensure that similar criteria are applied to all of them and that the assessment of their merits is made objectively and is not colored by interinstitutional rivalry. These institutes should also be responsible for work on the presentation and popularization of the products made. It may be that an extension of normal agriculture will meet the world's food needs for a few more years, but ultimately more radical research will be needed. It would be prudent to start it before the need is even more pressing than it is at present.

ROPE COVERED WITH MUSSELS is lifted from the water after mussels are mature.

BIOGRAPHICAL NOTES AND BIBLIOGRAPHIES

I AGRICULTURAL BEGINNINGS

1. The Agricultural Revolution

The Author

ROBERT J. BRAIDWOOD is professor in the Oriental Institute and in the department of anthropology at the University of Chicago. He was born in Detroit, Mich., in 1907 and acquired his B.A. and M.A. at the University of Michigan in 1932 and 1933, respectively. He obtained a Ph.D. in archeology at Chicago in 1942, joining the faculty there two years later. Since 1947 he has been field director of a series of archeological expeditions to Iraq and Iran, the results of which are discussed in his article for this issue. This work has been partly financed by the National Science Foundation, the American Philosophical Society and the Wenner-Gren Foundation for Anthropological Research.

Bibliography

ANIMAL DOMESTICATION IN THE PREHISTORIC NEAR EAST. Charles A. Reed in *Science,* Vol. 130, No. 3,389, pages 1,629–1,639; December 11, 1959.

DOMESTICATION OF FOOD PLANTS IN THE OLD WORLD. Hans Helbaek in *Science,* Vol. 130, No. 3,372, pages 365–372; August 14, 1959.

NEAR EASTERN PREHISTORY. Robert J. Braidwood in *Science,* Vol. 127, No. 3,312, pages 1,419–1,430; June 20, 1958.

PREHISTORIC INVESTIGATIONS IN IRAQI KURDISTAN. Robert J. Braidwood and Bruce Howe in *Studies in Ancient Civilization,* No. 31. University of Chicago Press, 1960.

2. The Origins of New World Civilization

The Author

RICHARD S. MACNEISH is director of the Robert S. Peabody Foundation for Archaeology at Andover, Massachusetts. A native of New York City and a former Golden Gloves boxing champion, MacNeish was graduated from the University of Chicago in 1940. He went on to acquire an M.A. in anthropology and a Ph.D. in archaeology, physical anthropology and ethnology from Chicago in 1944 and 1949, respectively. Since 1936 MacNeish has participated in more than 45 archaeological field expeditions to various sites in the U.S., Canada and Mexico, and his books, articles and reviews on New World archaeology number more than 130. From 1949 to 1963 he was chief archaeologist for the National Museum of Canada, and from 1964 to 1968 he was head of the Department of Archaeology at the University of Calgary. He led four expeditions between 1960 and 1964 to the valley of Tehuacan in southern Mexico on behalf of the Robert S. Peabody Foundation for Archaeology; the results of these expeditions form the basis of the present article. He is now directing a research program concerned with origins of agriculture in highland Peru.

Bibliography

ANCIENT MESOAMERICAN CIVILIZATION. Richard S. MacNeish in *Science,* Vol. 143, No. 3606, pages 531–537; February 1964.

ORIGINS OF AGRICULTURE IN MIDDLE AMERICA. Paul C. Mangelsdorf, Richard S. MacNeish, and Gordon R. Willey in *Handbook of Middle American Indians: Vol. I* (Robert Wauchope, gen. ed.), edited by Robert C. West. University of Texas Press, 1964.

THE ORIGINS OF AMERICAN AGRICULTURE. Richard S. MacNeish in *Antiquity,* Vol. XXXIX, pages 87–94; 1965.

THE PREHISTORY OF THE TEHUACAN VALLEY, VOL I: ENVIRONMENT AND SUBSISTENCE. Edited by D. S. Byers. University of Texas Press, 1968. (See, particularly, Chapters 1 and 9–15.)

3. Forest Clearance in the Stone Age

The Author

JOHANNES IVERSEN is head of the Paleobotanical Laboratory of the Geological Survey of Denmark and lecturer at the University of Copenhagen. He took a Ph.D. in biology at the University of Copenhagen in 1936. His principal research has been in plant ecology and vegetational history. While tracing the factors governing the vegetational succession of the past as they are reflected in pollen records, he came upon evidence of large-scale clearances in the Danish forests of Neolithic times. He was a guest of Yale University on a Rockefeller Foundation grant during one of his visits to the U.S.

Bibliography

THE INFLUENCE OF PREHISTORIC MAN ON VEGETATION. Johs. Iversen in *Danmarks Geologiske Undersogelse,* Vol. 4, pages 1–25; 1949.

THE HISTORY OF THE BRITISH FLORA. H. Godwin. Cambridge University Press, 1956

4. The Chinampas of Mexico

The Author

MICHAEL D. COE is professor and chairman of the Department of Anthropology at Yale University. He is also adviser to Harvard University's Robert Woods Bliss Collection of pre-Columbian art at Dumbarton Oaks. A native of New York, Coe received an A.B. and a Ph.D. in anthropology from Harvard in 1950 and 1959 respectively. After teaching for a year at the University of Tennessee he went to Yale in 1960. Since 1957 Coe has concentrated on the development of early civilizations in Central America. He has excavated and traveled extensively in Mexico, British Honduras, Costa Rica and Guatemala. His interest in the chinampa problem, he writes, "was the by-product of a busman's holiday in Mexico City in the summer of 1962, when I became convinced that the 'floating garden' theory was all wet."

Bibliography

THE AGRICULTURAL BASIS OF URBAN CIVILIZATION IN MESOAMERICA. Angel Palerm in *Irrigation Civilizations: A Comparative Study,* Social Science Monograph I, pages 28–42, Pan American Union, 1955.

DAILY LIFE AMONG THE AZTECS. Jacques Soustelle. The Macmillan Company, 1962.

MEXICO. Michael D. Coe. Frederick A. Praeger, 1962.

II PLANT GROWTH AND DEVELOPMENT

5. The Role of Chlorophyll in Photosynthesis

The Authors

EUGENE I. RABINOWITCH and GOVINDJEE are, respectively, professor of chemistry at the State University of New York at Albany, and associate professor of botany and biophysics at the University of Illinois. Rabinowitch was born in Russia, received a Ph.D. in inorganic chemistry at the University of Berlin in 1926, and worked in Germany, Denmark, and Britain before coming to the U. S. in 1938. During World War II he worked on the Manhattan project. He joined the staff of the University of Illinois in 1947 and moved to the State University of New York at Albany in 1968. Rabinowitch is the author of many books and papers, editor of the *Bulletin of the Atomic Scientists,* and a translator of Russian poetry. A book of his poems written in Russian has been published in Paris. Govindjee, who has no other name, is a graduate of the University of Allahabad in India; he obtained a Ph.D. in biophysics at the University of Illinois in 1960. His father dropped the family name, Asthana, in an effort to wipe out caste distinctions; one can often judge the caste or subcaste of Indians by their family names. Rabinowitch says he has "often suggested to Govindjee that he should invent a first name to make life easier for abstracters and indexers, but he seems to enjoy the distinction." Govindjee is the author (or co-author) of many papers and a book on photosynthesis.

Bibliography

CHLOROPHYLL FLUORESCENCE AND PHOTOSYNTHESIS. Govindjee, G. Papageorgiou, and E. Rabinowitch in *Fluorescence: Theory, Instrumentation and Practice,* edited by G. G. Guilbault. Marcel Dekker, Inc., 1967.

MOLECULAR PHYSICS IN PHOTOSYNTHESIS. R. K. Clayton. Blaisdell Publishing Co., 1965.

PHOTOSYNTHESIS. G. E. Fogg. American Elsevier Publishing Co., 1968.

PHOTOSYNTHESIS. E. Rabinowitch and Govindjee. John Wiley and Sons, Inc., 1969.

PRIMARY PROCESSES IN PHOTOSYNTHESIS. M. D. Kamen. Academic Press, 1963.

TRANSFORMATION OF LIGHT ENERGY INTO CHEMICAL ENERGY. Govindjee in *Crop Science,* Vol. 7, No. 6, pages 551–560; November-December 1967.

6. The Control of Plant Growth

The Author

JOHANNES VAN OVERBEEK is director of the Institute of Life Science and of the department of biology at Texas A&M University. Born in Holland, he was graduated from the University of Leiden in 1928 and obtained a Ph.D. from the State University of Utrecht in 1933. He then spent nine years in research and teaching at the California Institute of Technology. From 1943 to 1947 he headed the work in plant physiology at the Institute of Tropical Agriculture in Puerto Rico. For the next 20 years, until he took up his present work, he was chief plant physiologist at the agricultural research laboratory of the Shell Development Company at Modesto, Calif. Van Overbeek is also a farmer; he owns a commercial vineyard in California and in that capacity he is a member of the Farm Bureau, the Allied Grape Growers and the United Vintners.

Bibliography

BIOCHEMISTRY AND PHYSIOLOGY OF PLANT GROWTH SUB-STANCES. Edited by F. Wightman and G. Setterfield. The Runge Press, 1968.

THE LORE OF LIVING PLANTS. Johannes van Overbeek and Harry K. Wong. Scholastic Book Services, 1964.

PAPERS ON PLANT GROWTH AND DEVELOPMENT. Watson M. Laetsch and Robert Cleland. Little, Brown and Company, 1967.

PLANT HORMONES AND REGULATORS. J. van Overbeek in *Science*, Vol. 152, No. 3723, pages 721–731; May 6, 1966.

7.   Germination

The Author

DOV KOLLER is professor of botany at the Hebrew University of Jerusalem. Born in Israel in 1925, he obtained his M.Sc. degree in 1950 and his Ph.D. in 1954, both from the Hebrew University. His career as a student was interrupted for six years of military service—four years as a Royal Air Force volunteer during World War II, and two as an infantryman in the Israeli army during Israel's war of independence. He is specially interested in developmental and environmental plant physiology. "The ability of plants to exist in the desert has fascinated me ever since my intimate contact with the deserts of North Africa and the Middle East in wartime," he says. "Germination is a crucial phase in the life cycle of the desert plant, since it is at this time that the plant commits itself, so to speak, to try weathering the hazards of its extreme environment." Koller spent two years at the California Institute of Technology as a Rockefeller Research Fellow. More recently, he spent a year as a visiting research plant ecologist at the University of California at Los Angeles.

Bibliography

GERMINATION INHIBITORS. Michael Evanari in *The Botanical Review*, Vol. 15, No. 3, pages 153–194; March 1949.

PHYSIOLOGY OF SEEDS. William Crocker and Lela V. Barton. Chronica Botanica Company, 1953.

PHYSIOLOGY OF SEED GERMINATION. Eben H. Toole, Sterling B. Hendricks, Harold A. Borthwick, and Vivian K. Toole in *Annual Review of Plant Physiology*, Vol. 7, pages 299–324; 1956.

SEED GERMINATION. Dov Koller, Alfred M. Mayer, Alexandra Poljakoff-Mayber, and Shimon Klein in *Annual Review of Plant Physiology*, Vol. 13, pages 437–464; 1962.

DORMANCY AND GERMINATION IN SEEDS. In *Encyclopedia of Plant Physiology: Vol. XV*, edited by W. Ruhland. Springer-Verlag New York, Inc., 1965.

A MODEL OF SEED DORMANCY. Ralph D. Amen in *The Botanical Review*, Vol. 34, pages 1–31; 1968.

THE PHYSIOLOGY OF DORMANCY AND SURVIVAL OF PLANTS IN DESERT ENVIRONMENTS. Dov Koller in *Dormancy and Survival: XXIII Symposium of the Society for Experimental Biology and Medicine*, pages 449–469; 1969.

8.   The Flowering Process

The Author

FRANK B. SALISBURY is professor and head of the department of plant science at Utah State University. Born in Provo, Utah, he graduated from the University of Utah in 1951. He received his M.A. from the University of Utah, and his Ph.D. from the California Institute of Technology, where he studied plant physiology under James Bonner and geochemistry under Harrison Brown. He has authored four books, over sixty technical papers, and many popular articles. He is also the father of seven children—five boys and two girls. His research at the present time—under the support of the National Science Foundation and the National Aeronautics and Space Administration—concerns the biological clock and the physiology of flowering; and various exobiological problems, including the responses of plants to ultraviolet radiation and other stress factors, and the possibilities of life on Mars.

Bibliography

THE DUAL ROLE OF AUXIN IN FLOWERING. Frank B. Salisbury in *Plant Physiology*, Vol. 30, No. 4, pages 327–334; July 1955.

PARTIAL REACTIONS IN THE FORMATION OF THE FLORAL STIMULUS IN XANTHIUM. James A. Lockhart and Karl C. Hamner in *Plant Physiology*, Vol. 29, No. 6, pages 509–513; November 1954.

THE REACTION CONTROLLING FLORAL INITIATION. H. A. Borthwick, S. B. Hendricks and M. W. Parker in *Proceedings of the National Academy of Sciences*, Vol. 38, No. 11, pages 929–934; November 1952.

THE REACTIONS OF THE PHOTOINDUCTIVE DARK PERIOD. Frank B. Salisbury and James Bonner in *Plant Physiology*, Vol. 31, No. 2, pages 141–147; March 1956.

THE FLOWERING PROCESS. Frank B. Salisbury. Pergamon Press, Inc., 1963.

THE INDUCTION OF FLOWERING: SOME CASE HISTORIES. Edited by L. T. Evans. Macmillan of Australia, 1969.

ASPECTS OF CLOCK RESETTING IN FLOWERING OF XANTHIUM. H. D. Papenfuss and Frank B. Salisbury in *Plant Physiology*, Vol. 42, No. 11, pages 1562–1568; November 1967.

THE PHYSIOLOGY OF FLOWERING. N. E. Searle in *Annual Review of Plant Physiology*, Vol. 16, pages 97–118; 1965.

9.   The Ripening of Fruit

The Author

J. B. BIALE started out to be a scientific farmer, but science proved more interesting than farming and he decided to devote his career to understanding plants rather than to growing them. He was born in Poland in 1908, came to the U.S. at the age of 20 and registered at the Agricultural School of the University of California. Knowing no English, he found a Polish-English dictionary "the most indispensable companion in freshman botany." He went on to take his Ph.D. in plant physiology and joined the faculty of the University of California at Los Angeles, where he is now professor of plant physiology. His chief research interests are in the fields of plant respiration and fruit physiology. Biale's spare time is divided between handball, tennis, skiing and his three children. Of his adopted country he says, "the U. S. A. in general and California in particular afforded me the opportunities denied elsewhere for a full and rich life, for which I am deeply grateful."

Bibliography

THE PHYSIOLOGY OF FRUIT GROWTH. J. P. Nitsch in *Annual Review of Plant Physiology*, Vol. 4, pages 199–236; 1953.

THE BIOCHEMISTRY OF FRUIT MATURATION. J. B. Biale and R. E. Young in *Endeavour*, Vol. XXI, No. 83-84, pages 164–174; October 1962.

GROWTH, MATURATION, AND SENESCENCE IN FRUITS. J. B. Biale in *Science*, Vol. 146, No. 3646, pages 880–888; November 13, 1964.

ETHYLENE ACTION AND THE RIPENING OF FRUITS. S. P. Burg and E. A. Burg in *Science*, Vol. 148, No. 3674, pages 1190–96; May 28, 1965.

III PLANT ENVIRONMENT

10. Light and Plant Development

The Authors

W. L. BUTLER and ROBERT J. DOWNS were until recently employed by the U.S. Department of Agriculture, the former as a biophysicist at the Instrumentation Research Laboratory, the latter as a plant physiologist at the Plant Industry Station. Butler acquired a B.A. in physics at Reed College in 1949. He did graduate work in photosynthesis at the University of Chicago under the Nobel laureate James Franck, receiving a Ph.D. in biophysics in 1955. He joined the Instrumentation Laboratory in 1956, and in 1964 became professor of biology at the University of California, San Diego. Downs took three degrees at George Washington University, receiving his Ph.D. in botany two years after he joined the Plant Industry Station in 1952. Since 1965 he has been professor of botany and horticulture and director of the Phytotron at North Carolina State University.

Bibliography

LIGHT AND PLANTS: A SERIES OF EXPERIMENTS DEMONSTRATING LIGHT EFFECTS ON SEED GERMINATION, PLANT GROWTH AND PLANT DEVELOPMENT. R. J. Downs, H. A. Borthwick, and A. A. Piringer in *United States Department of Agriculture Miscellaneous Bulletin 879*. U.S. Government Printing Office, 1966.

PHYTOCHROME ACTION AND ITS TIME DISPLAYS. H. A. Borthwick in *American Naturalist*, Vol. 98, No. 902, pages 347–355; September-October 1964.

PROPERTIES OF PHYTOCHROME. H. W. Siegelman and W. L. Butler in *Annual Review of Plant Physiology*, Vol. 16, pages 383–392; 1965.

A RAPID PROCEDURE FOR THE VISIBLE DETECTION OF PHYTOCHROME. C. O. Miller, R. J. Downs and H. W. Siegelman in *Bioscience*, Vol. 15, No. 9, pages 596–597; September 1965.

PURIFICATION AND PROPERTIES OF PHYTOCHROME: A CHROMOPROTEIN REGULATING PLANT GROWTH. H. W. Siegelman and S. B. Hendricks in *Federation Proceedings*, Vol. 24, No. 4, Part 1, pages 863–867; July-August 1965.

11. Soil

The Author

CHARLES E. KELLOGG is deputy administrator of the Soil Conservation Service in the U.S. Department of Agriculture. After obtaining a Ph.D. in soil science in 1929 from Michigan State College and serving as professor in charge of soils at North Dakota College, he went to the Department of Agriculture as chief of the division of soil surveys in 1935. Kellogg has been a guest scientist of the Academies of Science of the U.S.S.R., France, Algeria, Israel and India, and of the National Institute for the Study of Agronomy in the Belgian Congo; he has headed the agricultural division of the U.S. Mission on Soil and Water Uses to the U.S.S.R., and been secretary of the F.A.O. Agricultural Committee. He was given the Distinguished Service Award and gold medal by the U.S. Department of Agriculture in 1950.

Bibliography

MODERN SOIL SCIENCE. Charles E. Kellogg in *American Scientist*, Vol. 36, No. 4, pages 517–535; October 1948.

THE SOILS THAT SUPPORT US. Charles E. Kellogg. The Macmillan Co., 1941.

SOIL. The United States Department of Agriculture in *The Yearbook of Agriculture, 1957*. U.S. Government Printing Office.

12. Water

The Author

ROGER REVELLE is the director of the Center for Population Studies at Harvard University. He was formerly the university dean of research at the University of California and director of that university's Scripps Institution of Oceanography. Revelle began his long association with the Scripps Institution in 1931, two years after acquiring an A.B. in geology from Pomona College. He received a Ph.D. from Scripps in 1936 and was professor of oceanography there in 1951, when he became the first alumnus of that institution to be appointed its director. During World War II Revelle served as a commander in the U.S. Navy and immediately after the war joined the Office of Naval Research as head of the Geophysics Branch. In 1946 he organized the oceanographic expedition associated with the atomic bomb test in Bikini Lagoon, measuring the diffusion of radioactive waters and their effects on marine organisms. During the early 1950's he led several other expeditions to the central and southern Pacific, developing new methods for measuring the flow of heat out through the floor of the ocean. He served as president of the first International Oceanographic Congress held by the United Nations in 1959, and in 1961 he became the first man to hold the post of science advisor to the Secretary of the Interior. Revelle has been president of the Committee on Oceanographic Research of the International Council of Scientific Unions and a member of the U. S. Commission to the UNESCO Office of Oceanography.

Bibliography

DESIGN OF WATER RESOURCES SYSTEMS. Arthur Mass, Maynard Hufschmidt, Robert Dorfman, and Harold Thomas. Harvard University Press, 1962.

A HISTORY OF LAND USE IN ARID REGIONS: ARID ZONE RESEARCH, XVII. Edited by L. Dudley Stamp. UNESCO, 1961.

MISSION TO THE INDUS. Roger Revelle in *New Scientist,* Vol. 17, No. 326, pages 340–342; February 1963.

POSSIBILITIES OF INCREASING WORLD FOOD PRODUCTION:
BASIC STUDY NO. 10. Food and Agriculture Organization of the UN, 1963.

THE VALUE OF WATER IN ALTERNATIVE USES. Nathaniel Wollman. University of New Mexico Press, 1962.

WATER FACTS FOR THE NATION'S FUTURE. W. B. Langbein and W. G. Hoyt. The Ronald Press, 1959.

WATER RESOURCES: A REPORT TO THE COMMITTEE ON NATURAL RESOURCES PUBLICATION 1000-B. National Academy of Sciences–National Research Council, 1962.

13. Climate and Agriculture

The Author

FRITS W. WENT was professor of plant physiology at the California Institute of Technology from 1953–1958. He is now professor of botany at the University of Nevada. His father was professor of botany at the University of Utrecht. Went writes that "in spite of his concern that he would push me into his own field, I became a plant physiologist. It was my high school teacher in biology, who had no inhibitions in acquainting me with the endless problems of nature, who made me a biologist." For his Ph.D., which he took in 1927, Went investigated the plant growth hormone, for which he developed a quantitative assay.

Bibliography

THE EXPERIMENTAL CONTROL OF PLANT GROWTH. Frits W. Went. The Ronald Press, 1957.

CROP PRODUCTION AND ENVIRONMENT. R. O. Whyte. Faber and Faber Limited, 1946.

ENVIRONMENTAL CONTROL OF PLANT GROWTH. Edited by L. T. Evans. Academic Press, 1963.

IV PRODUCTION TECHNOLOGY

14. The Reclamation of a Man-Made Desert

The Author

WALTER C. LOWDERMILK is a leading conservationist who was associate chief of the Soil Conservation Service in the U. S. Department of Agriculture from 1933 until he retired in 1947. He was also chief of research for the Service from 1937 until 1947. Born in North Carolina in 1888, Lowdermilk acquired his B.A. at the University of Arizona in 1912 and then attended the University of Oxford as a Rhodes scholar. He was in forestry work for many years in this country and in China, where he was associated with the University of Nanking. In 1929 he took his Ph.D. at the University of California. Since 1943 Lowdermilk has been serving as consultant on conservation to many foreign governments as well as to presidential commissions and the Supreme Allied Command of Japan. From 1955 to 1957 he was visiting professor of

agricultural engineering at Technion, the Israeli technological institute of Haifa.

Bibliography

PALESTINE, LAND OF PROMISE. Walter Clay Lowdermilk. Harper & Brothers, 1944.

RIVERS IN THE DESERT. Nelson Glueck. Jewish Publication Society of America, 1959.

15. Hybrid Corn

The Author

PAUL C. MANGELSDORF, one of the leading workers in the U. S. on the genetics of crop plants, dates his interest in cultivated plants from childhood. His father was a Kansas florist and seedsman. Young Mangelsdorf studied wheat- and corn-breeding at Kansas State College and after graduation in 1921 went to the Connecticut Agricultural Experiment Station as an assistant geneticist. He worked there under Donald F. Jones and took graduate work at Harvard under Edward M. East—these two played major parts in the development of hybrid corn. From 1926 to 1940 Mangelsdorf was agronomist at the Texas Agricultural Experiment Station, where he produced hybrid corn strains for the Texas climate and developed new varieties of wheat, oats and barley. He went to Harvard in 1940 and was professor of botany and director of the Harvard Botanical Museum until his retirement in 1968. Since 1941 Mangelsdorf has been consultant for agriculture with the Rockefeller Foundation, helping underdeveloped countries produce more corn and wheat. He has written two other articles for SCIENTIFIC AMERICAN, "The Mystery of Corn," in July, 1950, and "Wheat," in July, 1953.

Bibliography

CORN AND CORN IMPROVEMENT. Edited by George F. Sprague. Academic Press, 1955.

A PROFESSOR'S STORY OF HYBRID CORN. Herbert K. Hayes. Burgess Publishing Company, 1963.

16. Hybrid Wheat

The Authors

BYRD C. CURTIS and DAVID R. JOHNSTON are both with Cargill, Inc.; Curtis is head of the hybrid wheat development program at Cargill Research Farms in Fort Collins, Colorado, and Johnston is a wheat breeder. Curtis received his bachelor's degree in agronomy at Oklahoma State University in 1950, his master's degree in agronomy at Kansas State University in 1951, and his Ph.D. in plant breeding and genetics from Oklahoma State University in 1959. He was a member of the staff at Oklahoma State until 1963, working on the breeding of wheat, oats and barley; he then was appointed associate professor—and later full professor—at Colorado State University, and director of wheat research for the state of Colorado. He joined Cargill in 1967. He also serves as an affiliate professor in Agronomy at Colorado State University. Johnston received his Bachelor of Science degree in agronomy from the University of Massachusetts in 1952. He commenced post graduate studies in plant breeding and genetics at the University of Minnesota in 1956. In 1958, he was appointed to the staff as a research fellow in wheat breeding. He held that position until 1967, at which time he took his present position.

Bibliography

SUBSTITUTION OF NUCLEUS AND ITS EFFECTS ON GENOME MANIFESTATIONS. H. Kihara in Cytologia, Vol. 16, No. 2, pages 177–193; August 1951.

STUDIES ON RESTORATION AND SUBSTITUTION OF NUCLEUS (GENOME) IN AEGILOTRICUM, IV: GENOME EXCHANGE BETWEEN DURUM AND OVATA CYTOPLASM AND ITS THEORETICAL CONSIDERATION FOR MALE-STERILITY. H. Fukasawa in Cytologia, Vol. 22, No. 1, pages 30–39; March 1957.

MALE-STERILITY INTERACTION OF THE TRITICUM AESTIVUM NUCLEUS AND TRITICUM TIMOPHEEVI CYTOPLASM. J. A. Wilson and W. M. Ross in Wheat Information Service No. 14, pages 20–31; 1962.

CROSS-BREEDING IN WHEAT, TRITICUM AESTIVUM, I: FREQUENCY OF THE POLLEN-RESTORING CHARACTER IN HYBRID WHEATS HAVING AEGILOPS OVATA CYTOPLASM. J. A. Wilson and W. M. Ross in Crop Science, Vol. 1, No. 3, pages 191–193; May-June 1961.

CROSS BREEDING IN WHEAT, TRITICUM AESTIVUM L., II: HYBRID SEED SET ON A CYTOPLASMIC MALE-STERILE WINTER WHEAT COMPOSITE SUBJECTED TO CROSS-POLLINATION. J. A. Wilson and W. M. Ross in Crop Science, Vol. 2, No. 5, pages 415–417; September-October 1962.

HYBRID WHEATS: THEIR DEVELOPMENT AND FOOD POTENTIAL. Ricardo Rodríguez, Marco A. Quiñones L., Norman E. Borlaug, and Ignacio Narváez in Research Bulletin No. 3, International Maize and Wheat Improvement Center, Mexico; July 1967.

HYBRID WHEAT. V. A. Johnson and J. W. Schmidt in Advances in Agronomy, Vol. 20, pages 199–233; 1968.

HETEROSIS IN WHEAT: A REVIEW. L. W. Briggle in Crop Science, Vol. 3, No. 5, pages 407–412; September-October 1963.

17. Chemical Fertilizers

The Author

CHRISTOPHER J. PRATT is manager of international agricultural chemical development for the Mobil Chemical Company. He was born and educated in England, studying industrial chemistry at the Rutherford College of Technology and psychology and music at Goldsmiths' College. For several years before coming to the U.S. in 1953 he was a management consultant to British companies in the field of heavy chemical processing. In his spare time he is a free-lance writer, in which capacity he has written several articles (including the present one) on heavy chemicals and on fertilizer production, and he indulges an interest in baroque music by playing the contrabass—he did so professionally for several years in England—and the organ.

Bibliography

THE CHEMISTRY AND TECHNOLOGY OF FERTILIZERS. Edited by Vincent Sauchelli. Reinhold Publishing Corporation, 1960.

COMMERCIAL FERTILIZERS. G. H. Collings. McGraw-Hill Book Company, Inc., 1955.

FERTILIZER NITROGEN: ITS CHEMISTRY AND TECHNOLOGY. Edited by Vincent Sauchelli. Reinhold Publishing Corporation, 1964.

18. Third-Generation Pesticides

The Author

CARROLL M. WILLIAMS is professor of biology at Harvard University. After his graduation from the University of

Richmond in 1937 he went to Harvard and successively obtained master's and doctor's degrees in biology and, in 1946, an M.D. He joined the Harvard faculty in 1946 and became full professor in 1953. From 1959 to 1961 he was chairman of the biology department. He has been a member of the National Academy of Sciences since 1961. Williams' studies of insects have won a number of awards; in 1967 he received the George Ledlie Prize of $1,500, which is given every two years to the member of the Harvard faculty who has made "the most valuable contribution to science, or in any way for the benefit of mankind."

Bibliography

THE HORMONAL REGULATION OF GROWTH AND REPRODUCTION IN INSECTS. V. B. Wigglesworth in *Advances in Insect Physiology: Vol. II*, edited by J. W. L. Bement, J. E. Treherne, and V. B. Wigglesworth. Academic Press Inc., 1964.

HORMONAL INTERACTIONS BETWEEN PLANTS AND INSECTS. C. M. Williams in *Perspectives in Chemical Ecology*, edited by E. Sondheimer. Academic Press Inc., 1970.

THE CHEMISTRY AND BIOLOGY OF JUVENILE HORMONE. H. Roller and K. H. Dahm in *Recent Progress in Hormone Research: Vol. 24*, edited by Gregory Pincus. Academic Press Inc., 1968.

19. Mechanical Harvesting

The Author

CLARENCE F. KELLY is director of the University of California's Agricultural Experiment Station. He received his bachelor's and master's degrees in agricultural engineering from North Dakota State University, then joined the U. S. Department of Agriculture in 1935, where he remained for 15 years except for three years naval service during World War II. In 1950 he became a member of the agricultural engineering staff of the University of California at Davis, and was chairman of the department after 1961. In 1963 he became associate director of the university's statewide Agricultural Experiment Station, and was made director in 1965. He was elected a Fellow in the American Society of Agricultural Engineers in 1958 and received that society's Cyrus Hall McCormick Medal in 1963. In 1968 he was elected to the National Academy of Engineering.

Bibliography

CYRUS HALL MCCORMICK; HARVEST, 1856–1884. William T. Hutchinson. D. Appleton-Century Company Incorporated, 1935.

POWER TO PRODUCE. The United States Department of Agriculture in *The Yearbook of Agriculture, 1960*. Government Printing Office.

ROOTS OF THE FARM PROBLEM. Earl O. Heady, Edwin O. Haroldsen, Leo V. Mayer, and Luther G. Tweeten. The Iowa State University Press, 1965.

20. Wine

The Author

MAYNARD A. AMERINE is professor of enology at the University of California at Davis. Amerine was graduated from the University of California at Berkeley in 1932 and obtained a Ph.D. in plant physiology there in 1936. He joined the Davis faculty in 1935; from 1957 to 1962 he was chairman of the department of viticulture and enology at Davis. During World War II Amerine served with the Chemical Warfare Service of the U.S. Army in Algeria and India. He is a past president of the American Society of Enologists and the author of several books on wine making.

Bibliography

AMERICAN WINES AND WINE-MAKING. Philip M. Wagner. Alfred A. Knopf, 1956.

GENERAL VITICULTURE. A. J. Winkler. University of California Press, 1962.

THE NOBLE GRAPES AND THE GREAT WINES OF FRANCE. A. L. Simon. McGraw-Hill Book Company, Inc., 1957.

THE TECHNOLOGY OF WINE MAKING. M. A. Amerine, H. W. Berg, and W. V. Cruess. The Avi Publishing Company, Inc., 1967.

WINES OF FRANCE. Alexis Lichine. Alfred A Knopf, 1965.

WINES OF GERMANY, Frank Schoonmaker. Hastings House, Publishers, Inc., 1956.

DESSERT WINES. Maynard A. Joslyn and M. A. Amerine. University of California, Division of Agricultural Sciences, 1962.

TABLE WINES: THE TECHNOLOGY OF THEIR PRODUCTION. Maynard A. Amerine and Maynard A. Joslyn. University of California Press, 1969.

V FOOD NEEDS AND POTENTIALS

21. The Ecosphere

The Author

LA MONT C. COLE is professor of zoology at Cornell University. Despite a boyhood passion for snakes, he graduated from the University of Chicago as a physicist. His return to the animal kingdom resulted from a trip down the Colorado River with A. M. Woodbury of the University of Utah, who inspired him to study ecology. Cole's chief interest is now in natural populations. He has taught at Cornell since 1948. Before that he occupied the late Alfred Kinsey's post in entomology at Indiana University, which he had taken over when Kinsey "turned to the study of bigger and better things."

Bibliography

THE ECOLOGY OF ANIMALS. Charles Elton. John Wiley & Sons, Inc.,1950.

ENERGY IN THE FUTURE. Palmer Cosslett Putnam. D. Van Nostrand Company, Inc., 1953.

ELEMENTS OF PHYSICAL BIOLOGY. Alfred J. Lotka. Williams & Wilkins Company, 1925.

FUNDAMENTALS OF ECOLOGY. Eugene P. Odum. W. B. Saunders Company, 1953.

GEOCHEMISTRY. Kalervo Rankama and Th. G. Sahama. The University of Chicago Press, 1950.

THE WEB OF LIFE: A FIRST BOOK OF ECOLOGY. John H. Storer. The Devin-Adair Company, 1953.

22. The Human Population

The Author

EDWARD S. DEEVEY, JR., is now professor of biology at Dalhousie University, Halifax, Nova Scotia; a member of the Fisheries Research Board of Canada; and president of the Ecological Society of America. For many years he was professor of biology at Yale University, and in 1967–68 he was the section head of environmental and systematic biology for the National Science Foundation. Though his researches have centered on paleoecology, he notes that the ecology of the recent past, especially of the last ten thousand years when human disturbance became the major environmental process, "is getting more relevant every day." For example, his Ph.D. dissertation—done under the direction of G. E. Hutchinson in 1938—was on *eutrophication*, though there was no such word in the English language prior to that time. He is building a new home on Shag Bay, Halifax County, Nova Scotia, confident that "Canada still has time."

Bibliography

THE NEXT HUNDRED YEARS: MAN'S NATURAL AND TECHNICAL RESOURCES. Harrison Brown, James Bonner and John Weir. Viking Press, Inc., 1957.

POPULATION AHEAD. Roy Gustaf Francis. University of Minnesota Press, 1958.

SCIENCE AND ECONOMIC DEVELOPMENT: NEW PATTERNS OF LIVING. Richard L. Meier. John Wiley & Sons, Inc., 1956.

WORLD POPULATION AND PRODUCTION: TRENDS AND OUTLOOK. W. S. Woytinsky and E. S. Woytinsky. Twentieth Century Fund, 1953.

23. Food

The Author

NEVIN S. SCRIMSHAW is professor of nutrition and head of the Department of Nutrition and Food Science at the Massachusetts Institute of Technology. A graduate of Ohio Wesleyan University, Scrimshaw received an M.A. and a Ph.D. from Harvard University in 1939 and 1941, respectively. He did postdoctoral work in nutrition and endocrinology at the University of Rochester and received an M.D. from that university's medical school in 1945. After interning at Gorgas Hospital in the Canal Zone he returned to the University of Rochester in 1946 to do research in the department of obstetrics and gynecology. In 1948 he went back to Panama to do field research on nutrition and pregnancy and shortly thereafter became chief of the Nutrition Section of the World Health Organization's Pan American Sanitary Bureau. From 1949 to 1961 Scrimshaw served as director of the Institute of Nutrition of Central America and Panama. He acquired a degree in public health from Harvard in 1959 and was adjunct professor of public health nutrition at the Columbia University College of Physicians and Surgeons from 1959 until 1961, when he took up his present post. Scrimshaw has served as adviser on world nutrition problems to various Government and United Nations agencies. At M.I.T. he is currently engaged in studies involving the effect of stress on nutritional requirements and the effect of nutrition on resistance to infection and on mental development.

Bibliography

SCIENCE, TECHNOLOGY, AND DEVELOPMENT, VOLUME III: AGRICULTURE. United States Papers Prepared for the United Nations Conference on the Application of Science and Technology for the Benefit of the Less Developed Areas. U.S. Government Printing Office, 1962.

SCIENCE, TECHNOLOGY, AND DEVELOPMENT, VOLUME VI: HEALTH AND NUTRITION. United States Papers Prepared for the United Nations Conference on the Application of Science and Technology for the Benefit of the Less Developed Areas. U.S. Government Printing Office, 1963.

PROSPECTS OF THE WORLD FOOD SUPPLY: PROCEEDINGS OF A SYMPOSIUM. National Academy of Sciences, 1966.

THE WORLD FOOD PROBLEM: A REPORT, VOLS. I, II, III. The President's Science Advisory Committee. U.S. Government Printing Office, 1967.

24. Peasant Markets

The Author

SIDNEY W. MINTZ is professor of anthropology at Yale University. He acquired his B.A. in psychology at Brooklyn College in 1943, served three years with the Air Force, then took his Ph.D. in anthropology at Columbia University in 1951. He has done field work in Puerto Rico, Jamaica, Haiti and Iran. Mintz has been at Yale since 1951, where he is editor of the Yale University Press's Caribbean Series and chairman of the Faculty Committee for Afro-American Studies. His publications include a study of a Puerto Rican sugarcane plantation, a life-history of a plantation worker, and numerous articles. He is now at work on a social history of the Antilles.

Bibliography

TRADITIONAL EXCHANGE AND MODERN MARKETS. Cyril S. Belshaw. Prentice-Hall, 1965.

INTERNAL MARKET SYSTEMS AS MECHANISMS OF SOCIAL ARTICULATION. Sidney W. Mintz in *The Proceedings of the Annual Meeting of the American Ethnological Society,* pages 20–30; 1959.

TRADE AND MARKETS. Richard Salisbury in *International Encyclopedia of the Social Sciences,* Vol. 16, pages 118–122; 1968.

PEASANT MARKET PLACES AND ECONOMIC DEVELOPMENT IN LATIN AMERICA. Sidney W. Mintz in *Vanderbilt University Graduate Center for Latin American Studies Occasional Paper No. 4,* pages 1–9; 1964.

THE ECONOMICS OF MARKETING REFORM. P. T. Bauer and B. S. Yamey in *The Journal of Political Economy,* Vol. LXII, No. 3, pages 210–235; 1954.

WEST AFRICAN TRADE. Peter Tamas Bauer. Cambridge University Press, 1954.

25. Orthodox and Unorthodox Methods of Meeting World Food Needs

The Author

N. W. PIRIE is head of the department of biochemistry at the Rothamsted Experimental Station in England. A Fellow of the Royal Society, he has interested himself not only in biochemical matters but also in social issues. Among the former are the separation and the properties of several plant viruses; factors controlling the infectability of plants by viruses, and the preparation of edible proteins from leaves. Pirie writes: "That is recounted in 100-plus papers. Another 100-plus deal in a general way with fractionation of macromolecules and the criteria of purity; classification of viruses and similar entities; contraception and the population problem; protein sources and other aspects of world food supplies, and strictures on government policies on such issues as air raid precautions, nuclear weapon testing and the planning of scientific research." Pirie was graduated from the University of Cambridge and served in the biochemical laboratory there from 1929 to 1940, when he went to the Rothamsted Station.

Bibliography

INTERNATIONAL ACTION TO AVERT THE IMPENDING PROTEIN CRISIS. Advisory Committee on the Application of Science and Technology to Development. United Nations, 1968.

LEAF PROTEIN AS A HUMAN FOOD. N. W. Pirie in *Science,* Vol. 152, No. 3730, pages 1701–1705; June 24, 1966.

THE PRODUCTION AND USE OF LEAF PROTEIN. N. W. Pirie in *Proceedings of the Nutrition Society,* Vol. 28, page 85; 1969.

COMPLEMENTARY WAYS OF MEETING THE WORLD'S PROTEIN NEED. N. W. Pirie in *Proceedings of the Nutrition Society,* Vol. 28, page 255; 1969.

FOOD RESOURCES: CONVENTIONAL AND NOVEL. N. W. Pirie. Penguin Books Inc., 1969.

THE FOOD RESOURCES OF THE OCEANS. S. J. Holt in *Scientific American,* Vol. 221, No. 3, pages 178–194; September 1969.

INDEX

ATP, 44
Abscisin, 53
Abscisic acid, 54
Acidity of soil, 89, 153
Adenine, 53
Aegilops caudata. See Goat grass
Agricultural revolution, 4–12
Agriculture, 4, 13, 96, 108, 123–132, 159, 163, 206–214
  chinampa system of, 28–36
  economics of, 206–214, 215–222
  efficiency of, 56
  expansion of, 106, 123–132, 152–162, 206–214
  in Israel, 123–126, 128
  origins of, 7, 8, 14, 22, 28, 223
  regions of development of, 4, 15, 105–106
  in West Pakistan, 105
  slash and burn, 24
Agroclimatology, 117
Ajuereado phase, 20
Alcohol in wine, 185
Aleurone, 51
Algae, 46
Alkalinity of soil, 89
Amaranth, 13, 20, 33
Amino acid in wine, 182
Ammonia for fertilizer, 155–157
Amouq, 12
Anau, 12
Animals and livestock, 107, 131, 133, 141, 195, 205, 209, 226
  domestic, 8, 10, 13, 141
  prehistoric, 13
  wild, 9, 11
Anthocyanin, 80, 81
Apazote, 13, 20
Archaic period, 7
Archeological records, 3ff, 12, 33
Arsenate of lead as pesticide, 163
Artifacts, 8, 11, 16
Artificial lighting, 113
Asia, agricultural revolution in, 8
Aswan High Dam, 105
Auxin, 50–52, 69–70
Axe, hafted, 23
Azonal soils, 88
Aztec empire, 28ff, 31
B-Nine, 56
Banana, 70ff

Barley, 7, 8, 10, 23
Bean, 13, 20, 33
Beet, sugar, 173
Benzyladenine, 54, 55
Biosphere, 192
Boron, 86
Bracken, 25, 26
Buds, 64, 68, 116
Burning, 23–26
  energy dissipation by, 193
  influence on flora of, 26

CCC, 56
Calcium, 86
Canal, 30ff
  drainage, 128
  irrigation, 105
Capillary action, 104
Capullis, 34
Carbon, reduction of, 44
Carnivore, efficiency of, 193
Cereals, 8, 23, 25
  See also Corn, Grains, Wheat
Chapínes, 32, 33
Chayote, 13
Chernozem, 93
Chinampas, 28–36
Chinampero, 28, 31–34
Chive, 13
Chlorophyll, 41, 47–48
Chloroplast, 41
Chocolate, 13
Chromosome, linkage of, 136
Civilization, origin of, 13, 21, 28, 203
Clay, 89
Climate, 8
  control of, 112, 117
  of Israel, 124
  for plant growth, 108
  See also Agroclimatology
Climacteric, 72, 73
Coa (cultivating stick), 32ff
Cocklebur, 64
Coconut milk, 52–53
Coleoptile, 51, 56
Combine
  operation of, 170
  wheat, 168
Communities, human, 7, 8, 198ff, 215ff
Compositae, 27

Compost, 155
Copper, 86
Coprolites, 8
Corn, 14–20, 133–141, 157, 162
  hybrid, 133–141
Corncobs, ancient, 15
Cotton, 173–174
Coxcatlán Cave, 14–17
Crop breeding, 133–141, 143–151, 170, 225
Crop yields, influence of fertilization on, 152
Crops, 7, 33, 108–117, 131, 133, 143, 168–178, 209
Cucurbita mixta. See Squash
Culture, human, 4, 8, 10, 13, 20, 21, 106, 215ff
Cytochrome, 46, 48
Cytokinin, 51–56
Cytoplasmic sterility, 145

DDT, metabolism of, 163
DNA, 53
Dam
  multipurpose, 97
  Aswan High, 105
Dark period for flowering, 66
Desalination. *See* Salinity
Desert
  land, 130
  reclamation of, 123
  soils, 94, 95
Detasseling, 133, 138–139
Diet, 193
Dispersion of seed, 59
Domestication of plants and animals, 4, 9, 10, 11, 13, 15, 20, 133, 141, 143
Dormancy of seeds, 57
Dormin, 54
Double cross of corn, 135
Drainage, 30, 95, 103, 104, 124, 128
Dunes, 132

Earth, heating of, 193
East, Edward M., 135
Ecosphere, 192–198
Ecosystem, 192ff
Einkorn wheat, 25
Electrochemical potential, 44
Electron transfer, 44ff